A SPECIAL ISSUE OF
COGNITIVE NEUROPSYCHOLOGY

Perception and Action: Recent Advances in Cognitive Neuropsychology

edited by

Jean Decety

INSERM Unit 280, Lyon, France

Copyright © 1998 by Psychology Press Ltd, a member of the Taylor & Francis group
   All rights reserved. No part of this publication may be reproduced, stored in a retrieval system, or transmitted in any form or by any means, electronic, mechanical, photocopying, recording or otherwise, without permission in writing from the publisher.

Psychology Press Ltd, Publishers
27 Church Road
Hove
East Sussex
BN3 2FA
UK

British Library Cataloguing-in-Publication Data

A catalogue record for this book is available from the British Library

   ISBN: 0-86377-600-0

Typeset by Quorum Technical Services Ltd, Cheltenham
Printed and bound in the United Kingdom by Henry Ling Ltd, Dorchester

Contents

481 Jean Decety, *Preface and Acknowledgements* *

483 Marjan Jahanshahi and Christopher D. Frith, *Willed Action and its Impairments*

535 Ian M. Thornton, Jeannine Pinto, and Maggie Shiffrar, *The Visual Perception of Human Locomotion*

553 Julie Grèzes, Nicolas Costes, and Jean Decety, *Top-down Effect of Strategy on the Perception of Human Biological Motion: A PET Investigation*

583 Lawrence M. Parsons and Peter T. Fox, *The Neural Basis of Implicit Movements Used in Recognising Hand Shape*

617 Laurel J. Buxbaum, Myrna F. Schwartz, and Michael W. Montgomery, *Ideational Apraxia and Naturalistic Action*

645 M. Jane Riddoch, Martin G. Edwards, Glyn W. Humphreys, Richard West, and Tom Heafield, *Visual Affordances Direct Action: Neuropsychological Evidence from Manual Interference*

685 Anna M. Barrett, Ronald L. Schwartz, Anastasia L. Raymer, Gregory P. Crucian, Leslie Gonzalez Rothi, and Kenneth M. Heilman, *Dyssynchronous Apraxia: Failure to Combine Simultaneous Preprogrammed Movements*

---

* This book is also a special issue of the journal *Cognitive Neuropsychology* which forms Issues 6, 7 and 8 of Volume 15 (1998). The page numbers used here are taken from the journal and so begin on p. 481.

705 Kelly J. Murphy, David P. Carey, and Melvyn A. Goodale, *The Perception of Spatial Relations in a Patient with Visual Form Agnosia*

723 Catherine L. Reed and Ian M. Franks, *Evidence for Movement Preprogramming and On-line Control in Differentially Impaired Patients with Parkinson's Disease*

747 Isabel M. Smith and Susan E. Bryson, *Gesture Imitation in Autism I: Nonsymbolic Postures and Sequences*

771 Glyn W. Humphreys and E.M.E. Forde, *Disordered Action Schema and Action Disorganisation Syndrome*

813 *Subject Index for Special Issue*

Preface and Acknowledgements

I was honoured when Professor Alfonso Caramazza asked me to act as a Guest Editor for the publication of a Special Issue of *Cognitive Neuropsychology* on perception and action. My duty was to invite a limited number of authors (consequently, it is selective and arbitrary) and to propose that they submit an original paper to be refereed as if for a regular issue. My intention has been, on the one hand, to gather within a single volume the contributions from cognitive neuropsychologists whose interests deal with perception and action in the normal brain as well as in neurological patients; and, on the other hand, to illustrate the current—interdisciplinary and converging—approaches to this topic. Indeed, within the past 10 years, the field of perception and action has gained much from cross-fertilisation between experimental psychology, neurophysiology and neuropsychology. For example, the division of labour within the visual system (ventral versus dorsal stream) has had a great impact on neuropsychology, although most (of the first) experimental evidence comes from investigations in the monkey (see Milner, 1998).

The present volume contains 11 papers, which provide an excellent illustration of the cognitive neuropsychology of perception and action. About one third of the contributions are studies of patients, whilst a quarter, unsurprisingly, report neuropsychological disorders in action-perception with a strong emphasis on apraxic disorders, which are still poorly understood in neurology. Three articles report experiments in healthy subjects, two of which use neuroimaging techniques. Their results are of great interest for the validation of cognitive models in neuropsychology. Since the cognitive neuropsychological approach is now also invading the field of psychiatry (Halligan & Marshall, 1996), one contribution addresses imitation in autistic patients and one reviews neurophysiological evidence of willed action and its impairments both in neurological and psychiatric patients.

I am grateful to the following for their help with the preparation of this issue. First, I would like to thank all the authors and referees who contributed. Second, I wish to acknowledge Kate Moysen (Psychology Press) and Deborah Edelman (Harvard University) who provided efficient assistance with the production of the issue.

Jean Decety
Plateau d'Albion, France
August, 1998

REFERENCES

Halligan, P.W., & Marshall, J.C. (Eds.) (1996). *Method in madness: Case studies in cognitive neuropsychiatry*. Hove, UK: Psychology Press.

Milner, A.D. (Ed.) (1998). *Comparative neuropsychology*. Oxford, UK: Oxford University Press.

WILLED ACTION AND ITS IMPAIRMENTS

Marjan Jahanshahi
Institute of Neurology and National Hospital for Neurology and Neurosurgery, London, UK

Christopher D. Frith
Institute of Neurology, London, UK

Actions are goal-directed behaviours that usually involve movement. There is evidence that intentional self-generated actions (willed actions) are controlled differently from routine, stereotyped actions that are externally triggered by environmental stimuli. We review evidence from investigations using positron emission tomography (PET), recordings of movement-related cortical potentials (MRCPs) or transcranial magnetic stimulation (TMS), and conclude that willed actions are controlled by a network of frontal cortical (dorsolateral prefrontal cortex, supplementary motor area, anterior cingulate) and subcortical (thalamus and basal ganglia) areas. We also consider evidence suggesting that some of the cognitive and motor deficits of patients with frontal lesions, Parkinson's disease, or schizophrenia as well as apathy and abulia and rarer phenomena such as primary obsessional slowness can be considered as reflecting impairment of willed actions. We propose that the concept of a willed action system based on the frontostriatal circuits provides a useful framework for integrating the cognitive, motor, and motivational deficits found in these disorders. Problems remaining to be resolved include: identification of the component processes of willed actions; the specific and differential role played by each of the frontal cortical and subcortical areas in the control of willed actions; the specific mechanisms of impairment of willed actions in Parkinson's disease, schizophrenia, and frontal damage; and the precise role of the neurotransmitter dopamine in the willed action system.

INTRODUCTION

In this article our aim is to put forward a number of proposals about the physiological basis of willed action and provide some supporting evidence from a selective and non-exhaustive review of the literature. First, we suggest that intentional self-generated aspects of behaviour (willed action) are controlled differently from routine externally triggered

---

Requests for reprints should be addressed to Dr. M. Jahanshahi, MRC Human Movement & Balance Unit, The National Hospital for Neurology and Neurosurgery, Queen Square, London WC1N 3BG, UK (Tel: 071 829 8759; Fax: 071 278 9836; E-mail: m.jahanshahi@ion.ucl.ac.uk).

The financial support of the Wellcome Trust is gratefully acknowledged.

actions. Second, that the frontostriatal circuits may be the anatomical substrates of such a willed action system. Third, that akinesia in Parkinson's disease, poverty of action in schizophrenia, and disorganised behaviour following frontal lesions are impairments of willed actions. Fourth, that some of the cognitive deficits of patients with these disorders can also be considered to reflect impairment of willed actions. Finally, we suggest that the concept of a willed action system based on the frontostriatal network allows an integrated approach to the study of the impairments of cognitive and motor function as well as the motivational deficits in these disorders, which so far have largely been investigated separately. We will start by clarifying what we mean by willed actions.

WHAT ARE WILLED ACTIONS?

The study of motor control has been primarily concerned with the mechanistic aspects of motor function. Even with the central control of movement through motor programming (Keele, 1968), the focus has been on the nature of the physical parameters (e.g. direction, force) that may be represented in such motor programmes. However, even the simplest nonreflexive movements are preceded by goal formulation, giving rise to an intention to act. A scheme for action involves a number of processes: from formulation of a goal and an intention to act, to more basic processes such as response selection, programming, and initiation that precede the production and execution of an action. Contemporary theories of action (for example, Norman & Shallice, 1986) include higher-order supervisory processing structures, which oversee and control the basic motor processes in nonroutine circumstances, and exert "executive" control over action, prior to its production by effector mechanisms. The concept of willed action allows the establishment of links between cognitive psychological models of control of action and the field of motor control.

Actions are purposeful, goal-directed behaviours, usually involving movement. While all actions are voluntary, not all are willed. For example, in most situations walking is an automatic act, but when walking on ice this normally automatic action requires our attention and becomes willed. William James (1890) made a distinction between "ideo-motor" and "willed" acts. According to James (1890, p. 522) "wherever movement follows unhesitatingly and immediately from the notion of it in the mind, we have ideo-motor action. We are then aware of nothing between the conception and the execution". In contrast, willed acts were considered to be those where "an additional conscious element in the shape of a fiat, mandate, or expressed consent" is involved. James also suggested that, "Effort of attention is thus the essential phenomenon of will", emphasising the central role of attention in willed action. Besides attention, other processes are also relevant to willed actions. Let us consider an example. This manuscript is being written using a series of willed cognitive and motor actions. As I write each sentence, I am attending to my thoughts, that is, what I want to write and communicate. Since I am not a skilled typist, I also attend to the movements of my hands over the keyboard as well as what appears on the computer screen. I feel that I am able to choose what I write and how I write it. My actions are intentional, the immediate intention being to communicate my thoughts about willed actions in order to fulfil the more long-term goal of meeting the deadline for submission of the article to the editor. These actions clearly fit in the category of willed actions as they meet criteria of involving (1) attention and conscious awareness, (2) choice and control, and (3) intentionality.

Until such time when physiological criteria for defining willed actions become available, the above criteria based on common-sense psychology are the only means of doing so. A major problem concerns how we can use these criteria to design experimental tasks that engage willed action. One commonly used strategy is to ask subjects to produce random sequences of responses (e.g. Friston, Liddle, & Frackowiak, 1991). In such tasks each response has to be consciously selected. Another strategy is to ask subjects performing a routine task to attend to their actions (e.g. Jueptner, Stephan, et al., 1997). Another strategy is based on internal vs. external locus of control. Willed actions are defined as those that are not elicited directly by environmental stimuli but are mainly controlled by interoceptive stimuli (Frith, 1992). However, even when there is an external trigger for the initiation of an action there may be component processes that depend upon internal control. Two examples of this are found in reaction time tasks. In a simple reaction time (SRT) task, where presentation of one particular stimulus (S) elicits a specific response (R), the invariant S–R relationship across trials allows preparation of the response prior to the presentation of the stimulus. This optional (willed) preparation of the response in advance confers a speed advantage to SRT over choice reaction time (CRT) tasks. The attention-demanding nature of this preprogramming has been demonstrated by the finding that concurrent performance of a secondary task prolongs SRT but not CRT (Frith & Done, 1986; Goodrich, Henderson, & Kennard, 1989). A second such volitional process prior to an externally triggered action is the S–R decoding necessary for response selection in CRT tasks with low S–R compatability resulting from the use of arbitrary or novel S–R pairings that lack a conceptual link.

Closely related to the concept of willed action are high-level control processes such as those supplied by the "Supervisory Attentional System" (SAS) of Norman and Shallice (1986). Tasks that require the operation of the SAS are those that (1) involve planning or decision making, (2) involve components of trouble-shooting, (3) are ill-learned or contain novel sequences of actions, (4) are judged to be dangerous or technically difficult, (5) require the overcoming of a strong habitual response or resisting temptation. Many complex tasks have been developed which engage these various high-level control processes (e.g. The Tower of London task: Shallice, 1982).

The Anatomy of the Frontostriatal Circuits

There is substantial evidence that willed actions depend upon regions of frontal cortex and associated subcortical structures. Before reviewing this evidence we shall briefly describe the anatomy of the frontostriatal circuits. Anatomical and functional reciprocity of the frontal cortex and the basal ganglia is implied by the multiple circuits between them (Alexander, DeLong, & Strick, 1986). Five such circuits have been distinguished. Each circuit originates from a discrete area of the frontal cortex, passes through specific portions of the basal ganglia, and then projects back to the original frontal cortical site via distinct thalamic nuclei (Fig. 1). Briefly, the five circuits and their main cortical and subcortical components are as follows. First, the "motor" loop between the SMA and putamen, second, the "complex" loop between the dorsolateral prefrontal cortex and the caudate nucleus, third, the "anterior cingulate" loop between this structure and the ventral striatum, fourth, the "oculomotor" loop between the frontal eye fields and the caudate nucleus, and finally, the "lateral orbitofrontal" loop between this section of the prefrontal cortex and the caudate nucleus.

Besides the fundamental similarity in their cortico-subcortical-thalamic-cortical arrangement, the various frontostriatal circuits share a number of other features (Alexander & Crutcher, 1990). As shown in Fig. 2, each circuit has a direct and an indirect pathway from the striatum to the output nuclei of the basal ganglia, with different neurotransmitters being involved. The direct pathway from the striatum to the internal segment of the globus pallidus (GPi), which also projects to the substantia nigra pars reticulata (SNr), projects to the cortex via the thalamus. Activation of the direct pathway results in disinhibition of thalamic nuclei, which in turn facilitate cortically initiated activity. The indirect pathway passes first to the external segment of the globus pallidus (GPe), then from GPe to the subthalamic nucleus (STN), and then to GPi. The net effect of activity in the indirect pathway is increased inhibition of thalamic targets and consequently reduced thalamic input to cortical areas. Therefore, the direct and indirect pathways within each circuit are considered to have opposing effects on the basal ganglia output nuclei and the thalamic targets of basal ganglia outflow. Consequently, the direct and indirect pathways respectively facilitate or suppress cortically initiated activity. Such a configuration may be ideally suited for the mediation of willed actions as it allows either a magnification and reinforcement or suppression and braking of activity initiated at the cortical level. This model of frontostriatal functioning is supported by direct recordings of neuronal activity (for example, Miller & DeLong, 1987) and PET studies of cortical and subcortical activation in patients with Parkinson's disease (Jahanshahi et al., 1995a; Playford et al., 1992; for review see Brooks, 1995).

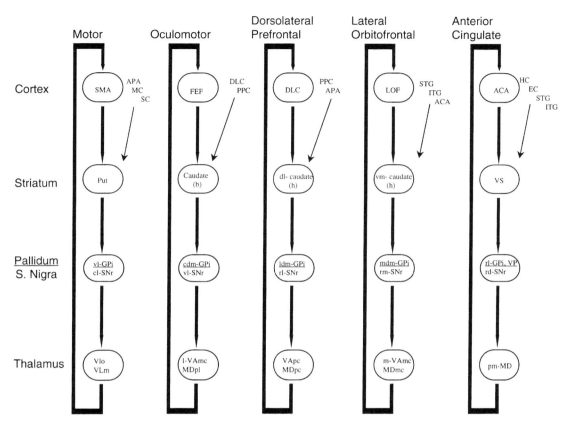

Fig. 1. The frontostriatal network, adapted with permission from Alexander, DeLong, & Strick, (1986). The five circuits showing projections from specific areas of the frontal cortex to discrete areas of the striatum which project back to the originating frontal areas via distinct output sections of the basal ganglia and the thalamus. Abbreviations are as follows: ACA: anterior cingulate; APA: acruate premotor area; caudate, (b) body, (h) head: DLC: dorsolateral prefrontal cortex; EC: entorhinal cortex, FEF: frontal eye fields; GPi: internal segment of the globus pallidus; HC: hippocampal cortex; ITG: inferior temporal gyrus; LOF: lateral orbitofrontal cortex; MC: motor cortex; MDpl: medialis dorsalis pars paralamellaris; MDmc: medialis dorsalis pars magnocellularis; MDpc: medialis dorsalis pars parvocellularis: PPC: posterior parietal cortex; PUT: putamen; SC: somatosensory cortex; SMA: supplementary motor area; SNr: substantia nigra pars reticulata; STG: superior temporal gyrus; VAmc: ventralis anterior pars magnocellularis; VApc: ventralis anterior pars parvocellularis; VLm: ventralis lateralist pars medialis; Vlo: ventralis lateralis pars oralis; VP: ventral pallidum; VS: ventral striatum; cl-: caudolateral; cdm-: caudal dorsomedial; dl-: dorsolateral; l-: lateral; ldm-: lateral dorsomedial; m-: medial; mdm-: medial dorsomedial; pm-: posteromedial; rd-: rostrodorsal; rl-: rostralateral; rm-: rostromedial; vm-: ventromedial; vl-: ventrolateral.

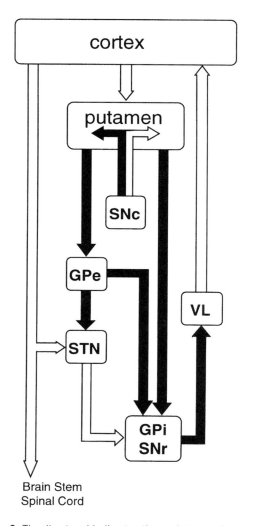

Fig. 2. The direct and indirect pathways between the basal ganglia and the frontal cortex for the motor circuit, adapted with permission from Wichmann & DeLong (1993). The direct pathway from the putamen to internal segment of the globus pallidus (GPi)/substantia nigra pars reticulata (SNr) and the indirect pathway via the external pallidus (GPe) and subthalamic nucleus (STN). White arrows are excitatory connections, black arrows are inhibitory connections. SNc: substantia nigra pars compacta. VL: ventral lateral nucleus of the thalamus.

The Medial and Lateral Motor Systems and Self-generated Actions

Since the mid 1980s, components of the frontostriatal network have been implicated in the control of self-generated actions. Based on experimental evidence from animal lesion studies, Passingham (1985) made a distinction between two premotor systems. The first, centred around the SMA, was considered to be necessary for movements generated in the absence of external stimuli. The second premotor system, involving the lateral premotor cortex, was proposed to be engaged in selecting movements on the basis of external stimuli. A similar distinction between medial and lateral premotor systems was proposed by Goldberg (1985). From an extensive review of the literature, Goldberg (1985, p. 568) presented an anatomico-physiological model of the medial premotor system. "In this model the SMA is considered to be a key element in a medial bilaterally organized system which operates in concert with a variety of other cortical and subcortical structures to perform context dependent selection, linkage, initiation, and anticipatory control of a set of 'precompiled' motor subroutines each of which corresponds to a particular component perceptual-motor strategy or schema of complete action." In contrast, Goldberg (1985, p. 568) noted that "The lateral system is part of a system responsible for recognizing and associating motivational significance with external objects and, in relation to action, operates in a responsive mode in which each action is dependent upon an explicit external input." Goldberg (1985, p. 567) also suggested that "the SMA has an important role to play in the intentional process whereby internal context influences the elaboration of action." This function was proposed to be achieved though the medial premotor system, as part of which the SMA and anterior cingulate formed loops with distinct areas of the basal ganglia via thalamic connections. More recently, on the basis of imaging studies, Passingham (1997) has suggested that, rather than self-generated movements, SMA and the medial motor system is concerned with "open loop" performance, that is in controlling movement sequences that have become overlearned and can be performed without reference to external information. The lateral system is engaged when performance depends upon feedback (closed loop), whether the movements are externally or internally triggered.

THE PHYSIOLOGICAL BASIS OF WILLED ACTION

Our criteria for an action being "willed" depend in large part on reports of being aware of selecting or rejecting possible responses. In consequence, studies of willed action are largely restricted to humans. However, we will mention here a few relevant studies in animals. These studies involve internally generated movements and tasks in which prepotent movements must be suppressed. Single cell recording studies have revealed the existence of cells in the SMA that selectively fire several hundred milliseconds prior to a self-paced movement of the upper limb (Okano & Tanji, 1987). Animal lesion studies have shown that monkeys with lesions of the SMA have difficulty with generating self-paced arm movements to obtain a reward (Thaler, Rolls, & Passingham, 1988), whereas animals with premotor lesions that spare the SMA are not impaired on self-initiated movements (Passingham, Thaler, & Chen, 1989). This evidence suggests that SMA is essential for movements based on an internal "what to do" decision. Monkeys with lesions of the cingulate cortex also make fewer self-initiated movements (Thaler, Chen, Nixon, Stern, & Passingham, 1995). The anti-saccade task requires inhibition of strong response tendencies, the subject being required to suppress responses in the direction of a visual stimulus and instead direct gaze to the opposite direction. In monkeys, performance on the anti-saccade task is associated with neuronal activity in the dorsolateral prefrontal cortex, which is considered to hold this information "on-line" to either suppress or produce a response (Funahashi, Chafee, & Goldman-Rakic, 1993). From single cell recording studies in monkeys, Brotchie, Ianesk, & Horne (1991) suggested that phasic activity in the basal ganglia prior to movement onset provides an internal cue terminating sustained activity in the SMA for the impending movement and signalling the preparatory phase for the next movement in a sequence.

Evidence from Investigations with Healthy Subjects

Studies Using Functional Imaging with Motor Tasks

Random response selection. The tasks used in these studies share the common feature of requiring an internally driven decision for response selection or timing of responses on each trial. Frith et al. (1991) used PET to study a motor task which required the subject to move the first or second finger of the right hand at will in a random sequence paced by touches to the fingers made by the experimenter. This was compared to a control condition, when subjects lifted the finger touched by the experimenter. Compared to the externally determined actions, random finger lifting was associated with greater activation of the dorsolateral prefrontal cortex and the anterior cingulate. Deiber et al. (1991) and Playford et al. (1992) required subjects to make random movements of a joystick in one of four possible directions: forward, backward, left, or right. This task necessitated self-selection of movements or decisions about "what to do" on each trial. Compared to rest, performance of willed random joystick movements was associated with activation of components of the frontostriatal network including the right dorsolateral prefrontal cortex, the anterior cingulate, the SMA, the left lateral premotor cortex, and the left putamen among other areas (Playford et al., 1992). In order to produce a random sequence, a subject must not only select a new response on each trial, but also keep in mind previous responses in order to maintain the randomness of the sequence. Spence, Hirsch, Brooks, & Grasby (in press) attempted to control for this memory component by comparing random joystick movements to the production of a prespecified sequence. Both these conditions have a memory component, but only the random condition involves free selection. The random generation condition was associated with greater activity in left prefrontal cortex and right premotor cortex among other areas. Spence et al. also compared two random generation tasks using different response modalities. In one task responses were four joystick movements, while in the other the responses were speaking four different words. A common area of activation was observed in left dorsolateral prefrontal cortex (BA 9).

Attention to action. The function of practice is to facilitate skilled performance of tasks such that they can be performed effortlessly without the need to specify the component elements volitionally or to attend consciously to their serial integration and execution. Therefore, it is not surprising that practice is associated with changes in the pattern of activation of the components of the willed action system. Seitz, Roland, Bohm, Greitz, and Stone-Elander (1990) used PET to measure rCBF during initial learning, advanced learning, and skilled performance of a finger-to-thumb opposition task. There were rCBF increases in the cerebellum throughout all phases of learning and decreases in the striatum that changed to striatal increased as the motor skill was learned. Inferior frontal areas that were activated in the initial phases of learning were no longer activated when performance had become skilled. However, in this study, as the task became more skilled, the rate and number of movements performed increased, which confounded the changes in rCBF observed with practice. To overcome this rate problem, Jenkins, Brooks, Nixon, Frackowiak, and Passingham (1994) and Jueptner and colleagues (Jueptner, Frith, Brooks, Frackowiak, & Passingham, 1997; Jueptner, Stephan, et al., 1997), kept the rate of performance constant by using an auditory pacing tone. Subjects learned, through trial and error, the order of pressing keys to produce a sequence of eight key presses, with the rate of movement controlled by a pacing stimulus. In one condition subjects were scanned while learning a new sequence, while in a second condition they performed a prelearned sequence. The dorsolateral prefrontal cortex and the anterior cingulate were only activated during new learning. The lateral premotor cortex and the parietal cortex were significantly more activated during new learning, whereas the SMA was more activated during the prelearned sequence (Jenkins et al., 1994). In addition, Jueptner et al. (1997a) demonstrated that the striatum, particularly the caudate nucleus, is differentially active during learning of new sequences compared to performance of familiar or repetitive sequences of button presses. Conversely, Jueptner, Stephan, et al. (1997) showed that when subjects were instructed to attend to what they were doing during performance of the prelearned sequence, there was reactivation of the dorsolateral prefrontal cortex and the anterior cingulate, thus demonstrating that attention to action was partly responsible for the activations in these areas. However, the activations of the dorsolateral prefrontal cortex and anterior cingulate were greater when subjects learned a new sequence than when they attended to performance of a prelearned sequence, thus highlighting the association of these areas with new learning. The results of these studies suggest that with practice, when tasks can be performed more automatically and with less attention to their performance, some of the frontal components of the willed action system are no longer engaged, which liberates them for involvement in other tasks. A similar effect has been shown with a verb generation task. This task is initially associated with activation of the dorsolateral prefrontal cortex and the anterior cingulate, but after practice, activity in these regions is no longer significant (Raichle et al., 1994).

Selecting when to respond. Another aspect of willed actions, decisions about the precise time of initiation of actions, was investigated by Jahanshahi et al. (1995a, b). These studies used PET to measure rCBF changes associated with lifting of the right index finger in three conditions: first, when these movements were self-initiated (i.e. the subject chose precisely when to move) and performed at an average rate of once every 3sec (Jahanshahi et al., 1995a) or at a more variable rate of 2 to 7sec (Jahanshahi et al., 1995b), second, when the same movements were externally triggered by a stimulus at a rate yoked to that of the self-initiated movements, and third, a rest condition during which subjects passively listened to tones presented at the same yoked rates but did nothing. Compared to rest, the regular self-initiated actions, which required the participant to decide the precise timing of movements as well as engage motor preparation, were associated with significant activation of the right dorsolateral prefrontal cortex, the anterior cingulate cortex, the anterior SMA bilaterally, the lateral premotor cortex bilaterally, the left putamen, the left thalamus, the left primary motor cortex, inferior parietal cortex bilaterally, the insular cortex bilaterally, and the right superior temporal cortex (Jahanshahi et al., 1995a). The more variable self-initiated movements in the second study activated the same components of the "motor" and "complex" and "anterior cingulate" frontostriatal circuits, with the exception that the dorsolateral prefrontal activation was bilateral and there was also activation of the right superior parietal cortex (Jahanshahi et al., 1995b). When the externally triggered actions were performed in response to a tone presented at a regular rate, which allowed stimulus anticipation and motor preparation, there was significant activation of the SMA bilaterally, the left lateral premotor cortex, the anterior cingulate, the inferior parietal cortex bilaterally, the left insular cortex, the left putamen, and the left sensorimotor cortex, compared to rest. Thus, the regular rate externally triggered actions, which engage motor preparation but do not involve decision-making about precise initiation time, were associated with activation of components of the "motor" and "anterior cingulate" circuits but not the "complex" circuit. In contrast, compared to rest, the variable rate externally triggered actions, which engaged neither motor preparation nor decision-making about the precise timing of actions, only activated the left sensorimotor cortex, the left putamen, and the insular cortex bilaterally. This suggests that when the volitional requirements of action are low, as in the case of variable rate externally triggered actions, the frontal components of the willed action system are not engaged and performance of these more "reflexive" actions are perhaps mediated through the more direct route between the putamen and the sensorimotor cortex via the thalamus (Holsapple, Preston, & Strick, 1991).

Evidence from a recent imaging study using fMRI has also shown that willed control of the precise timing of movements is associated with activation of the loop between the putamen, ventrolateral thalamus, and the SMA. Rao et al. (1997) used a synchronisation-continuation paradigm for the study of timing of repetitive right index finger tapping with inter-response intervals of 300 or 600msec. In this task, the subject is instructed to synchronise finger tapping with a tone presented at a particular frequency (synchronisation phase) and to continue tapping at the same frequency even after cessation of the external stimulus (continuation phase). While both the synchronisation and continuation tasks were associated with significant activation foci in the left sensorimotor cortex, the right cerebellum and the right superior temporal gyrus (BA 22), the continuation phase also activated the left putamen, the left ventrolateral thalamus, and the SMA as well as the right inferior frontal gyrus (BA 44). These results suggest that the motor circuit involving the SMA, the putamen, and the ventrolateral thalamus is associated with precise timing of willed actions.

Advance preparation. The willed nature of motor preparation and its association with activation of the SMA, lateral premotor cortex, and anterior cingulate found by Jahanshahi et al. (1995a) has also been reported by Deiber, Ibanez, Sadato, & Hallett (1996). These investigators compared blood flow during fully cued, partially cued, and uncued choice RT tasks, in which a precue provided full, partial, or no information about the finger (index or little) to be moved or the direction of movement (abduction or elevation) in advance or a second go signal, with a rest condition involving similar visual stimuli but no responses. They found motor preparation to be associated with activation of the SMA, lateral premotor cortex, and the anterior cingulate in addition to the sensorimotor cortex, the parietal association cortex, the striatum, thalamus, and the cerebellum. Another finding of this study is also of interest in relation to willed actions. A "free" movement condition was also included, which required subjects to decide which finger to move and the direction of movement at will on every trial in order to produce a random sequence of the four possible movements. Compared to the fully cued condition (externally/triggered), the free movement condition (self-generated actions) was associated with significantly greater activation of the dorsolateral prefrontal cortex bilaterally (BA 9), the anterior cingulate (BA 32), the anterior section of the SMA, the left prefrontal area 10, the premotor area 8 bilaterally, and the left parietal association cortex (BA 40). This suggests that internally drive decisions about which response to make are associated with activation of the dorsolateral prefrontal in addition to the anterior cingulate and SMA which were activated with motor preparation (see also Passingham, 1997).

Novel S–R pairings. Deiber et al. (1991) studied the effects of novel S–R pairings using a variant of the paradigm in which a joystick must be moved in four different directions. In the "fixed" condition subjects moved the joystick repeatedly in the same direction. In the "conditional" condition they performed a well practised four-choice RT task with overlearned S–R relationships. In the "opposite" condition they had to use novel S–R pairings to perform the four-choice RT task. The left superior parietal cortex was significantly activated with both the "conditional" and "opposite" conditions relative to the control task. The "opposite" condition with novel S–R pairings also significantly activated the right dorsolateral prefrontal cortex and the medial superior parietal cortex.

Tasks with overlearned or novel S–R paring were directly compared in a study by Paus, Petrides, Evans, and Meyer (1993). They studied a series of tasks using different response modalities (speech, finger movements, eye movements) in which subjects performed two choice RT tasks with well practised or novel S–R relationships. Across all response modalities, activity of the anterior cingulate cortex and, to a lesser extent, DLPFC, was observed when the S–R pairings were novel. The task-specific activations were obtained in specific subregions of the anterior cingulate depending on the output modality. The rostral and caudal regions of the anterior cingulate cortex were respectively activated by the oculomotor and manual tasks, whereas the task involving verbal output activated the intermediate dorsal and rostral areas of this structure.

Response suppression. In a "go no-go" RT task, subjects have to prepare a response that is activated and executed when a particular stimulus is presented (go trials) but suppressed when a different stimulus occurs (no-go trials). Both activation and suppression of a response require trial-by-trial decision making and willed control. Using functional magnetic resonance imaging (fMRI), Humberstone et al. (1997) found that the anterior part of the SMA was activated prior to both go and no-go trials, whereas the caudal SMA was activated only in association with motor execution on go trials. This result supports the findings of Ikeda et al. (1996), who used subdural recordings of MRCPs and found negativity, considered to reflect SMA activation, prior to both go and no-go trials.

Conclusions. Although the precise nature of the tasks differ across the functional imaging studies reviewed here, the common feature is that the tasks involved willed actions. Compared to externally triggered or stereotyped actions, a number of cortical (dorsolateral prefrontal cortex, anterior cingulate cortex, SMA) and subcortical (thalamus, basal ganglia) areas are activated with self-generated actions involving nonroutine decision making, regardless of whether the nature of the decision making is related to "what to do" (selection of one of a set of possible responses) (Deiber et al., 1991, 1996; Frith, Friston, Liddle, & Frackowiak, 1991; Jenkins et al., 1994; Jueptner, Frith, et al., 1997; Playford et al., 1992; Jueptner, Stephan, et al., 1997) or "when to do" (precise time of initiation of response) (Jahanshahi et al., 1995a, b; Rao et al., 1997), or whether or not to act (activate a response on go trials and suppress it on no-go trials) (Humberstone et al., 1997). The similarity in patterns of brain activation across these studies supports the concept of a willed action system, with the dorsolateral prefrontal cortex, anterior cingulate, and the SMA as its cortical components receiving subcortical input from the striatum via the thalamus.

Studies Using Functional Imaging with Cognitive Tasks

So far we have discussed willed actions reflected in motor responses. However, more abstract entities such as words and ideas can also be selected by the willed action system. Petersen, Fox, Posner, Nintus, and Riachle (1988) studied a word generation task that required subjects to respond with a use for the word presented (e.g. stimulus word "cake"—response word "eat"). In this task, the response is not completely specified by the stimulus word and requires the willed action of searching for an appropriate response word on each trial. Compared to a control word reading task, the word generation task was associated with activation of the dorsolateral prefrontal cortex and the anterior cingulate. The study by Frith et al. (1991), described earlier, also included a "willed" word generation task, paced first letter verbal fluency. This test requires subjects to generate words beginning with a particular letter of the alphabet. It involves a self-directed search of word associative networks to retrieve appropriate responses, as well as suppression of inappropriate responses and repetition of already retrieved exemplars. This was compared to a control task where the subject repeated the words uttered by the experimenter. As in the willed motor task, verbal fluency was associated with significant activation of the left dorsolateral prefrontal cortex and the anterior cingulate. Compared to the control task, verbal fluency was also associated with inhibition of the left superior temporal gyrus, suggesting that willed generation of words may require low activity in this area where the word associative network is presumably distributed (Friston, Frith, Liddle, & Frackowiak, 1991).

We (Jahanshahi, Dirnberger, Fuller, & Frith, 1997) have found a similar pattern of frontotemporal connectivity with another willed action task, random number generation (RNG). Performance of RNG requires selection and maintenance of strategies, holding information "on line", suppression of habitual counting, internally driven response generation, and monitoring of responses. Thus RNG involves willed actions as well as high-level executive functions. Compared to counting, RNG is associated with significant activation of the left dorsolateral prefrontal cortex (BA 9), the anterior cingulate (BA 32) and right lateral premotor cortex (BA 6), the precuneus and the superior parietal cortex (BA 7) bilaterally, the right inferior frontal cortex (BA 47), and the right and left cerebellar hemispheres. RNG is also associated with widespread deactivations in the right and left temporal cortex relative to counting. Thus, the component processes of RNG that necessitate willed actions were associated with activation of the frontostriatal network. Relative to counting, Petrides, Alivisatos, Meyer, and Evans (1993) also found activations of the dorsolateral prefrontal cortex, anterior cingulate, the premotor cortex, and the superior and inferior parietal cortex when subjects performed a task requiring generation of self-ordered random sequences of 10 numbers or when holding in working memory a random sequence of 9 experimenter-generated numbers to identify the missing number. Both the self-ordered and externally ordered tasks involve willed action, as on each trial the required response is not specified and necessitates internally driven action.

Among many other components such as planning and working memory, complex cognitive tasks such as the Wisconsin Card Sorting test and the Tower of London task inevitably involve willed action. For example, inappropriate strategies must be suppressed and new ones initiated. Performance of such tasks is associated with activation in the willed action system, especially the dorsolateral prefrontal cortex (e.g. Baker et al., 1996; Weinberger, Berma, & Zinc, 1986). We shall not discuss such tasks in detail here because of the difficulty of isolating components specifically associated with willed action.

Studies with Transcranial Magnetic Stimulation

Transcranial magnetic stimulation (TMS) is a technique that allows temporary disruption of processing during fairly focal stimulation of specific brain regions; an effect which is equivalent to a short-lasting temporary "lesion". As a result TMS can reveal areas that are essential for performance of a task. Support for the role of the dorsolateral prefrontal cortex and SMA in component processes of willed actions such as response selection, response preparation, suppression of habitual responses, and sequencing and timing of responses comes from studies using this technique.

To date, the effects of TMS over the prefrontal cortex have been examined only in a handful of studies, and we will consider those relevant to willed performance of actions. Ammon and Gandevia (1990) reported that in nine right-handed normal subjects, TMS over Fz, which is over the prefrontal cortex, resulted in a bias in hand selection. The direction of current flow determined the hand selected: the left and right hands being preferentially selected respectively with anticlockwise and clockwise current flow in the stimulating coil. This bias was specific to stimulation over the prefrontal cortex as it was not produced by stimulation over the occipital cortex.

Willed action is one of the component processes of the random number generation task (RNG) described above. In a recent study with 11 normal subjects, Jahanshahi et al. (1998) found that rapid rate TMS over the left dorsolateral prefrontal cortex resulted in a reversal of the subjects' usual counting bias: Increasing the most habitual response of "counting in ones" and decreasing the less stereotyped "counting in twos". This alteration of the counting bias was not obtained with TMS over the right dorsolateral prefrontal cortex or over the medial frontal cortex. This suggests that the left dorsolateral prefrontal cortex plays a role in suppression of habitual counting, a process that is likely to require considerable allocation of attention and volitional control. Indeed, use of dual-task methodology has shown that in normal subjects concurrent performance of a visual tracking task with RNG results in reversal of the usual counting bias and increases habitual counting in ones compared to when RNG is performed alone (Brown, Soliveri, & Jahanshahi, 1998).

Electrical stimulation of the SMA during surgery for epilepsy results in inhibition of ongoing voluntary activity (Penfield & Welch, 1949). Fried et al. (1991) placed subdural electrode grids on the SMA in 13 patients with epilepsy undergoing investigation for surgery. In some cases electrical stimulation of the SMA evoked a sensation or urge to perform a movement or anticipation that a movement was about to be performed. Electrical stimulation of the cingulate cortex has been reported to give rise to stimulus-driven behaviour. For example, on presentation of a banana the patient started eating it, a behaviour that was abandoned if electrical stimulation was interrupted (Damasio & Van Hoesen, 1983). Stimulation of these deeper medial structures with TMS is not as easily achieved as stimulation of the lateral prefrontal areas, which lie closer to the surface and are more accessible. Nevertheless, a number of studies have reported disruptive effects of TMS over the SMA on finger movements and saccadic eye movements.

TMS over the SMA has been shown to interfere with the performance of periodic sequential finger movements. Amassian, Cracco, Maccabee, Bigland-Ritchie, and Cracco (1991) required subjects to perform self-timed sequential finger movements in a periodic and rhythmic manner. TMS over the SMA increased the interval between sequential movements of the first to the fifth fingers, particularly prolonging the last interval, and also resulted in repetition/perseveration of a previously made movement. In a more recent study, Gerloff, Corwell, Chen, Hallett, and Cohen (1997) used high-frequency repetitive TMS to investigate the effects of stimulation over the SMA on the performance of overlearned sequential finger movements paced by a metronome. There were three conditions, "simple" (16 repeated presses of the index finger); "scale" (4 repetitions of sequential presses of the little, ring, middle, and index fingers); and "complex" (sequence of 16 presses involving use of the 4 fingers in a nonconsecutive nonrepetitive order). TMS over the SMA induced errors only for the "complex sequences", and interfered with the organisation of elements in the sequence. TMS over the motor cortex increased errors for both the "scale" and "complex" sequences, while stimulation at prefrontal or parietal sites produced no effects. The errors induced by TMS over the SMA occurred about 1 second later than with TMS over the motor cortex. These results support the role of the SMA in control of the sequential order of willed actions.

TMS over the SMA or the prefrontal cortex have also been reported to disrupt nonreflexive saccadic eye movements. Memory-guided saccades require production of a nonreflexive saccade to the remembered position of a target. TMS over the SMA disrupted the accuracy of memory-guided saccades, when stimulation was given during the target presentation (learning phase) but not when it was delivered during the memorisation or execution phase of such saccades (Muri, Rivaud, Vermersch, Leger, & Pierrot-Deseilligny, 1995). TMS over the occipital cortex produced no such effects. This suggests that the SMA plays an essential role in the learning phase of such memory-guided saccades. Ro, Henik, Machado, and Rafal (1997) found that TMS over the superior prefrontal cortex increased the latency of volitional saccades made to a central arrowhead (endogenous go signal) that indicated the location of the required response in the right or left visual field. TMS over the superior prefrontal cortex had no effect on the saccades triggered by a peripheral asterisk (exogenous go signal) that marked the hemifield where a response was required. TMS over the parietal cortex had no effect on either the volitional or triggered saccades. Therefore, the critical role of the SMA and prefrontal cortex in controlling memory-guided and endogenous saccades that require willed actions is confirmed.

Studies Recording Movement-related Cortical Potential

Another strand of evidence concerning components of the willed action system comes from the study of movement-related cortical potentials (MRCPs). The bereitschaftspotential (BP), first described by Kornhuber and Deecke (1965), is a negative cortical potential that is associated with the preparation and initiation of a movement. This MRCP develops from 1 to 1.5sec prior to self-initiated movement. An early, a late, and a peak component are usually distinguished. The exact generators of these BP components are not known. However, as the early part of the BP is bilaterally symmetrical and is maximal at Cz and FCz, which are over the SMA, the latter structure has been considered as a major source of the early component. In contrast, "late" and "peak" BP become maximal over the hemisphere contralateral to the responding hand suggesting that the primary motor cortex is the source of these BP components (Deecke, Scheid, & Kornhuber, 1969; Shibasaki, Barrett, Halliday, & Halliday, 1980). Animal studies using microstimulation and unit recording techniques with subdural or depth electrodes (Arezzo & Vaughan, 1975; Hashimoto, Gembra, & Sasaki, 1979; Sasaki & Gemba, 1981; Tanji & Kurata, 1982, 1985), and studies that recorded the BP with intracranial electrodes in patients with epilepsy (Ikeda, Luders, Burgess, & Shibasaki, 1992; Neshige, Luders, & Shibasaki, 1988; Rektor, Feve, Buser, Bathien, & Lamarche, 1994) agree that both the SMA and primary motor cortex are active prior to movement and contribute to the BP. Dipole source analysis has also suggested that the SMA contributes to the BP (McKinnon et al., 1996; Praamstra, Stegeman, Horstink, & Cools, 1996). Other evidence of the contribution of the SMA to the BP is provided from studies that have recorded MRCPs and measured changes in regional cerebral blood flow using the same paradigms in different sessions (Jahanshahi et al., 1995a) or simultaneously in the same session (Mackinnon et al., 1996). The weight of existing evidence supports the idea that, as well as the motor cortex, the SMA also contributes to the generation of the BP.

Timing. When making self-paced responses the subject has to prepare to move and make a decision about the precise time of initiation of the response. When manual movements are self-paced, both the early and late components of the BP can be discerned. When movements are made in response to an external stimulus that occurs relatively regularly and predictably, anticipation of the stimulus and motor preparation are possible, and this is reflected by the presence of premovement negativity, although there is no late accentuation of this negativity. In contrast to the MRCP obtained prior to predictably triggered movements, when the triggering stimulus occurs more irregularly, there is no build-up of negativity prior to the movement (Aminoff, & Goodin, 1993; Jahanshahi et al., 1995a; Kutas & Donchin, 1980; Libet, Wright, & Gleason, 1982; Papa, Artieda, & Obeso, 1991; Thickbroom, & Mastaglia, 1985). MRCPs prior to self-initiated and triggered saccades have also been investigated and have demonstrated a similar context-sensitivity to the internal vs. external generation of movements and relative timing of the triggering stimuli (Moster & Goldberg, 1990; Klosterman et al., 1994; Kurtzberg & Vaughan, 1982; Shimizu & Okiyama, 1993; Thickbroom & Mastalgia, 1985).

Response selection. In the above investigations, the tasks differed in terms of whether the precise time of initiation of the action was determined by the subject (selfpaced/self-initiated) or by external stimuli (triggered conditions). In a number of more recent studies, the effect of "what to do" decisions on MRCPs have been assessed. MRCPs were recorded in conditions where subjects decided which one of a set of possible key presses or joystick movements to make on each trial or they performed a stereotyped and predetermined movement repeatedly across trials. The amplitude of MRCPs were significantly higher prior to the self-selected movements (Dirnberger, Fickel, Lindinger, Lang, & Jahanshahi, 1998; Praamstra, Stegeman, Horstink, Brunia, & Cools, 1995; Touge, Werhahn, Rothwell, & Marsden, 1995).

Response suppression. The results of these studies, demonstrating the context-sensitivity of MRCPs recorded prior to manual or saccadic eye movements, support Libet et al.'s (1982) suggestion that two volitional processes contribute to the BP; the first process is associated with the development of preparation to act in the near future, while the second process is associated with voluntary choice and with the endogenous "urge" or intention to act. The effect of deciding whether or not to activate or suppress a prepared response has been examined by Ikeda et al. (1996). They used subdural electrodes to record MRCPs in a go no-go task with an S1-S2 paradigm in five patients with epilepsy. They reported a slow potential prior to S2 in orbitofrontal and mesial frontal areas, which was considered to reflect anticipatory and preparatory processes. In addition, a bilateral transient potential was recorded over the mesial frontal areas on presentation of S2 for both go and no-go trials, which was related to the decision-making process of whether to initiate (go trials) or suppress the response (no-go trials). As noted earlier, the study by Humberstone et al. (1997) with fMRI has confirmed that willed decision-making, which is required for both response activation and suppression, is associated with activation of the anterior section of the SMA.

Conclusions. This evidence suggests that certain components of MRCPs, which are considered to reflect the activation of the SMA, are affected by internal vs. external context, that is whether decisions about the nature or precise timing of movements are taken by the subject or triggered by external stimuli. Decisions about whether or not to act affect the amplitude of these MRCP components. Study of these components can also provide important information about the timing as well as the location of events in the willed action system.

Impairment of Willed Actions in Patients with Frontal Damage

Impairments of willed action are frequently seen after damage to the various anatomical components of the system and especially in patients with prefrontal lesions. Cummings (1993) has suggested that specific behavioural syndromes are associated with dysfunction of particular frontostriatal circuits. He proposed that injury to the anterior cingulate circuit gives rise to problems in response initiation (apathy), damage to the dorsolateral prefrontal circuit results in deficits in executive function, and lesions of the orbitofrontal circuit give rise to problems in response suppression (disinhibition). Damage to prefrontal cortex is associated with perseverance of well-learned responses, increased distractibility, inability to plan and maintain a sequence of goals and actions, combined with stimulus-driven, behaviour. This pattern would be expected from impairment of an executive system mediating flexible, self-generated planning and maintenance and shifting of action (Norman & Shallice, 1986). Utilisation behaviour is perhaps the best example of failure to inhibit prepotent, but inappropriate responses seen in patients with frontal lesions and is clearly a failure in the willed action system. Patients will make stereotyped responses to objects placed in front of them such as combing the hair if a comb is placed in the hand or putting on three pairs of glasses one on top of another (Lhermitte, 1983; Shallice, Burgess, Schon, & Baxter, 1989). According to Lhermitte, the parietal cortex creates a link between external stimuli and behaviour, which is normally kept in check by the frontal cortex. With damage to the frontal cortex, the parietal cortex is released from this inhibitory influence and the patient's behaviour becomes driven by external stimuli. The existence of utilisation behaviour suggests that patients with prefrontal lesions have problems suppressing prepotent tendencies such as behaviours evoked by environmental stimuli. According to Frith (1992), impairment of willed action after prefrontal lesions is a result of an imbalance between stimulus intentions and willed intentions. Thus these patients are unable to formulate a plan of action and their behaviour is largely guided by stimulus intentions even when these are inappropriate.

A "pseudodepressive" syndrome can be associated with damage to the dorsolateral prefrontal cortex, with passivity, flattening of affect, reduction of verbal output, and slowness to respond as it features (Blumer & Benson, 1975). In humans, damage to the SMA is associated with a decrease of spontaneous movements of the limbs and partial mutism (Damasio & Van Hoesen, 1980; Laplane, Talairach, Meininger, Bancaud, & Orgogozo, 1977).

Studies of Cognitive Function

Patients with lesions of the frontal cortex are impaired on performance of tests of verbal and design fluency (Jones-Gotman & Milner, 1977; Perret, 1974) that involve internally generated response selection coupled with suppression of inappropriate responses.

Patients with frontal lesions also perform poorly on random number generation and show higher habitual counting than normals, because they persist with an inappropriate production strategy (Spatt & Goldenberg, 1993). They are also impaired in the generation of self-ordered random sequences of numbers between 1 and 10, which requires internal generation and subjective ordering of responses (Wiegersma, Van der Scheer, & Hijman, 1990). In contrast, these patients are not impaired on a test of working memory, the missing-scan test, which requires identification of the digit missing from a sequence of numbers generated by the experimenter and does not involve internal generation or active ordering of items (Wiegersma et al., 1990).

Movement Initiation and Execution

The willed process of motor preparation appears to be impaired in patients with prefrontal lesions, particularly those with lesions in the right hemisphere. These patients do not use advance information to speed up RTs with precued CRT compared to uncued CRT tasks (Alivisatos & Milner, 1989; Verfaellie & Heilman, 1987). Compared to the RT performance of normals and patients with lesions of the temporal cortex, the error rates of patients with frontal lesions were preferentially affected by S–R compatibility, associative complexity, and number of stimuli per response (Decary & Richer, 1995), suggesting deficits in the process of response selection following damage to frontal cortex.

Patients with damage to the medial frontal cortex are also impaired on go no-go RT tasks and make more errors of commission, which indicate a failure to inhibit the response on the no-go trials (Drewe, 1975). Guitton, Buchtel, and Douglas (1985) found that on the anti-saccade task, patients with frontal lesions, particularly those with lesions in the dorsolateral prefrontal cortex, were unable to suppress reflexive glances and to generate goal-directed saccades in the direction opposite to that of the target presented.

Patients with frontal lesions are impaired on tasks requiring coordinating hand and arm movements, particularly when different movements have to be performed simultaneously. Following SMA lesions, the only long-term deficits are in alternating movements of the two hands (Laplane et al., 1977). In a patient with an infarction of the right SMA, Dick, Benecke, Rothwell, Day, and Marsden (1986) found performance of complex simultaneous and sequential movements to be impaired while simple movements were relatively intact. Leonard, Milner, and Jones (1988) compared simple unimanual tapping, spatially ordered unimanual tapping, and bimanual tapping of out-of-phase movements in patients with frontal or temporal lesions and normal controls. While spatially ordered unimanual tapping was impaired for all patients, only those with frontal lesions performed poorly on bimanual tapping involving out-of-phase movements. Halsband, Ito, Tanji, and Freund (1993) reported that patients with lesions of the lateral or medial premotor cortex had a severe impairment in reproduction of rhythmic tapping when required to use both hands in an alternating manner. In addition, patients with left SMA lesions had difficulty in reproducing rhythmic tapping at particular frequencies from memory but were unable to produce them with auditory pacing. This means that the patients could not generate the sequence using internal cues.

Studies of Movement-related Cortical Potentials

As described in previous sections, evidence from a variety of sources has shown that the SMA as well as the primary motor cortex is active prior to self-initiated movements. Therefore, it is not surprising that a reduced amplitude of BP has been found in patients with lesions of the SMA (Deecke, Lang, Heller, Hufnagl, & Kornhuber, 1987; Singh & Knight, 1990). Using magnetoencephalography and EEG recording in a patient with a lesion of the right SMA, Lang et al. (1991) have provided evidence for the contribution of the intact SMA to the "early" and the motor cortex to the "late" BP. A lower amplitude of the BP has also been found in patients with lesions of the dorsolateral prefrontal cortex (Singh & Knight, 1990).

The contingent negative variation (CNV) is a brain potential that develops in the interval between warning and go stimuli, and is considered to reflect processes of stimulus anticipation and motor preparation (Walter, Copper, Aldridge, McCallum, & Winter, 1964). Lesions of the prefrontal cortex reduce the amplitude of the late component of the CNV from about 1000msec prior to the go stimulus (Rosahl & Knight, 1995).

Impairment of Willed Actions in Parkinson's Disease

Parkinson's disease is a movement disorder which presents with a range of motor as well as cognitive deficits. The appearance of the symptoms relate to degeneration of dopamine-producing neurons in the substantia nigra, resulting in dopaminergic deficiency in the striatum, and particularly the putamen (Brooks et al., 1990). Bradykinesia (slowness of movement) and akinesia (poverty of spontaneous movements), together with tremor and rigidity, are the cardinal signs of the disorder. Kinnier Wilson (1925) referred to akinesia as "paralysis of the will" and believed that the akinetic patient had a defect at the "highest level" of motor control, which failed to transmit the "normal impulses to movement, both of the 'voluntary' and the 'spontaneous' kind". The coexistence of normal and bradykinetic performance on the two sides of the body in hemi-parkinsonism, together with the reports of the patients, suggest that the intention to act is unimpaired in these patients and that the deficit is in the translation of intentions of will into action. Patients with this disorder know what they want to do, but cannot do it (Frith, 1992, see Fig. 3). Similarly, the phenomenon of "paradoxical kinesis", whereby an ordinarily immobile patient can walk or run normally in situations of risk such as in a fire, suggests that the mechanical machinery for performance of actions is in no way impaired, but rather the

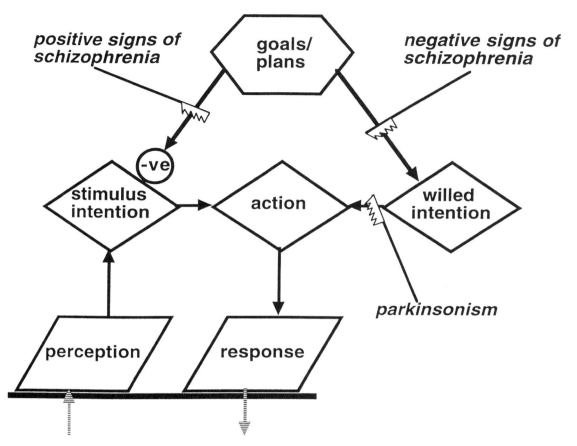

Fig. 3. Two routes to action, a stimulus-driven route and a willed action route. Stimulus-drive route: perception → stimulus intention → action → response; Willed action route: goals/plans → willed intention → action → response. Three disconnections are shown. First goals fail to generate willed intentions, which gives rise to negative signs in schizophrenia. Second, goals fail to inhibit stimulus-driven actions, which gives rise to the positive signs of schizophrenia. Third, willed intentions fail to generate actions, which results in akinesia in parkinsonism.

problem in everyday circumstances is one of "willing" that machinery into action. A further clinical observation that goes back to Purdue Martin (1967) is that provision of external guidance can improve mobility in Parkinson's disease. For example, the shuffling gait of patients with Parkinson's disease can be improved by provision of lines on the floor over which the patients step. Taken together, these observations are consistent with the proposal that bradykinesia and akinesia represent impairment of willed actions in Parkinson's disease, which are to some extent overcome when guidance by external stimuli are provided.

This proposal is supported by empirical evidence from a range of investigations, which we will consider next.

Studies Using Functional Imaging

Current opinion (Alexander, Crutcher, & DeLong, 1990) considers akinesia and bradykinesia in Parkinson's disease to result from overactivity of the GPi, which results in excessive inhibition of the thalamus and the cortical targets of thalamic projections (Fig. 4b). Reduction of dopamine projections in the nigroputaminal pathway increases tonic and phasic GPi activity, mainly due to overactivity in the "indirect" pathway via the input from GPe and STN to GPi. The subsequent excessive inhibition of the thalamocortical circuits gives rise to akinesia and bradykinesia. In support of this model, the results of a number of PET activation studies have shown that in Parkinson's disease performance of willed actions is associated with underactivation of the SMA, the dorsolateral prefrontal cortex, the anterior cingulate, and the putamen. These are components of the "motor", "complex", and "anterior cingulate" circuits between the frontal cortex and the basal ganglia. Playford et al. (1992) used PET to study performance of the random joystick movements in Parkinson's disease. Compared to a reset condition, age-matched normal participants showed significant activation of the right dorsolateral prefrontal cortex, the anterior cingulate, the SMA, the left lateral premotor cortex, the left primary sensorimotor cortex, the left putamen, and the parietal association areas bilaterally. For the patients with Parkinson's disease who were tested off medication, compared to rest, the random joystick movements were associated with underactivation of the left putamen, the anterior cingulate, the SMA, and the dorsolateral prefrontal cortex. The study by Jahanshahi et al. (1995a) found a similar pattern of cortical and subcortical underactivation during willed self-initiated actions and in addition demonstrated that patients with Parkinson's disease did not differ from age-matched normals in terms of rCBF change during externally triggered actions.

Two recent functional imaging studies have suggested that in addition to underactivation of the cortical targets of the striatal-frontal circuits, in Parkinson's disease there may be overactivation of the parietal-lateral premotor-cerebellar circuits during performance of willed sequential finger movements. Using 133 Xe-SPECT, Rascol et al. (1997) found that when tested off medication patients with Parkinson's disease showed significant underactivation of the SMA together with significant overactivation of the ipsilateral cerebellar hemisphere during a finger-to-thumb opposition task. Samuel, Ceballos-Baumann, Blin, et al. (1997a) used PET to measure rCBF change during performance of unimanual or bimanual sequential tapping of index to little fingers. They found that relative to matched healthy controls, for Parkinson's disease patients tested off medication, in addition to underactivation of the SMA and the prefrontal cortex, the parietal cortex and the lateral PMC were overactive. The results of these two studies suggest that when tested off medication, patients with Parkinson's disease compensate for the dysfunction in their basal ganglia-fron-

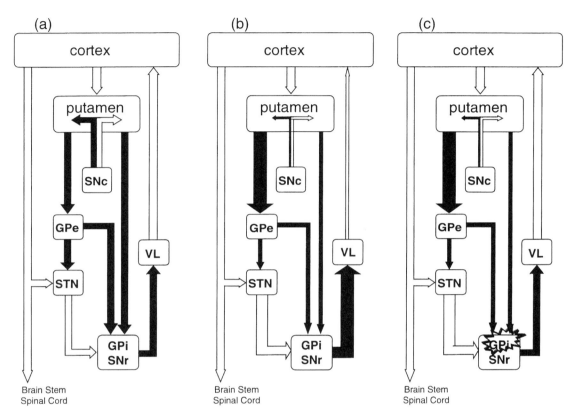

Fig. 4. The current model of akinesia and bradykinesia in Parkinson's disease and the effects of surgical intervention, adapted with permission from Wichmann & DeLong (1993). Compared to normal (a), in Parkinson's disease (b), degeneration of the dopamine projection from the substantia nigra pars compacta (SNc) leads to underactivity of the inhibitory direct input to GPi and overactivity of the excitatory indirect input to GPi. The result is excessive activity of output neurons of the GPi and a reduction of the final excitatory output from ventrolateral (VL) thalamus to cortex. White arrows are excitatory connections, black arrows are inhibitory connections. In Parkinson's disease, changes in the thickness of the arrows relative to normal represents increased or decreased activity.

The effects of surgical intervention on the internal segment of the globus pallidus (GPi) as in posteroventral pallidotomy and deep brain stimulation of the GPi or subthalamic nucleus (STN) are also indicated (c). Abbreviations: GPi: internal segment of the globus pallidus, GPe: external segment of the globus pallidus, SNr: substantia nigra pars reticulata.

tal cortical loops by reliance on alternative pathways such as the cerebellar-parietal-lateral premotor circuits, which have been implicated in the control of actions in the presence of external cues (Goldberg, 1985; Passingham, 1985).

Further evidence suggests that the patterns of mesial frontal and prefrontal under-activation and ipsilateral cerebellar overactivation found in Parkinson's disease when these patients are assessed off medication are to a large extent ameliorated by dopaminergic medication. Using PET, Jenkins et al. (1992) reported that injections of apomorphine resulted in increased rCBF in the SMA and the dorsolateral prefrontal cortex in patients with Parkinson's disease during performance of joystick movements. In the study by Rascol et al. (1997) mentioned earlier, a second group of Parkinson's disease patients tested on dopaminergic medication showed significant activation of the SMA and the contralateral sensorimotor cortex similar to healthy controls. These findings, together with evidence from other sources that will be considered later, suggest that dopamine plays a role in modulating the activity of the frontostriatal circuits during willed actions.

Studies of Cognitive Function

Patients with Parkinson's disease tend to perform poorly on tasks requiring high-level executive function such as the Wisconsin Card Sorting test (Taylor, Saint-Cyr, & Lang, 1986) and the Tower of London task (Owen et al., 1993), on which patients with frontal damage also perform poorly (Milner, 1964; Owen, Downes, Sahakian, Polkey, & Robbins, 1990). Patients with Parkinson's disease are also bad at tasks more closely linked to willed action, such as verbal fluency (Taylor et al., 1986) and random number generation (Spatt & Goldenberg, 1993).

Direct evidence for the value of the internal/external distinction in the problems of patients with Parkinson's disease was provided by Brown and Marsden (1988). They used a computerised version of the Stroop task, for example with the word RED printed in green ink, requiring subjects to either name the colour of ink (INK) or to read the words (WORD). In alternating blocks, subjects had to switch between INK and WORD as the relevant stimulus dimension. In the uncued condition, stimuli were preceded by a warning signal (READY) and subjects had to rely on internal information to focus attention on the currently relevant stimulus attribute, whereas in the cued condition each stimulus was preceded by an explicit external cue, INK or WORD. While the controls showed no difference between the cued and uncued conditions, the patients with Parkinson's disease were impaired on the uncued but not the cued condition, and the impairment on the uncued internally driven condition was greatest for the trials where the subject had to switch between the two attributes.

Studies of Movement Initiation and Execution

In Parkinson's disease, slowness of movement initiation and execution is present for all actions, routine and nonroutine. Routine actions such as walking, which are normally performed in an automatic fashion, are slowed. However, more complex nonroutine actions (for example, getting up from a chair and simultaneously shaking hands) are particularly affected. There is experimental evidence from a number of paradigms supporting the idea that willed actions are more impaired than routine actions.

Timing and sequencing movements. In Parkinson's disease, simultaneous and particularly sequential movements that require greater involvement of the SMA are more impaired than simple movements (Agostino, Berardelli, Fromica, Accornero, & Manfredi, 1992; Benecke, Rothwell, Dick, Day, & Marsden, 1986, 1987a; Harrington & Haaland, 1991; Roy, Saint-Cyr, Taylor, & Lang, 1993; Weiss, Stelmach, & Hefter, 1997). With sequential movements, patients with Parkinson's disease take longer to perform each component movement and show an increased pause between the first and second movement (Benecke et al., 1987a) or marked hesitation between movement segments (Weiss et al., 1997), suggesting that they have difficulty in switching between motor programmes. Levodopa medication results in differentially greater improvement of movement times of complex sequential than simple movements (Benecke, Rothwell, Dick, Day, & Marsden, 1987b).

Concurrent performance of two motor tasks requires greater volitional control than when each task is carried out on its own. Patients with Parkinson's disease have particular difficulty in carrying out bimanual tasks, especially when the tasks are different for each hand (Benecke et al., 1986; Brown, Jahanshahi, & Marsden, 1993a; Schwab, Chafter, & Walker, 1954) or employ different time elements (Stelmach & Worringham, 1988). Performance of a normally skilled action such as doing up buttons is slowed, particularly when performed concurrently with foot tapping, with the slowing under dual-task conditions being differentially greater for the patients than for normals (Soliveri, Brown, Jahanshahi, & Marsden, 1992).

Response preparation. The results of RT studies have also indicated deficits in Parkinson's disease in willed action such as the optional preprogramming in simple RT which confers its speed advantage over choice RT (Bloxham, Mindel, & Frith, 1984; Goodrich et al., 1989; Sheridan, Flowers, & Hurrell, 1987), or decoding S–R relations in choice RT tasks with low S–R compatibility and high uncertainty (Brown, Jahanshahi, & Marsden, 1993b). Patients with Parkinson's disease require a longer interval between a precue and a go signal to use advance information for motor preparation (Jahanshahi, Brown & Marsden, 1992), and these patients show additional slowness with increased choice complexity in choice RT tasks (Cooper, Sagar, Tidswell, & Jordan, 1994). Goodrich et al. (1989) have also shown that performance of an attention-demanding secondary task such as oral reading significantly prolongs SRTs of normal subjects by preventing volitional preprogramming but has little effect on SRTs of patients with Parkinson's disease. The effects of a secondary task on CRT are comparable for patients and normals. These results suggest that under standard conditions, the SRT slowing found in Parkinson's disease is due to impairment of the attention-demanding process of volitional preprogramming. Filipovic et al. (1997) found that the difference between choice and simple RT was significantly and negatively associated with the amplitude of the early and peak BP components in Parkinson's disease. This confirms that the failure to preprogramme in simple RT is associated with less motor preparation as indexed by lower amplitude of the early BP component.

Dependence on external cues. Patients with Parkinson's disease are more dependent on visual cues for the control of movements (Cooke, Brown, & Brooks, 1978) and show greater slowness in the initiation and execution of internally generated sequential movements performed in conditions where external cues are reduced or absent (Georgiou et al., 1993, 1994). Instead of fast, preprogrammed ballistic or "open-loop" movements, these patients executive movements in a "closed-loop" fashion relying on visual guidance (Flowers, 1976). Visual guidance improves motor performance in Parkinson's disease (Flowers, 1976; Georgiou et al., 1994). Memory-guided goal-directed prehension movements that require reliance on internal representation of the target have been shown to be significantly impaired in this disorder (Jackson, Jackson, Harrison, Henderson, & Kennard, 1995).

Parkinson's disease patients also have deficits in memory-guided and self-initiated saccadic eye movements requiring volitional control, but not in visually guided saccades (Crawford, Henderson, & Kennard, 1989; Shimizu & Okiyama, 1993). Interestingly, they do not have problems with anti-saccades (Lueck et al., 1990), a finding which has been replicated in other studies (Fukushima, Fukushima, Miyasak, & Yamashita, 1994).

Movement-related Cortical Potentials

There is now considerable support for the finding originally reported by Deecke, Englitz, Kornhuber, and Schmitt (1977), that BPs recorded prior to self-paced movements are abnormal in Parkinson's disease (Cunnington, Iansek, Bradshaw, & Phillips, 1995; Dick et al., 1989; Jahanshahi et al., 1995a; Shibasaki, Shima, & Kuroiwa, 1978; Simpson & Khurabeit, 1987; Vidailhet et al., 1993). A number of studies (Cunnington et al., 1995; Dick et al., 1989; Jahanshahi et al., 1995a; Vidailhet et al., 1993) have further revealed that it is the early BP component, considered to reflect motor preparation associated with SMA activation, which has a smaller amplitude in patients with Parkinson's disease. Furthermore, the amplitude of the early BP component appears sensitive to dopamine. Administration of dopaminergic medication increases the amplitude of the "early" but not the "late" BP in these patients (Dick et al., 1987). The amplitude of the CNV is also significantly increased by levodopa therapy in Parkinson's disease (Amabile et al., 1986).

More recent studies have examined the influence of other factors on MRCPs in Parkinson's disease. In a study in which MRCP recordings and rCBF measurements were obtained using the same task (Jahanshahi et al., 1995a), underactivation of the SMA and a reduced amplitude of the early BP were demonstrated in a group of patients with Parkinson's disease compared to age-matched normals. In the same study, it was shown that while the amplitude of the early BP component was significantly lower for patients with Parkinson's disease compared to normals prior to self-initiated movements, the two groups did not differ in terms of MRCPs recorded prior to movements that were externally triggered by presentation of a regularly occurring tone. To maximise volitional motor preparation and to distinguish it from movement execution, Cunnington, Iansek, Johnson, and Bradshaw (1997) recorded MRCPs prior to imagined versus actual performance of sequential movements on a tapping board or while simply watching visual cues presented on the tapping board. By subtraction of MRCPs in the imagined and "watching cues" conditions, they found that MRCP components relating to motor preparatory processes were impaired in patients with Parkinson's disease, whereas MRCPs relating to movement execution obtained by subtracting MRCPs for imaged from actual movements were not. As described earlier, in normal subjects, the mode of movement selection affects the amplitude of MRCPs, which is higher prior to self-selected random movements compared to repetitive movements (Dirnberger et al., 1998; Praamstra, et al., 1995; Touge et al., 1995). In their study, Touge et al. (1995) found that in contrast to the age-matched normal group, for the patients with Parkinson's disease, the amplitudes of MRCPs were no different when self-paced and self-selected joystick movements were compared to repetitive joystick movements in a single direction. This was considered to indicate that processes involved in the self-selection of movement are impaired in Parkinson's disease.

Effects of Pallidotomy of Deep Brain Stimulation (DBS) of the STN or GPi on Motor and Cognitive Function in Parkinson's disease

In light of current models of akinesia and bradykinesia in Parkinson's disease, there has been renewed interest in surgical approaches to the treatment of the disorder. The focus has been on three particular surgical procedures. The first, posteroventral pallidotomy (PVP), involves lesioning of the posteroventral section of the internal segment of the globus pallidus (GPi). The two other procedures involve deep brain stimulation (DBS), which uses implanted electrodes for high-frequency stimulation of the GPi or the subthalamic nucleus, which has an input to GPi via the indirect pathway. The effects of these various procedures are illustrated in Fig. 4c. The aim of all three surgical procedures is to reduce overactivation of the GPi, which is considered responsible for the excessive inhibition of the thalamocortical targets of this area, thus "releasing the brake" open cortical activity in order to reduce bradykinesia and akinesia.

All three procedures are effective in reducing rigidity, tremor, and drug-induced dyskinesia, and improving bradykinesia (Baron et al., 1996; Limousin et al., 1995; Lozano et al., 1995; Siegfried & Lippitz, 1994). Davis et al. (1997) reported that stimulation of the GPi led to significant clinical benefits including reduction of bradykinesia and increased activation of the pre-SMA as measured with PET. Increased activation of the SMA has also been found during DBS of the STN while patients performed random joystick movements (Limousin et al.,1997). PVP is also associated with improvement of bradykinesia coupled with increased activation of the SMA during performance of random joystick movements (Ceballos-Baumann et al., 1994; Samuel, Ceballos-Baumann, Turjanski, et al., 1997b) or visually cued arm movements (Grafton, Waters, Sutton, Lew, & Couldwell, 1995). Decreases in pallidal and thalamic metabolism following PVP surgery have been shown to covary with bilateral increases in lateral premotor cortex activation as well as with clinical improvement in limb akinesia (Eidelberg et al., 1996). These data clearly indicate that the effects of surgery in overactive subcortical regions (GPi or STN) has the desired effect of "releasing the brake" on the cortical motor target of GPi, namely the SMA, with consequent reduction of bradykinesia and akinesia. These results constitute strong evidence for the role of the motor circuit in mediating the deficits in willed actions in Parkinson's disease. Nevertheless, the very same results have cast doubt over the validity of current models of frontostriatal functioning and akinesia/bradykinesia in Parkinson's disease. For example, on the basis of these models some deficit in voluntary movement, particularly in novel situations, would be expected following pallidotomy or DBS of the GPi, whereas using standard clinical assessment of movement, no such deficits are reported following surgery (Marsden & Obeso, 1994).

An associated finding from these studies is of interest. The ventral (sensorimotor) region of GPi through which the motor circuit between the putamen and the SMA passes is the target of pallidal surgery. The dorsal region of the GPi, which is part of the complex loop between the caudate nucleus and the dorsolateral prefrontal cortex is not involved. Nevertheless, there is evidence from PET studies suggesting that activity of the other frontostriatal circuits is also affected by PVP or DBS. The activation of the dorsolateral prefrontal cortex as well as the SMA is significantly increased following PVP (Ceballos-Baumann et al., 1994; Samuel, Ceballos-Baumann, Turjanski et al., 1997b) and DBS of the STN significantly increases activation of the dorsolateral prefrontal cortex and the anterior cingulate in addition to the SMA (Limousin et al., 1997). These may constitute an inadvertent effect of the surgical procedure. As the trajectories of the recording, lesioning, or stimulating electrodes pass through the dorsal pallidum to reach the ventral section, the functioning of the "complex" circuit may in some way be affected by the lesioning or simulation. Another possibility is that functionally the frontostriatal circuits are not as segregated as is assumed, a possibility that we will return to later.

Impairment of Willed Actions in Schizophrenia

Schizophrenia is a severe psychiatric disorder. The precise physiological basis of the disorder is unknown, but treatment with drugs that modulate the dopamine system has marked effects on symptoms in most cases. The signs and symptoms can be divided into two classes, positive symptoms and negative signs (Crow, 1980). Positive symptoms such as thought disorder, hallucinations, or delusions are abnormal by their presence, whereas negative signs such as flattening of affect, poverty of action, and social withdrawal reflect an abnormal absence of behaviour. In an attempt to explain the behavioural signs of schizophrenia, Frith (1987) proposed a cognitive model involving two routes to action (Fig. 3), which have parallels to the distinction between the medial vs. lateral premotor systems. In the first stimulus-driven route, perception of the stimulus triggers an action and leads to production of a response. In the second willed action route, goals and plans lead to formulation of intentions, which result in initiation of appropriate actions (and suppression of inappropriate ones) and production of a response. Disconnections at different points of these two routes to action can be considered responsible for various negative or positive signs in schizophrenia as well as akinesia in Parkinson's disease. The negative signs of schizophrenia such as poverty of action, stereotyped action, and incoherent action reflect a dysfunction of "willed" action, whereas the processes involved in "stimulus-driven" actions remain largely intact (Frith, 1987; see Fig. 3). As in Shallice's (1988) model of the effect of frontal lesions, the defect in willed action in schizophrenia can lead to failure to act or to inappropriate or stereotyped actions. Robbins (1990, 1991) also suggested that the cognitive and motor deficits in patients with schizophrenia may reflect dysfunction of the frontostriatal network, in which dopamine plays a crucial role.

Studies Using Functional Imaging

Since the study of Ingvar and Franzen (1974), many, but not all, functional imaging studies have reported lower frontal activation in patients with schizophrenia scanned at rest. Hypofrontality has also been observed when patients are engaged in tasks that activate the frontal regions in normal subjects (e.g. Weinberger et al., 1986). Furthermore, this "hypofrontality" has been related to the presence of negative signs (Andreasen et al., 1992; Liddle, Friston, Frith, Hirsch, Jones, & Frackowiak, 1992; Wolkin et al., 1992) or cognitive deficits on tests such as the Wisconsin (Seidman et al., 1994). These findings suggest that in schizophrenia, the impairment of willed action, as reflected by negative behavioural signs or poor performance on frontal tasks, is associated with underactivation of the dorsolateral prefrontal cortex.

A problem for the interpretation of these studies is that the performance of the patients is often worse than that of the controls. Frith et al. (1995) matched performance by using a paced verbal fluency task. They found that the groups did not differ in terms of activation of the left dorsolateral prefrontal cortex or anterior cingulate when word generation was compared to control tasks. A later study of drug-free patients (Dolan et al., 1995) using the same paradigm replicated the observation of normal activation in the dorsolateral prefrontal cortex, but observed reduced activity in the anterior cingulate cortex. In both these studies the patients failed to show the normal decrease of activation in the left superior temporal gyrus associated with word generation. This may indicate lack of functional connectivity between the left superior temporal gyrus and the left dorsolateral prefrontal cortex in schizophrenia (Friston, Liddle, Frith, Hirsch, & Frackowiak, 1995). In the study of drug-free patients, injections of apomorphine normalised the activation of the anterior cingulate cortex (Fletcher, Frith, Grasby, Friston, & Donlan, 1996).

Studies of Cognitive Function

There is now a considerable literature on the cognitive deficits associated with schizophrenia. Patients are impaired on executive tasks such as the Wisconsin Card Sorting task (Malmo, 1974) and the Tower of London task (Pantellis et al., 1997), especially when showing negative signs. Here we will focus on tasks that are more specific to the impairment of willed actions. Patients with schizophrenia are impaired on verbal fluency (Allen, Liddle, & Frith, 1993). Furthermore, impaired performance is significantly associated with negative symptoms considered indicative of poverty of willed action (Allen et al., 1993; Brown & White, 1991; Frith, Leary, Cahill, & Johnstone, 1991; Liddle & Morris, 1991).

Frith and Done (1983) tested patients with schizophrenia and matched controls on a two-choice guessing task, where subjects have to choose whether a tossed coin will land heads or tails. This task requires selection of responses in the absence of relevant stimuli. Normal individuals produced a random sequence of guesses. Patients with schizophrenia and negative features produced stereotyped sequences marked by alternations or perseverations. Similar results were obtained by Lyon, Mejsholm, and Lyon (1986). Random generation of sequences of numbers or letters are related tasks that require intrinsic generation of responses. Patients with schizophrenia perform poorly on random number generation tasks (Horne et al., 1982; Rosenberg, Weber, Crecq, Duval, & Macher, 1990).

Movement Initiation and Execution

A consistent finding from RT studies is that schizophrenic patients have significantly slower RTs than normal controls (for review, see Nuechterlein, 1997). A second consistent finding is the "cross-over effect". At longer intervals between warning and go stimuli, patients with schizophrenia fail to benefit from the temporal predictability of stimuli when the S1-S2 interval is constant or blocked rather than varies randomly across trials. Schizophrenics show a cross-over effect and have slower RTs for blocked than the random S1-S2 intervals. In the cross-over effect, patients are failing to use the advance information provided by the warning signal about the temporal predictability of the imperative stimulus to speed up the response. The effect of other types of advance information on RTs in schizophrenia has also been investigated. Carnahan, Chua, Elliot, Velamoor, and Carnaham (1994) compared uncued and precued CRTs in leukotomised and unleukotomised schizophrenic patients compared to normals. The data show that the unleukotomised schizophrenics performed as slowly on the fully cued CRT task as on the uncued CRT, suggesting that they were not using the advance information provided by the precue to volitionally preprogramme the response in the fully cued CRT. Performance of SRT tasks concurrently with a second attention-demanding task, which introduces a capacity load and requires greater volitional control, has also been shown to be particularly detrimental to the performance of patients with schizophrenia (Schwartz et al., 1989). Under conditions of low S-R compatibility, patients with schizophrenia are particularly slowed by increases in task complexity in CRT tasks (Slade, 1971; Venables, 1958). These findings are in agreement with the proposal that in schizophrenia RT deficits become more evident as tasks require greater volitional control. Of interest is the finding that slowness of RT in schizophrenia is significantly associated with presence and severity of negative symptoms (Schwartz et al., 1991).

A task that has frequently been used to assess the ability to sustain attention in schizophrenia is the Continuous Performance Test. On this task, similar to a go no-go RT task, subjects only respond to specific target stimuli and withhold responding to other stimuli. Patients with schizophrenia perform poorly particularly on the version where they are required to respond to a letter only if it matched the immediately preceding letter (Cornblatt, Lensanweger, & Erlenmeyer-Kimling, 1989). Cohen and Servan-Schreiber (1992) have interpreted this deficit as indicating an inability to maintain an internal representation of context, that is an inability to hold information in mind to control subsequent responses.

Using saccadic RT tasks that involved self-initiation (SIS) or automatic initiation (AIS) of saccadic eye movements, Done and Frith (1989) found that drug-free patients with schizophrenia or psychotic depression were significantly slower than normals on the SIS condition requiring strategic volitional initiation of saccades but not on the AIS condition. The inability to benefit from temporal predictability to speed up saccade initiation on the SIS task differentiated the patients with schizophrenia from those with psychotic depression. In agreement with these findings, several other studies have found impairment of anticipatory saccades (Hommer, Clem, Litman, & Ackar, 1991), and suppression deficits on the antisaccade task (Fukushima et al., 1994) or other oculomotor tasks requiring suppression of reflexive saccades (Paus, 1991) or holding information about the position of target "on line" during a delay period (Park & Holzman, 1992), whereas visually guided saccades are intact. Such deficits in suppression of reflexive saccades correlate with poor performance on cognitive tests of executive function (Paus, 1991) and are associated with decreased rCBF in the anterior cingulate, insula, and striatum (Crawford et al., 1996). Furthermore, Park and Holzman (1992) reported that patients showed similar deficits when performing a tactile or an oculomotor version of the delayed response task, thus demonstrating the generality of the problem.

Studies of Movement-related Cortical Potentials

In schizophrenia, impairment of motor preparatory processes prior to self-paced movements is reflected by the lower amplitude of the BP in medicated (Fuller, Nathaniel-James, Ron, & Jahanshahi, 1996; Singh et al., 1992) and unmedicated patients (Chiarenza, Papakostopoulos, Dini, & Cazzullo, 1985). In contrast, patients with schizophrenia and tardive dyskinesia have significantly larger amplitudes of the BP relative to normal controls and the amplitude of the BP significantly correlated with the severity of tardive dyskinesia (Adler, Pecevich, & Nagamoto, 1989), suggesting a role for dopaminergic influences on the BP.

The amplitude of the CNV is reduced in schizophrenia suggesting impairment of anticipatory and preparatory processes prior to movement (for reviews see Cohen, 1990; Pritchard, 1986).

Spectral analysis of the EEG also reveal that while untreated patients with schizophrenia do not differ from normals at rest (Wesetphal et al., 1990a), the resting EEG-spectra of treated patients (Westphal et al., 1990b) and the movement-related spectra of untreated patients (Westphal, Grozinger, Becker, Diekmann, & Kornhuber, 1992) differ from those of normal controls. In contrast to the normals, the patients showed a clear decrease of power density in the theta, alpha, and beta bands during movement execution and lacked the enhancement of theta power normally observed during movement preparation compared to rest. Since the normal enhancement of theta activity during movement preparation is considered to reflect the involvement of frontal areas in decision making and voluntary action (Kornhuber, Lang, Lang, Kure, & Kornhuber, 1990), the absence of theta enhancement in schizophrenia has been interpreted to reflect impairment of decision making, volition, and movement generation (Westphal et al., 1992).

Impairment of Willed Actions in Other Disorders

Certain symptoms of other disorders can also be considered to reflect deficits in willed action. One example of this is psychomotor retardation in depressive illness and the associated symptoms of lethargy and lack of interest, social withdrawal, flatness of speech, and an unchanging facial expression. PET studies of steady-state rCBF change in patients with depression (Bench et al., 1992) reported an association between decreased rCBF in the dorsolateral prefrontal cortex and psychomotor retardation. "Primary obsessional slowness" (POS) is a variant of obsessive-compulsive neurosis. POS can be considered another example of the breakdown of the willed action system. These mainly male patients are overmeticulous and take an excessively long time to perform most actions, particularly self-care behaviours and other activities of daily living. These patients have difficulty in making decisions, initiating goal-directed action, and suppressing intrusive and perseverative behaviour (Hymas, Lees, Bolton, Epps, & Head, 1991). Treatment of POS by pacing, prompting, shaping, and modelling all aim to use external means to make up for the deficiency in volitional control of timing the duration of action. These patients show bilaterally increased oxygen metabolism at rest in various areas of the frontal cortex (BA 6, 8, 44, 46), which includes the SMA and dorsolateral prefrontal cortex (Sawle, Hymas, Lees, & Frackowiak, 1991), while ^{18}F-dopa uptake into caudate and putamen is normal.

Another example of the dissociation between intention and action is the "alien hand" phenomenon. In patients who show this sign one hand will perform actions, like grasping a door knob, which are not intended by the patient. The alien hand sign can occur after lesions of the anterior cingulate or SMA (Goldberg & Bloom, 1990; Goldberg, Mayer, & Toglia, 1981). It can also be present in corticobasal degeneration, one of the akineto-rigid syndromes (Thompson & Marsden, 1992). PET studies (Sawle, Brooks, Marsden, & Frackowiak, 1991) have shown that this disorder is associated with an asymmetric reduction in uptake of ^{18}F-dopa in the caudate and putamen and in the medial frontal cortex. The "alien hand" phenomenon suggests that disturbances of volitional control of action can exist in isolation. Another example of this is "gait ignition" failure, reported by Atchison, Thompson, Frackowiak, and Marsden (1993) in six cases who had difficulty in initiating walking, although gait was relatively normal once it was started. These patients had no other motor abnormalities. One of the cases who had a PET scan showed hypometabolism of the inferomedial portions of the frontal cortex.

Abulia and Apathy

The term abulia comes from the Greek world boul, which means will. Thus abulia refers to a "lack of will". So far we have concentrated on studies that demonstrate abnormal task performance linked to problems with willed action. However, the most obvious consequence of a defect of willed action will be a failure to produce spontaneous behaviour. Demonstration of this lack of will depends upon clinical observation rather than experimental investigation. We shall now consider the motivational deficits that can be observed in various patient groups.

Bhatia and Marsden (1994) reviewed 240 cases with lesions of the basal ganglia reported in the literature and concluded that "The commonest behavioural disorder was abulia, which manifested in 30 of the 240 (13%) cases". These authors defined abulia or "psychic akinesia" as a "loss of drive or apathy including loss of spontaneous motor activity, loss of emotional affective expression and reduction of spontaneous thought content and initiative". Marin (1977a) considers abulia to be a more severe form of apathy and suggests that apathy has validity as a symptom, a syndrome, or a behavioural dimension, and introduced the term apathy and disorders of diminished motivation (ADDM). According to Marin, the diagnosis of ADDM depends on detecting "simultaneous diminution in the overt behavioural, cognitive and emotional concomitants of goal-directed behaviour". Inclusion of the reduction of "goal-directed behaviour" as a diagnostic criterion makes it clear that the investigation of apathy and the motivational deficits which are its core feature is relevant to understanding of willed actions. Furthermore, certain types of apathy are considered to be due to dysfunction of frontostriatal circuits (Cummings, 1993; Marin, 1997a).

Patients with frontal lesions may show abulia as a loss of spontaneity, curiosity, and initiative, and presence of apathy, and blunting of affect, drive, and mentation (Mesulam, 1986). Such impairments, which result in patients being described as apathetic and poorly motivated (Stuss & Benson, 1984) or "sluggish" and introverted (Grafman, Vance, Weingartner, Salazar, & Amin, 1986) are found after dorsolateral prefrontal lesions. However, lack of motivation and goal-directed behaviour, difficulty in decision-making, obsessional behaviour, and poor social judgment in real-life situations have also been reported in a patient, EVR, with a lesion predominantly affecting the orbitofrontal areas but extending to mesial and dorsolateral cortex (Eslinger & Damasio, 1985). From a review of the literature, Cummings (1993) has suggested that apathy is particularly associated with damage to the anterior cingulate.

Apathy and loss of interest can be features of Parkinson's disease and carers sometimes report that, if left to their own devices, patients just sit inactive for long periods and express no interest in doing anything. In Parkinson's disease, akinesia and apathy, that is the motor and the motivational deficits in willed action, can be loosely coupled. As noted by Marin (1997b), a mildly akinetic patient can show severe apathy, and a severely akinetic patient may be highly motivated. In a study of a consecutive series of 50 patients with Parkinson's disease, Starkstein et al. (1992) reported that 12% of the patients showed apathy and 30% suffered from both apathy and depression. The presence of apathy was found to be significantly associated with impaired performance on time-limited tests requiring willed action such as verbal fluency and version B of the Reitan Trail Making.

Apathy is also dissociable from depression and is a clinically distinct state (Marin, 1997b). Negative thoughts about the self and the future and dysphoric mood, which are core features of depression, are absent in apathy. Instead, apathy is characterised by lack of effort, lack of concern and interest, and lack of affect. Nevertheless, apathy, a purely motivational deficit of lack of interest and concern, is also a feature of depression and other psychiatric disorders such as schizophrenia (Marin, 1977a, b). Therefore, apathy and abulia, purely motivational deficits, are likely to contribute to the impairment of willed action such as the poverty of action in schizophrenia. There is some support for this from a study by Schmand et al. (1994), who operationalised motivational deficits in terms of "time on task" during an RT task. Significantly more psychotic than control patients showed such motivational deficits, which significantly correlated with negative but not positive symptoms.

Abulia and apathy, the motivationally based deficits of willed action, may be mediated by those frontostriatal circuits that have connections with the limbic systems, such as the anterior cingulate and the orbitofrontal circuits. Involvement of the orbitofrontal circuit in processing of feedback information that can influence motivational aspects of behaviour was demonstrated in a recent PET study. Elliot, Frith, and Dolan (1997) found that, compared to a no-feedback condition, processing of task-independent performance feedback during a Tower of London and a matched guessing task was associated with greater activation of the medial caudate nucleus. In addition, for the more uncontrollable guessing task where information has to be assimilated across a number of trials to assess performance, presence of feedback was associated with greater activation of the orbitofrontal cortex. In light of the growing body of evidence implicating the mesolimbic and the mesocortical dopamine system in motivation and responsiveness to reward (Schultz, 1997), abulia and apathy may be dopamine-related dysfunctions. Some support for this comes from the study of Powell, Al-Adawi, Morgan, and Greenwood (1996). In 11 patients with abulia following head injury, they reported that bromocriptine, a post-synaptic dopamine agonist, significantly improved scores on experimentally developed measures of motivation as well as on tests requiring willed action, such as verbal fluency.

IS THE CONCEPT OF WILLED ACTIONS USEFUL?

Tables 1 and 2 present in summary form the main evidence on the cognitive and motor deficits of patients with frontal damage, Parkinson's disease, or schizophrenia. We draw three main conclusions on the basis of the evidence from normal and abnormal groups. First, we conclude that a clear distinction can be made between willed action and routine action. The two kinds of action are associated with different kinds of task and willed action can be impaired while routine action remains intact. Second, we conclude that the frontostriatal circuits play a role in the mediation of willed cognitive and motor actions. More specifically, the dorsolateral prefrontal cortex, the anterior cingulate, and the SMA may be the cortical components whereas the thalamus and basal ganglia are the subcortical components of this system. Third, the data summarised in Tables 1 and 2 highlight the parallels between the cognitive and motor deficits in these different disorders. A similar story is told by brain imaging studies. For example, Frith et al. (1991) demonstrated the similarity in patterns of brain activation associated with willed cognitive (verbal fluency) and motor (willed random finger lifting) actions. We would like to suggest, therefore, that the concept of willed action provides an important common thread that links not only cognition, motivation, and motor functions, but also cognitive, motivational, and motor impairments in disorders associated with the frontostriatal system such as Parkinson's disease and schizophrenia.

Table 1. Tests of Cognitive Function on Which Impaired Performance Has Been Found for Patients with Parkinson's Disease, Schizophrenia, or Frontal Lesions of Relevance to Willed Actions. For Each Test, One Example of a Study Reporting Such Deficits is Presented in the Body of the Table

Test	Frontal Lesions	Parkinson's Disease	Schizophrenia
Verbal fluency	Milner (1964)	Taylor et al. (1986)	Allen et al. (1993)
Random number generation	Spatt & Goldenberg (1993)	Spatt & Goldenberg (1993)	Horne et al. (1982)
Wisconsin Card Sorting	Milner (1964)	Taylor et al. (1986)	Malmo (1974)
Planning—Tower of London	Owen et al. (1990)	Owen et al. (1993)	Pantelis et al. (1997)

Table 2. Aspects of Motor Function for Which Impaired Performance Has Been Found for Patients with Parkinson's Disease, Schizophrenia, or Frontal Lesions. For Each an Example of a Study Reporting Such Deficits is Presented in the Body of the Table

Impaired Function	Frontal Lesions	Parkinson's Disease	Schizophrenia
Failure to preprogramme in SRT CRT–SRT difference less than normal	?	Bloxham et al. (1984)	?
Failure/slowness to use advance information in precued CRT	Alivisatos & Milner (1989)	Jahanshahi et al. (1992)	Carnahan et al. (1994)
Increased RT slowness with S–R incompatibility	Decary & Richer (1995)	Brown et al. (1993b)	?
Increased slowness with choice complexity in CRT	Decary & Richer (1995)	Cooper et al. (1994)	Slade (1971)
Impaired go no–go RTs	Drewe (1975)	Cooper et al. (1994)	Cornblatt et al. (1989)
Differentially greater impairment of bimanual than unimanual movements	Leonard et al. (1988)	Brown et al. (1993a)	?
Differentially greater impairment of complex than simple movements	Dick et al. (1986)	Benecke et al. (1986, 1987b)	?
Impaired antisaccades but not visually guided saccades	Guitton et al. (1985)	Lueck et al. (1990) (not impaired)	Fukushima et al. (1994)
Impaired self-initiated or memory guided or anticipatory saccades but not visually guided	Pierrot-Deseilligny et al. (1991)	Crawford et al. (1989)	Done & Frith (1989)
Lower amplitude of BP	Singh & Knight (1990)	Dick et al. (1986)	Chiarenza et al. (1985)
Impaired CNV	Rosahl & Knight (1995)	Amabile et al. (1986)	Pritchard (1986)

RT: reaction time, SRT: simple RT, CRT: choice RT, BP: bereitschaftspotential, CNV: contingent negative variation, ?: studies not identified.

What Are the Functional Components of the Willed Action System?

If the concept of willed action is to be more than simply a useful heuristic for organising a disparate body of research, then we must be able to specify the components of the willed action system in terms of underlying cognitive processes and neurophysiology. In the last part of this review we shall consider a number of pertinent questions and different approaches for identifying these components.

Willed Action and the Frontostriatal Circuits

The proposal that the cognitive, motivational and motor deficits in various disorders may be manifestations of a common underlying dysfunction of a willed action system centred around the frontostriatal network appears at first sight contrary to traditional views of these circuits as parallel and segregated. As noted earlier, it is possible that the frontostriatal loops do not function in a fully segregated manner. On the basis of more recent evidence, the strict anatomical segregation of the frontostriatal circuits has been questioned (Joel & Weiner, 1994). A common feature of the five frontostriatal circuits is that each circuit has both closed- and open-loop elements. Each circuit has neurons that remain anatomically segregated from parallel neurons in the other circuits, thus forming a closed loop. On the other hand, inputs from regions outside the closed loop modulate the activity of each circuit and each circuit in turn projects to areas outside the closed loop. These afferent and efferent connections constitute the "open-loop" element of each circuit (Alexander & Crutcher, 1990). Furthermore, anatomical studies of dendritic arborization in the striatopallidal circuits led Percheron and Filion (1991) to conclude that "convergence is the probable essential device of the striatopallidal nigral system" ... and that "the same pallidal or nigral neuron might respond to signals from different parts of the body, and might integrate 'motor', 'oculomotor', 'limbic' and two types of 'prefrontal' signal according to context". Functional integration could also be achieved through temporal coincidence of processing within the different circuits, for example through simultaneous processing of information relating to coordinated hand and eye movements in the motor and oculomotor circuits (Alexander & Crutcher, 1990). Therefore, current opinion considers segregated parallelism to coexist with convergence in the frontostriatal network (Joel & Weiner, 1994; Marsden & Obeso, 1994; Wichmann & DeLong, 1996).

A related question is: Do each of the "motor", "complex", and "anterior cingulate" circuits play a specific and distinct role in the control of willed actions? Evidence is beginning to emerge that relates to the function of the different loops. There is increasing evidence that anterior SMA is associated with motor preparation, the precise timing of actions, and response inhibition in go no-go tasks (Deiber et al., 1996; Humberstone et al., 1997; Jahanshahi et al., 1995a; Kornhuber & Deecke, 1965; Rao et al., 1997). In contrast, the dorsolateral prefrontal cortex is primarily engaged by tasks that require decisions about what response to make or when to make it on each trial (Deiber et al., 1991, 1996; Frith et al., 1991; Jahanshahi et al., 1995a, b, 1997; Petersen et al., 1988; Petrides et al., 1993). From a review of the literature, Cummings (1993) suggests that motivational deficits such as apathy are associated with dysfunction of the anterior cingulate circuit, regardless of whether the lesion is at the frontal cortical (anterior cingulate) or subcortical (nucleus accumbens) end of the circuit. Recent evidence has shown the medial caudate nucleus and the orbitofrontal cortex to be activated when subjects process performance feedback, which can influence motivational aspects of behaviour (Elliott et al., 1997).

How Does the Frontostriatal Network Interact with Other Cortical and Subcortical Regions of the Brain to Control Willed Actions?

Successful actions cannot be generated from a frontostriatal system working in isolation. Relevant sensory information from posterior brain regions must be utilized. In willed action tasks involving simple movements the parietal cortex (Jahanshahi et al., 1995a, b, 1997; Jenkins et al., 1994; Playford et al., 1992) and the cerebellum (Jahanshahi et al., 1997; Jenkins et al., 1994; Jueptner, Stephan, et al., 1997; Playford et al., 1992) become activated in addition to the frontal cortex. For example, the activation of the cerebellum with these willed action tasks may relate to the timing component of these tasks (Ivry & Keele, 1989; Jueptner et al., 1995) or may reflect rehearsal of responses in the articulatory loop in order to maintain an "on line" record of the last few (Awh et al., 1996; Paulesu, Frith, & Frackowiak, 1993). Recent PET evidence suggests that in addition to underactivation of the SMA and dorsolateral prefrontal cortex, performance of willed actions in Parkinson's disease is associated with overactivation of the ipsilateral cerebellum (Rascol et al., 1997; Samuel, Ceballos-Baumann, Turjanski, et al., 1997b). This may reflect the need to recruit regions normally involved in routine performance of the task. Spence et al. (1997) observed excessive activation of the parietal cortex during the performance of a willed action task requiring joystick movements in schizophrenic patients who perceived their actions as being under the control of outside forces (delusions of control). Patients with schizophrenia also show excessive activation of the left superior temporal cortex during performance of a willed action task requiring word generation (Frith et al., 1995). Investigations of how the frontostriatal circuits interact with posterior brain systems during willed actions may provide information relevant to understanding physiological abnormalities in Parkinson's disease or schizophrenia.

What Are the Specific Mechanisms of Impairment of Willed Actions in Parkinson's Disease, Schizophrenia, and Frontal Damage?

Despite the surface similarity of the impairment of willed actions in various disorders, the precise nature and mechanisms of breakdown of volitional control of action are likely to be different in each disorder. These differences should provide clues to the role of different components of the willed action system. Frith (1992) has proposed that while damage to the prefrontal cortex leads to a failure to develop a plan for action, the poverty of action in chronic schizophrenia is associated with a failure to generate the specific willed intentions appropriate to the plan. In Parkinson's disease willed intentions are generated, but cannot be converted into overt actions. However, it is not clear how these differences could be distinguished experimentally. Direct comparisons of the three types of patient on various cognitive tasks have revealed subtle differences (Owen et al., 1993; Pantelis et al., 1997). For example, of relevance to willed action is the observation that patients with frontal damage or schizophrenia have problems with the antisaccade task (Fukushima et al., 1994; Guitton et al., 1985), while patients with Parkinson's disease do not (Fukushima et al., 1994; Lueck, Tanyeri, Crawford, Hudson, & Kennard, 1990).

What Role Does Dopamine Play in the Volitional Control of Action?

Within the framework of current models of basal ganglia function, release of dopamine from the nigrostriatal projections is considered to have a facilitatory effect in the direct pathway and an inhibitory effect in the indirect pathway (Gerfen, 1995). The net effect of striatal dopamine release is to reduce the output from the basal ganglia, which in turn results in increased activity of thalamocortical projections and facilitation of cortically initiated action. It has been proposed that for selection of action dopamine may play a role in setting the threshold for triggering of schemas (Robbins, 1991). There is dysfunction of dopaminergic systems in Parkinson's disease and schizophrenia that contributes to their deficits of willed actions. Thus it seems that dopamine may play a fundamental role in modulation of willed actions, an issue that requires further investigation to elucidate the precise nature of this modulatory influence.

Conclusions

These questions can only be adequately addressed by using a variety of investigatory techniques, such as lesion studies or single cell recordings in animals, functional imaging in human subjects, comparative clinical studies, recording of movement and cognitive-related cortical potentials, transcranial magnetic stimulation to briefly disrupt brain function, or connectionist modelling to simulate normal and abnormal performance of willed actions. The degree to which each of these techniques is successful in addressing the pertinent questions depends to a large extent on our ability as neuroscientists to develop and use novel tasks and paradigms to dissect the component processes of willed actions. A process-oriented approach and the study of both normal and abnormal willed actions would also be essential to move beyond mere description of performance, to understand the mechanisms of control of willed actions and its breakdown in disease. Such a focus will establish links between cognitive psychological models of control of action and the field of motor control, which have co-existed largely independently. It will also allow a more integrated approach to the study of normal cognitive and motor function and motivational determinants of these and their breakdown in disease.

REFERENCES

Adler, L.E., Pecevich, M., & Nagamoto, H. (1989). Bereitschaftspotential in tardive dyskinesia. *Movement Disorders, 4(2)*, 105–112.

Agostino, R., Berardelli, A., Fromica, A., Accornero, N., & Manfredi, M. (1992). Sequential arm movements in patients with Parkinson's disease, Huntington's disease and dystonia. *Brain, 115*, 1481–1495.

Alexander, G.E., & Crutcher, M.D. (1990). Functional architecture of basal ganglia circuits: Neural substrates of parallel processing. *Trends in Neurosciences, 13*, 266–271.

Alexander, G.E., Crutcher, M.D., DeLong, M.R. (1990). Basal ganglia thalamocortical circuits: Parallel substrates for motor, oculomotor, 'Prefrontal' and 'limbic' functions. *Progressive Brain Research, 85*, 119–146.

Alexander, G.E., DeLong, M.R., & Strick, P.L. (1986). Parallel organization of functionally segregated circuits linking basal ganglia and cortex. *Annual Review of Neuroscience, 9*, 357–381.

Alivisatos, B., & Milner, B. (1989). Effects of frontal or temporal lobectomy on the use of advance information in a choice reaction time task. *Neuropsychologia, 27(4)*, 495–503.

Allen, H.A., Liddle, P.F., & Frith, C.D. (1993). Negative features, retrieval process and verbal fluency in schizophrenia. *British Journal of Psychiatry, 163*, 769–775.

Amabile, G., Fattapposta, F., Poxxessere, G., Albani, G., Sanarelli, L., Rizzo, P.A., & Morocutti, C. (1986). Parkinson's disease: Electrophysiological (CNV) analysis related to pharmacological treatment. *EEG Clinical Electroencephalography, 64*, 521–524.

Amassian, V.E., Cracco, R.Q., Maccabee, P.J., Bigland-Ritchie, B., & Cracco, J.B. (1991). Matching focal and non-focal magnetic coil stimulation to properties of human nervous system: Mapping motor unit fields in motor cortex contrasted with altering sequential digit movements by premotor-SMA stimulation. In W.J. Levy, R.Q. Cracco, A.T. Barker, & J. Rothwell (Eds.), *Magnetic motor stimulation: Basic principles and clinical experience (EEG suppl. 43)*. Amsterdam: Elsevier.

Aminoff, M.J., & Goodin, D.S. (1993). Premovement cerebral potentials in simple reaction-time tasks. *Neurology, 43*, A268.

Ammon, K., & Gandevia, S.C. (1990). Transcranial magnetic stimulation can influence the selection of motor programmes. *Journal of Neurology, Neurosurgery and Psychiatry, 53(8)*, 705–777.

Andreasen, N.C., Rezai, K., Alligen, R., Swayze, V.W., Flaum, M., Kirchner, P., Cohen, G., & O'Leary, D. (1992). Hypofrontality in neuroleptic-naive patients and in patients with chronic schizophrenia. *Archives of General Psychiatry, 49*, 943–958.

Arezzo, J., & Vaughan, H.G.J. (1975). Cortical potentials associated with voluntary movements in the monkey. *Brain Research, 88(1)*, 99–104.

Atchison, P.R., Thompson, P.D., Frackowiak, R.S., & Marsden, C.D. (1993). The syndrome of gait ignition failure: A report of six cases. *Movement Disorders, 8*, 285–292.

Awh, E., Jonides, J., Smith, E.E., Schumacher, E.H., Koeppe, R.A., & Katz, S. (1996). Dissociation of storage and rehearsal in verbal working memory: Evidence from positron emission tomography. *Psychological Science, 7(1)*, 25–31.

Baker, S.C., Rogers, R.D., Owen, A.M., Frith, C.D., Dolan, R.J., Frackowiak, R.S.J., & Robbins, T.W. (1996). Neural systems engaged by planning: A PET study of the Tower of London task. *Neuropsychologia, 34(6)*, 515–526.

Baron, M.S., Vitek, J.L., Bakay, R.A.E., Green, J., Kaneoke, Y., Hashimoto, T., Turner, R.S., Woodard, J.L., Cole, S.A., McDonald, W.M., & DeLong, M.R. (1996). Treatment of advanced Parkinson's disease by posterior GPi pallidotomy: 1-year results of a pilot study. *Annals of Neurology, 40*, 355–366.

Bench, C.J., Friston, K.J., Brown, R.G., Scott, L.C., Frackowiak, R.S.J., & Donlan, R.J. (1992). The anatomy of melancholia—focal abnormalities of cerebral blood flow in major depression. *Psychological Medicine, 22*, 607–615.

Benecke, R., Rothwell, J.C., Dick, J.P., Day, B.L., & Marsden, C.D. (1986). Performance of simultaneous movements in patients with Parkinson's disease. *Brain, 109*, 739–757.

Benecke, R., Rothwell, J.C., Dick, J.P., Day, B.L., & Marsden, C.D. (1987a). Disturbance of sequential movements in patients with Parkinson's disease. *Brain, 110*, 361–379.

Benecke, R., Rothwell, J.C., Dick, J.P., Day, B.L., & Marsden, C.D. (1987b). Simple and complex movements off and on treatment in patients with Parkinson's disease. *Journal of Neurology, Neurosurgery, and Psychiatry, 50*, 296, 303.

Bhatia, K.P., & Marsden, C.D. (1994). The behavioural and motor consequences of focal lesions of the basal ganglia in man. *Brain, 117*, 859–876.

Bloxham, C.A., Mindel, T.A., & Frith, C.D. (1984). Initiation and execution of predictable and unpredictable movements in Parkinson's disease. *Brain, 107*, 371–384.

Blumer, D., & Benson, D. (1975). Personality changes with frontal and temporal lobe lesions. In D.F. Benson & D. Blumer (Eds.), *Psychiatric aspects of neurologic disease* (pp. 151–170). New York: Grune & Stratton.

Brooks, D.J. (1995). The role of the basal ganglia in motor control: Contributions from PET. *Journal of the Neurological Sciences, 128*, 1–13.

Brooks, D.J., Ibanez, V., Sawles, G.V., Quinn, N., Lees, A.J., Mathias, C.J., Banniseter, R., Marsden, C.D., & Frackowiak, R.S.J. (1990). Differing patterns of striatal ^{18}F-dopa uptake in Parkinson's disease, multiple system atrophy, and progressive supranuclear palsy (see comments). *Annals of Neurology, 28*, 547–555.

Brotchie, P., Iansek, R., & Horne, M.K. (1991). Motor function in the monkey globus pallidus II. Cognitive aspects of movement and phasic neuronal activity. *Brain, 114*, 1685–1702.

Brown, K., & White, T. (1991). The association among negative symptoms, movement disorders and frontal lobe psychological deficits in schizophrenic patients. *Biological Psychiatry, 30*, 1182–1190.

Brown, R.G., Jahanshahi, M., & Marsden, C.D. (1993a). The execution of bimanual movements in patients with Parkinson's, Huntington's and cerebellar disease. *Journal of Neurology, Neurosurgery, and Psychiatry, 56*, 295–297.

Brown, R.G., Jahanshahi, M., & Marsden, C.D. (1993b). Response choice in Parkinson's disease. The effects of uncertainty and stimulus-response compatibility. *Brain, 116*, 869–885.

Brown, R.G., & Marsden, C.D. (1988). Internal versus external cues and the control of attention in Parkinson's disease. *Brain, 111*, 323–345.

Brown, R.G., Soliveri, P., & Jahanshahi, M. (in press). Random response generation in Parkinson's disease and normal controls. Executive processes and the role of the prefrontal cortex. *Neuropsychologia*.

Carnahan, H., Chua, R., Elliot, D., Velamoor, V.R., & Carnahan, C.H. (1994). Effects of schizophrenia and prefrontal leukotomy on movement preparation and generation. *Journal of Clinical and Experimental Neuropsychology, 16(2)*, 253–260.

Ceballos-Baumann, A.O., Obesa, J.A., Vitek, J.L., DeLong, M.R., Bakay, R., Linazasoro, G., & Brooks, D.J. (1994). Restoration of the thalamocortical activity after posteroventral pallidotomy in Parkinson's disease. *Lancet, 344(8925)*, 814.

Chiarenza, G.A., Papakostopoulos, D., Dini, M., & Cazzullo, C. (1985). Neurophysiological correlates of psychomotor activity in chronic schizophrenics. *Electroencephalography and Clinical Neurophysiology, 61*, 218–228.

Cohen, R. (1990). Event-related potentials and cognitive dysfunction in schizophrenia. In H. Hafner & W.F. Gattaz (Eds.), *Search for the causes of schizophrenia, Vol. 2*. Berlin: Springer-Verlag.

Cohen, J.D., & Servan-Schreiber, D. (1992). Context, cortex and dopamine: A connectionist approach to behaviour and biology in schizophrenia. *Psychological Review, 99*, 45–77.

Cooke, J.D., Brown, J.D., & Brooks, V.B. (1978). Increased dependence on visual information for movement control in patients with Parkinson's disease. *Canadian Journal of Neurological Science, 5(4)*, 413–415.

Cooper, J.A., Sagar, H.J., Tidwell, P., & Jordan, N. (1994). Slowed central processing in simple and go/no-go reaction time tasks in Parkinson's disease. *Brain, 117*, 517–529.

Cornblatt, B., Lensenweger, M.F., & Erlenmeyer-Kimling, L. (1989). A continuous performance test, identical pairs version. II. Contrasting attentional profiles in schizophrenic and depressed patients. *Psychiatry Research, 29*, 65–85.

Crawford, T.J., Henderson, L., & Kennard, C. (1989). Abnormalitites of nonvisually guided eye movements in Parkinson's disease. *Brain, 112(6)*, 1573–1586.

Crawford, T.J., Puri, B., Nijan, K.S., Jones, B., Kennard, C., & Lewis, S.W. (1996). Abnormal saccadic distractibility in patients with schizophrenia: A 99m Tc-HMPAO SPET study. *Psychological Medicine, 26*, 265–277.

Crow, T.J. (1980). Molecular pathology of schizophrenia: More than one disease process? *British Medical Journal, 280*, 66–68.

Cummings, J.L. (1993). Frontal-subcortical circuits and human behaviour, *Archives of Neurology, 50*, 873–880.

Cunnington, R., Iansek, R., Bradshaw, J.L., & Phillips, J.G. (1995). Movement-related potentials in Parkinson's disease. Presence and predictability of temporal and spatial cues. *Brain, 118(4)*, 935–950.

Cunnington, R., Iansek, R., Johnson, K.A., & Bradshaw, J.L. (1997). Movement-related potentials in Parkinson's disease. Motor imagery and movement preparation. *Brain, 120(8)*, 1339–1353.

Damasio, A.R., & Van Hoesen, G.W. (1980). Structure and function of the supplementary motor area. *Neurology (NY), 30*, 359.

Damasio, A.R., & Van Hoesen, G.W. (1983). Emotional disturbance associated with focal lesions of the limbic frontal lobe. In K.M. Heilman & P. Satz (Eds.), *Neuropsychology of human emotion* (pp. 85–110). New York: Oxford University Press.

Davis, K.D., Taub, E., Houle, S., Langu, A.E., Dostovsky, J.O., Tasker, R.R., & Lozano, A.M. (1997). Globus pallidus stimulation activates the cortical motor system during alleviation of parkinsonian symptoms. *Nature Medicine, 3*, 671–674.

Decary, A., & Richer, F. (1995). Response selection deficits in frontal excisions. *Neuropsychologia, 33(10)*, 1243–1253.

Deecke, L., Englitz, H.G., Kornhuber, H.H., & Schmitt, G. (1977). Cerebral potentials preceding voluntary movement in patients with bilateral or unilateral Parkinson akinesia. *Prog. Clin Neurophysiol, 1*, 151–163.

Deecke, L., Lang, W., Heller, H.J., Hufnagel, M., & Kornhuber, H.H. (1987). Bereitschaftspotential in patients with unilateral lesions of the supplementary motor area. *Journal of Neurology, Neurosurgery and Psychiatry, 50*, 1430–1434.

Deecke, L., Scheid, P., & Kornhuber, H.H. (1969). Distribution of readiness potential, pre-motion positivity and motor potential of the human cerebral cortex preceding voluntary finger movements. *Experimental Brain Research, 7*, 158–168.

Deiber, M.P., Ibanez, V., Sadato, N., & Hallett, M. (1996). Cerebral structures participating in motor preparation in humans: A positron emission tomography study. *Journal of Neurophysiology, 75*, 233–247.

Deiber, M.P., Passingham, R.E., Colebatch, J.G., Friston, K.J., Nixon, P.D., & Frackowiak, R.S.J. (1991). Cortical areas and the selection of movement: A study with positron emission tomography. *Experimental Brain Research, 84*, 393–402.

Dick, J.P., Benecke, R., Rothwell, J.C., Day, B.L., & Marsden, C.D. (1986). Simple and complex movements in a patient with infarction of the right supplementary motor area. *Mov. Disord, 1*, 255–266.

Dick, J.P., Cantello, R., Buruma, O., Gioux, M., Benecke, R., Day, B.L., Rothwell, J.C., Thompson, P.D., & Marsden, C.D. (1987). The Bereitschaftspotential, L-DOPA and Parkinson's disease. *Electroencephalography and Clinical Neurophysiology, 66*, 263–274.

Dick, J.P., Rothwell, J.C., Day, B.L., Cantello, R., Buruma, O., Gioux, M., Benecke, R., Berardelli, A., Thompson, P.D., & Marsden, C.D. (1989). The Bereitschaftspotential is abnormal in Parkinson's disease. *Brain, 112*, 233–244.

Dirnberger, G., Fickel, I.U., Lindinger, G., Lang, W., & Jahanshahi, M. (1998). The mode of movement selection: Movement-related cortical potentials prior to random choice and repetitive movements. *Experimental Brain Research, 120,* 263–272.

Dolan, R.J., Fletcher, P., Frith, C.D., Friston, K.J., Frackowiak, R.S.J., & Grasby, P.M. (1995). Dopaminergic modulation of impaired cognitive activation in the anterior cingulate cortex in schizophrenia. *Nature, 378,* 180–183.

Done, D.J., & Frith, C.D. (1989). Automatic and strategic volitional saccadic eye movements in psychotic patients. *Eur. Arch. Psychiatry Neurol. Sci., 239,* 27–32.

Drewe, E.A. (1975). Go–no go learning after frontal lobe lesions in humans, *Cortex, 11,* 8–16.

Eidelberg, D., Moeller, J.R., Ishikawa, T., Dhawan, V., Spetsieris, D., Stern, E., Woods, R.P., Fazzini, E., Dogali, M., & Beric, A. (1996). Regional metabolic correlates of surgical outcome following unilateral pallidotomy in Parkinson's disease. *Annals of Neurology, 36,* 450–459.

Elliot, R., Frith, C.D., & Dolan, D.J. (1997). Differential neural response to positive and negative feedback in planning and guessing tasks. *Neuropsychologia, 35,* 1395–1404.

Eslinger, P.J., & Damasio, A.R. (1985). Severe disturbance of higher cognition after bilateral frontal lobe ablation: Patient EVR. *Neurology, 35,* 1731–1741.

Filipovic, S.R., Covickovic-Sternic, N., Radovic, V.M., Dragasevic, N., Stojanovic-Svetel, M., & Kostic, V.S. (1997). Correlation between Bereitschaftspotential and reaction time measurements in patients with Parkinson's disease. Measuring the impaired supplementary motor area function? *Journal of Neurological Sciences, 147,* 177–183.

Fletcher, P.C., Frith, C.D., Grasby, P.M., Friston, K.J., & Dolan, R.J. (1996). Local and distributed effects of apomorphine on fronto-temporal function in acute unmedicated schizophrenia. *The Journal of Neuroscience, 16(21),* 7055–7062.

Flowers, K.A. (1976). Visual "closed-loop" and "open-loop" characteristics of voluntary movement in patients with Parkinsonism and intentional tremor. *Brain, 99(2),* 269–310.

Fried, I., Katz, A., McCarthy, G., Sass, K.J., Williamson, P., Spencer, S.S., & Spender, D.D. (1991). Functional organisation of human supplementary motor cortex studied by electrical stimulation. *Journal of Neuroscience, 11(11),* 3656–3666.

Friston, K.J., & Frith, C.D. (1995). Schizophrenia: A disconnection syndrome? *Clinical Neuroscience, 3,* 89–97.

Friston, K.J., Frith, C.D., Liddle, P.F., & Frackowiak, R.S.J. (1991). Investigating a network model of word generation with positron emission tomography. *Proceedings of the Royal Society of London (Biological Sciences), 244,* 101–106.

Friston, K.J., Liddle, P.F., Frith, C.D., Hirsch, S.R., & Frackowiak, R.S.J (1995). The left medial temporal region in schizophrenia. A PET study, *Brain, 115,* 367–382.

Frith, C.D. (1987). The positive and negative symptoms of schizophrenia reflect impairments in the perception and initiation of action. *Psychological Medicine, 17,* 631–648.

Frith, C.D. (1992). *The cognitive neuropsychology of schizophrenia.* Hove, UK: Lawrence Erlbaum Associates Ltd.

Frith, C.D., & Done, D.J. (1983). Stereotyped responding by schizophrenic patients on a two-choice guessing task. *Psychological Medicine, 13(4),* 779–786.

Frith, C.D., & Done, D.J. (1986). Routes to action in reaction time tasks. *Psychol. Res., 48,* 169–177.

Frith, C.D., Friston, K.J., Herold, S., Silbersweig, D., Fletcher, P., Cahill, C., Dolan, R.J., Frackowiak, R.S.J, & Liddle, P.F. (1995). Regional brain activity in chronic schizophrenic patients during the performance of a verbal fluency task. *British Journal of Psychiatry, 167,* 343–349.

Frith, C.D., Friston, K.J., Liddle, P.F., & Frackowiak, R.S.J. (1991). Willed action and the prefrontal cortex in man: A study with PET. *Proceedings of the Royal Society of London (Biological Sciences), 244,* 241–246.

Frith, C.D., Leary, J., Cahill, C., & Johnstone, E. (1991). Performance on psychological tests. Demographic and clinical correlates of the results of these tests. *British Journal of Psychiatry, 159 (Suppl. 13),* 26–29.

Fukushima, J., Fukushima, K., Miyasak, K., & Yamashita, I. (1994). Voluntary control of saccadic eye movement in patients with frontal cortical lesions and Parkinsonian patients in comparison with that in schizophrenics. *Biol. Psychiatry, 36,* 21–30.

Fuller, R., Nathaniel-James, D., Ron, M., & Jahanshahi, M. (1997). Movement-related potentials preceding self-initiated vs externally-triggered movements in patients with schizophrenia, *Journal of Psychophysiology, 11,* 281–282.

Funahashi, S., Chaffee, M.V., & Goldman-Rakic, P.S. (1993). Prefrontal neuronal activity in rhesus monkeys performing a delayed anti-saccade task. Nature, 365(6448), 753–756.

Georgiou, N., Bradshaw, J.L., Iansek, R., Phillips, J.G., Mattingley, J.B., & Bradshaw, J.A. (1994). Reduction in external cues and movement sequencing in Parkinson's disease. *Journal of Neurology, Neurosurgery and Psychiatry, 57(3),* 368–370.

Georgiou, N., Iansek, R., Bradshaw, J.L., Phillips, J.G., Mattingley, J.B., & Bradshaw, J.A. (1993). An evaluation of the role of internal cues in the pathogenesis of parkinsonian hypokinesia. *Brain, 11(6),* 1575–1587.

Gerfen, C.R. (1992). The neostriatal mosaic: Multiple levels of compartmental organization. *Trends in Neuroscience, 15(4),* 133–139.

Gerloff, C., Corwell, B., Chen, R., Hallett, M., & Cohen, L.G. (1997). Stimulation over the human supplementary motor area interferes with the organization of future elements in complex motor sequences. *Brain, 120(9),* 1587–1602.

Goldberg, G. (1985). Supplementary motor area structure and function: Review and hypotheses. *The Behavioural and Brain Sciences, 8,* 567–616.

Goldberg, G., & Bloom, K.K. (1990). The alien hand sign. Localization, lateralization and recovery. *Am. J. Phys. Med. Rehabil., 69(5),* 228–238.

Goldberg, G., Mayer, N.H., & Togila, J.U. (1981). Medial frontal cortex infarction and the alien hand sign. *Archives of Neurology, 38(11),* 683–686.

Goodrich, S., Henderson, L., & Kennard, C. (1989). On the existence of an attention-demanding process peculiar to simple reaction time: Converging evidence from Parkinson's disease. *Cognitive Neuropsychology, 6,* 309–331.

Grafman, J., Vance, S.C., Weingartner, H., Salazar, A.M., & Amin, D. (1986). The effects of lateralized frontal lesions on mood regulation. *Brain, 109,* 1127–1148.

Grafton, W.T., Waters, C., Sutton, J., Lew, M.F., & Couldwell, W. (1995). Pallidotomy increases activity of motor association cortex in Parkinson's disease: A positron emission tomographic study. *Archives of Neurology, 37,* 776–783.

Guitton, D., Buchtel, H.A., & Douglas, R.M. (1985). Frontal lobe lesions in man cause difficulties in suppressing reflexive glances and in generating goal-directed saccades. *Experimental Brain Research, 58(3),* 455–472.

Halsband, U., & Freund, H.J. (1990). Premotor cortex and conditional motor learning in man. *Brain, 113,* 207–222.

Halsband, U., Ito, N., Tanji, J., & Freund, H.J. (1993). The role of premotor cortex and the supplementary motor area in the temporal control of movement in man. *Brain, 116,* 243–266.

Harrington, D.L., & Haaland, K.Y. (1991). Sequencing in Parkinson's disease. *Brain, 114,* 99–115.

Hashimoto, S., Gembra, H., & Sasaki, K. (1979). Analysis of slow cortical potentials preceding self-paced hand movements in the monkey. *Exp. Neurol., 65,* 218–229.

Holsapple, J.W., Preston, J.B., & Strick, P.L. (1991). The origin of thalamic inputs to the hand representation in the primary motor cortex. *Journal of Neuroscience, 11,* 2644–2654.

Hommer, D.W., Clem, T., Litman, R., & Pickar, D. (1991). Maladaptive anticipatory saccades in schizophrenia. *Biol. Psychiatry, 30(8),* 779–774.

Horne, R.L., Evans, F.J., & Orne, M.T. (1982). Random number generation, psychopathology and therapeutic change. *Archives of General Psychiatry, 39(6),* 680–683.

Humberstone, M., Sawle, G.V., Clare, S., Hykin, J., Coxon, R., Botwell, R., MacDonald, I.A., & Morris, P.G. (1997). Functional magnetic resonance imaging of single motor events reveals human presupplementary motor area. *Annual Neurology, 42,* 632–637.

Hymas, N., Lees, A., Bolton, D., Epps, K., & Head, D. (1991). The neurology of obsessional slowness. *Brain, 114(5),* 2203–2233.

Ikeda, A., Luders, H.O., Burgess, R.C., & Shibasaki, H. (1992). Movement-related potentials recorded from supplementary motor area and primary motor area. *Brain, 115*, 1017–1043.

Ikeda, A., Luders, H.O., Collura, T.F., Burgess, R.C., Morris, H.H., Toshiaki, H., & Shibasaki, H. (1996). Subdural potentials at orbitofrontal and mesial prefrontal areas accompanying anticipation and decision making in humans: A comparison with bereitschaftspotential. *EEG Clinical Electroencephalography, 98*, 206–212.

Ingvar, D.G., & Franzen, G. (1974). Abnormalities of cerebral blood flow distribution in patients with chronic schizophrenia. *Acta Psychiatr. Scand., 50(4)*, 425–462.

Ivry, R.B., Keele, S.E. (1989). Timing functions of the cerebellum. *Journal of Cognitive Neuroscience, 1*, 134–150.

Jackson, S.R., Jackson, G.M., Harrison, J., Henderson, L., & Kennard, C. (1995). The internal control of action and Parkinson's disease: A kinematic analysis of visually guided and memory-guided pretension movements. *Exp Brain Research, 50*, 1–16.

Jahanshahi, M., Brown, R.G., & Marsden, C.D. (1992). Simple and choice reaction time and the use of advance information for motor preparation in Parkinson's disease. *Brain, 115*, 539–564.

Jahanshahi, M., Dirnberger, G., Fuller, R., & Frith, C. (1997). The functional anatomy of random number generation studied with PET. *Journal of Cerebral Blood Flow and Metabolism, 17*, 643.

Jahanshahi, M., Jenkins, I.H., Brown, R.G., Marsden, C.D., Passingham, R.E., & Brooks, D.J. (1995a). Self-initiated versus externally triggered movements. I. An investigation using measurement of regional cerebral blood flow with PET and movement-related potentials in normal and Parkinson's disease subjects (see comments). *Brain, 118*, 913–933.

Jahanshahi, M., Jenkins, I.H., Brown, R.G., Marsden, C.D., Brooks, D.J., Passingham, R.E. (1995b). Self-initiated versus externally-triggered movements: Effects of stimulus predictability assessed with Positron Emission Tomography. *Journal of Psychophysiology, 9*, 177–178.

Jahanshahi, M., Profice, P., Brown, R.G., Ridding, M.C., Dirnberger, G., & Rothwell, J.C. (1998). The effects of transcranial magnetic stimulation over the dorsolateral prefrontal cortex on suppression of habitual counting during random number generation. *Brain, 121*, 1533–1544.

James, W. (1910). *The principles of psychology.* London: McMillan.

Jenkins, I.H., Brooks, D.J., Nixon, P.D., Frackowiak, R.S.J., & Passingham, R.E. (1994). Motor sequence learning: A study with positron emission tomography. *Journal of Neuroscience, 14*, 3775–3790.

Jenkins, I.H., Fernandez, W., Playford, D., Lees, A.J., Frackowiak, R.S.J., Passingham, R.E., & Brooks, D.J. (1992). Impaired activation of the supplementary motor area in Parkinson's disease is reversed when akinesia is treated with apomorphine. *Annals of Neurology, 32*, 749–757.

Joel, D., & Weiner, I. (1994). The organisation of the basal ganglia-thalamocortical circuits: Open interconnected rather than closed segregated. *Neuroscience, 63(2)*, 363–379.

Jones-Gotman, M., & Milner, B. (1977). Design fluency: The invention of nonsense drawings after focal cortical lesions. *Neuropsychologia, 15*, 653–674.

Jueptner, M., Frith, C.D., Brooks, D.J., Frackowiak, R.S.J., & Passingham, R.E. (1997). Anatomy of motor learning. II. Subcortical structures and learning by trial and error. *Journal of Neuroscience, 77(3)*, 1325–1337.

Jueptner, M., Rijntjes, M., Weiller, C., Faiss, J.H., Timmann, D., Mueller, S.P., & Diener, H.C. (1995). Localization of a cerebellar timed process using PET. *Neurology, 45*, 1540–1545.

Jueptner, M., Stephen, K.M., Frith, C.D., Brooks, D.J., Frackowiak, R.S.J., & Passingham, R.E. (1997a). Anatomy of motor learning. I. Frontal cortex and attention to action. *Journal of Neuroscience, 77(3)*, 1313–1324.

Keele, S.W. (1968). Movement control in skilled motor performance. *Psychological Bulletin, 70*, 387–248.

Klosterman, W., Kompf, D., Heide, W., Verleger, R., Wauschkuhn, B., & Seyfert, T. (1994). The presaccadic cortical negativity prior to self-paced saccades with and without visual guidance. *Electroencephalography and Clinical Neurophysiology, 91*, 219–228.

Kornhuber, A.W., Lang, M., Lang, W., Kure, W., & Kornhuber, H.H. (1990). Will and frontal theta activity. In C.H.M. Brunia, A.W.K. Gaillard, & A. Kok (Eds.), *Psychophysiological brain research, Vol. 2* (pp. 55–58). Tilburg, The Netherlans: Tilburg University Press.

Kornhuber, H.H., & Deecke, L. (1965). Hirnpotentialländerungen bei Willeitscaftspotential und reafferente Potentiale. *Pflügers Arch, 284*, 1–17.

Kurtzberg, D., & Vaughan, H.G. (1982). Topographic analysis of human cortical potentials preceding self-initiated and visually triggered saccades. *Brain Research, 243*, 1–9.

Kutas, M., & Donchin, E. (1980). Preparation to respond as manifested by movement-related brain potentials. *Brain Research, 202*, 95–115.

Lang, W., Cheyne, D., Kristeva, R., Beisteiner, R., Lindinger, G., & Deecke, L. (1991). Three-dimensional localization of SMA activity preceding voluntary movement. A study of electric and magnetic fields in a patient with infarction of the right supplementary motor area. *Experimental Brain Research, 87(3)*, 688–695.

Laplane, D., Talairach, J., Meininger, V., Bancaud, J., & Orgogozo, J.M. (1977). Clinical consequences of corticectomies involving the supplementary motor area in man. *Journal of Neurological Sciences, 34(3)*, 301–314.

Leonard, G., Milner, B., & Jones, L. (1988). Performance on unimanual and bimanual tapping masks by patients with lesion of the frontal or temporal lobe. *Neuropsychologia, 26(1)*, 79–81.

Lhermitte, F. (1983). "Utilization behaviour" and its relation to lesions of the frontal lobe. *Brain, 106(2)*, 137–255.

Libet, C.F., Wright, E.W., & Gleason, C.A. (1982). Readiness-potentials preceding restricted "spontaneous" vs. pre-planned voluntary acts. *Electroenceph. Clin. Neurophysiol. 54*, 322–335.

Liddle, P.F., Friston, K.J., Frith, C.D., Hirsch, S.R., Jones, T., & Frackowiak, R.S.F. (1992). Patterns of cerebral blood flow in schizophrenia. *British Journal of Psychiatry, 160*, 179–186.

Liddle, P.F., & Morris, D.L. (1991). Schizophrenic syndromes and frontal performance. *British Journal of Psychiatry, 158*, 340–345.

Limousin, P., Greene, J., Pollack, P., Rothwell, J.R., Benabid, A.L., & Frackowiak, R.S.J. (1997). Changes in cerebral activity pattern due to subthalamic nucleus or internal pallidum stimulation in Parkinson's disease. *Archives of Neurology, 42*, 283–391.

Limousin, P., Pollack, P., Benazzouz, A., Hoffman, D., Le Bas, J.F., Broussolle, E., Perret, J.E., & Benabid, A.L. (1995). Effect of Parkinsonian signs and symptoms of bilateral subthalamic nucleus stimulation. *Lancet, 345*, 91–95.

Lozano, A.M., Lang, A.E., Galvez-Jimenez, N., Miyasaki, J., Duff, J., Hutchinson, W.D., & Dostrovsky, J.O. (1995). Effects of GPi pallidotomy on motor functions in Parkinson's disease (see comments). *Lancet, 346(8987)*, 1383–1387.

Lueck, C.J., Tanyeri, S., Crawford, T.J., Hudson, L., & Kennard, C. (1990). Antisaccades and remembered saccades in Parkinson's Disease. *Journal of Neurology, Neurosurgery and Psychiatry, 53*, 284–288.

Lyon, N., Mejsholm, B., & Lyon, M. (1986). Stereotyped responding by schizophrenic outpatients: Cross-cultural confirmation of perseverative switching on a two-choice task. *Journal of Psychophysiology, 20(2)*, 137–150.

MacKinnon, C.D., Kapur, S., Hussey, D., Verrier, M.C., Houle, S., & Tatton, W.G. (1996). Contributions of the mesial frontal cortex to the premovement potentials associated with intermittent hand movements in humans. *Human Brain Mapping, 4*, 1–22.

Malmo, H.P. (1974). On frontal lobe functions: Psychiatric patient controls. *Cortex, 10*, 231.

Marin, R.S. (1997a). Apathy—Who cares? An introduction to apathy and related disorders of diminished motivation. *Psychiatric Annals, 27*, 18–23.

Marin, R.S. (1997b). Differential diagnosis of apathy and related disorders of diminished motivation. *Psychiatric Annals, 27*, 30–33.

Martin, J.P. (1967). *The basal ganglia and posture.* London: Pitman.

Marsden, C.D., & Obeso, J.A. (1994). The functions of the basal ganglia and the paradox of stereotaxic surgery in Parkinson's disease. *Brain, 117*, 877–897.

Mesulam, M.M. (1986). Frontal cortex and behaviour. *Archives of Neurology, 19(4)*, 320–325.

Miller, W.C., & DeLong, M.R. (1987). Altered tonic activity of neurons in the globus pallidus and subthalamic nucleus in the primate MPTP model of parkinsonism. In M.B. Carpenter, & A. Jayaraman (Eds.), *The basal ganglia II* (pp. 415–427). New York, Plenum Press.

Milner, B. (1964). Some effects of frontal lobectomy in man. In J.M. Warren & K. Akert (Eds.), *The frontal granular cortex and behaviour* (pp. 313–331). New York: McGraw Hill.

Moster, M.L., & Goldberg, G. (1990). Topography of scalp potentials preceding self-initiated saccades. *Neurology, 40*, 644–648.

Muri, R.M., Rivaud, S., Vermersch, A.I., Leger, J.M., & Pierrot-Deseilligny, C. (1995). Effects of transcranial magnetic stimulation over the region of the supplementary motor area during sequences of memory-guided saccades. *Experimental Brain Research, 104(1)*, 163–166.

Neshige, R., Luders, H., & Shibasaki, H. (1988). Recording of movement-related potentials from scalp and cortex in man. *Brain, 111*, 719–736.

Norman, D.A., & Shallice, T. (1986). Attention to action: Willed and automatic control of behavior. In R.J. Davidson, G.E. Schwartz, & D. Shapiro (Eds.), Consciousness and self-regulation. *Advances in research and theory, Vol. 4* (pp. 1–18). New York: Plenum Press.

Nuechterlein, K.H. (1977). Reaction time and attention in schizophrenia: A critical evaluation of the data and theories. *Schizophrenia Bulletin 3(3)*, 373–428.

Okano, K., Tanji, J. (1987). Neuronal activities in the primate motor fields of the agranular frontal cortex preceding visually triggered and self-paced movement. *Experimental Brain Research, 66*, 155–166.

Owen, A.M., Downes, J.J., Sahakian, B.J., Polkey, C.E., & Robbins, T.W. (1990). Planning and spatial working memory following frontal lobe lesions in man. *Neuropsychologia, 28*, 1021–1034.

Owen, A.M., Roberts, A.C., Hodges, J.R., Summers, B.A., Polkey, C.E., & Robbins, T.W. (1993). Contrasting mechanisms of impaired attentional set-shifting in patients with frontal lobe damage or Parkinson's disease. *Brain, 116*, 1159–1175.

Pantelis, C., Barnes, T.E., Nelson, H.E., Tanner, S., Weatherley, L., Owen, A.M., & Robbins, T.W. (1997). Frontal-striatal cognitive deficits in patients with chronic schizophrenia. *Brain, 120*, 1823–1843.

Papa, S.M., Artieda, J., & Obeso, J.A. (1991). Cortical activity preceding self-initiated and externally triggered voluntary movement. *Movement Disorders, 6*, 217–224.

Park, S., & Holzman, P. (1992). Schizophrenics show spatial working memory deficits. *Archives of General Psychiatry, 49(12)*, 975–982.

Passingham, R.E. (1985). Memory of monkeys (Macaca mulatta) with lesions in prefrontal cortex. *Behavioral Neuroscience, 99*, 3–21.

Passingham, R.E. (1997). Functional organisation of the motor system. In R.S.J. Frackowiak, K.J. Friston, C.D. Frith, R.J. Dolan, & J.C. Mazziotta (Ed.), *Human brain function*. San Diego, CA: Academic Press.

Passingham, R.E., Thaler, D.E., & Chen, Y. (1989). Supplementary motor cortex and self-initiated movement. In M. Ito (Ed.), *Neural programming* (pp. 13–24). Karger: Basel.

Paulesu, E., Frith, C.D., & Frackowiak, R.S.J. (1993). The neural correlates of the verbal component of working memory. *Nature, 362*, 342–345.

Paus, T. (1991). Two modes of central gaze fixation maintenance and oculomotor distractibility in schizophrenics. *Schizophrenia Research, 5*, 145–152.

Paus, T., Petrides, M., Evans, A.C., & Meyer, E. (1993). Role of the human anterior cingulate cortex in the control of oculomotor, manual, and speech responses: A positron emission tomography study. *Journal of Neuroscience, 70*, 453–469.

Penfield, W., & Welch, K. (1949). The supplementary motor area in other cerebral cortex of man. *Transactions of the American Neurological Association, 74*, 179–184.

Percheron, G., & Filion, M. (1991). Parallel processing in the basal ganglia: Up to a point (letter;

comment). *Trends in Neuroscience, 14*, 55–59. Comment on: Trends in Neuroscience, 1990. 13, 241–244; *Trends in Neuroscience, 1990, 13,* 266–271.

Perret, E. (1974). The left frontal lobe of man and the suppression of habitual responses in verbal categorical behaviour. *Neuropsychologia, 12(3),* 323–330.

Petersen, S.E., Fox, P.T., Posner, M.I., Nintus, M., & Raichle, M.E. (1988). Positron emission tomographic studies of the cortical anatomy of single word processing. *Nature, 331*, 585–589.

Petrides, M., Alivisatos, B., Meyer, R., & Evans, A.C. (1993). Functional activation of the human frontal cortex during the performance of verbal working memory tasks. *Proceedings of the National Academy of Science, 90*, 878–882.

Pierrot-Deseilligny, C., Rivaud, S., Gaymard, B., & Agid, Y, (1991). Cortical control of memory-guided saccades in man. *Experimental Brain Research, 83*, 607–617.

Playford, D., Jenkins, H., Passingham, R., Nutt, J., Frackowiak, R.S.J., & Brooks, D. (1992). Impaired mesial frontal and putamen activation in Parkinson's disease: A PET study. *Annals of Neurology, 32*, 151–161.

Powell, J.H., Al-Adawi, S., Morgan, J., & Greenwood, R.J. (1996). Motivational deficits after brain injury: Effects of bromocriptine in 11 patients. *Journal of Neurology, Neurosurgery and Psychiatry, 60*, 416–421.

Praamstra, P., Stegeman, D.F., Horstink, M.W., Brunia, C.H., & Cools, A.R. (1995). Movement-related potentials preceding voluntary movement are modulated by the mode of movement selection. *Experimental Brain Research, 103(3),* 429–439.

Praamstra, P., Stegeman, D.F., Horstink, M.W., & Cools, A.R. (1996). Dipole source analysis suggests selective modulation of the supplementary motor area contribution to the readiness potential. *Electroencephalography and Clin. Neurophysiology, 98(6),* 468–477.

Pritchard, W.S. (1986). Cognitive event-related potential correlates of schizophrenia. *Psychological Bulletin, 100(1),* 43–66.

Raichle, M.E., Fiez, J.A., Videen, T.O., MacLeod, A.M., Pardo, J.V., Fox, P.T., & Petersen, S.E. (1994). Practice-related changes in human brain functional anatomy during nonmotor learning. *Cerebral Cortex, 4(1),* 8–26.

Rao, S.S., Harrington, D.L., Haaland, K.Y., Bobholz, J.A., Cox, R.W., & Binder, J.R. (1997). Distributed neural systems underlying the timing of movements. *Journal of Neuroscience, 17(14),* 5528–5535.

Rascol, O., Sabatini, U., Fabre, N., Brefel, C., Loubnoux, I., Celsis, P., Senard, J.M., Montastruc, J.L., & Chollet, F. (1997). The ipsilateral hemisphere is overactive during hand movements in akinetic parkinsonian patients. *Brain, 120(1),* 103–110.

Rektor, I., Feve, A., Buser, P., Bathien, N., & Lamarche, M. (1994). Intracerebral recording of movement-related readiness potentials: An exploration in epileptic patients. *EEG & Clinical Neurophysiology, 90*, 273–283.

Ro, T., Henik, A., Machado, L., & Rafal, R.D. (1997). Transcranial magnetic stimulation of the prefrontal cortex delays contralateral endogenous saccades. *Journal of Cognitive Neuroscience, 9(4),* 433–440.

Robbins, T.W. (1990). The case of frontostriatal dysfunction in schizophrenia. *Schizophrenia Bulletin, 16*, 391–402.

Robbins, T.W. (1991). Cognitive deficits in schizophrenia and Parkinson's disease: Neural basis and the role of dopamine. In P. Wilner & J. Scheel-Kruger (Eds.), *The mesolimbic dopamine system: From motivation to action*. London: John Wiley.

Robertson, C., Hazlewood, R., & Rawson, M.D. (1996). The effects of Parkinson's disease on the capacity to generate information randomly. *Neuropsychologia, 34*, 1069–1078.

Rosahl, S.K., & Knight, R.T. (1995). Role of prefrontal cortex in generation of the contingent negative variation. *Cerebral Cortex, 5(2),* 123–134.

Rosenberg, S., Weber, N., Crocq, M., Duval, F., & Macher, J. (1990). Random number generation by normal, alcoholic and schizophrenic subjects. *Psychological Medicine, 20*, 953–960.

Roy, E.A., Saint-Cyr, J., Taylor, A., & Lang, A. (1993). Movement sequencing disorders in

Parkinson's disease. *International Journal of Neuroscience, 73*, 183–194.

Samuel, M., Ceballos-Baumann, A.O., Blin, J., Uema, T., Boecker, H., Passingham, R.E., & Brooks, D.J. (1997a). Evidence for lateral premotor and parietal overactivity in Parkinson's disease during sequential and bimanual movements: A PET study. *Brain, 120*, 963–976.

Samuel, M., Ceballos-Baumann, A.O., Turjanski, N., Boecker, H., Gorospe, A., Linazasoro, G., Holmes, A.P., DeLong, M., Vitek, J.L., Thomas, D.G.T., Quinn, N.P., Obeso, J.A., & Brooks, D.J. (1997b). Pallidotomy in Parkinson's disease increases supplementary motor area and prefrontal activation during performance of volitional movements. An $H_2\ ^{15}O$ PET study. *Brain, 120*, 1301–1313.

Sasaki, K., & Gemba, H. (1981). Cortical field potentials preceding self-paced and visually initiated movements in one and the same monkey and influences of cerebellar hemispheroctomy upon the potentials. *Neuroscience Letters, 25*, 287–292.

Sawle, G., Brooks, D.J., Marsden, C.D., & Frackowiak, R.S.J. (1991). Corticobasal degeneration. A unique pattern of regional cortical oxygen hypometabolism and striatal fluorodopa demonstrated by positron emission tomography. *Brain, 114(1b)*, 541–556.

Sawle, G., Hymas, N.F., Lees, A.J., & Frackowiak, R.S.J. (1991). Obsessional slowness. Functional studies with positron emission tomography. *Brain 114(5)*, 2191–2202.

Schmand, B., Kuipers, T., Van der Gaag, M., Bosveld, J., Bulthuis, F., & Jellema, M. (1994). Cognitive disorders and negative symptoms as correlates of motivational deficits in psychotic patients. *Psychological Medicine, 24(4)*, 869–884.

Schultz, W. (1997). Dopamine neurons and their role in reward mechanisms. *Current Opinion in Neurobiology, 7*, 191–197.

Schwab, R.S., Chafter, M.E., & Walker, S. (1954). Control of two simultaneous voluntary motor acts in Parkinsonism. *Archives of Neurology and Psychiatry, 72*, 591–598.

Schwartz, F., Carr, A.L., Munich, R.L., Glauber, S., Lesser, B., & Murray, J. (1989). Reaction time impairment in schizophrenia and affective illness: The role of attention. *Biological Psychiatry, 25*, 540–548.

Schwartz, F., Munich, R.F., Carr, A.F., Bartuch, E., Lesser, B., Rescigno, D., & Viegener, B. (1991). Negative symptoms and reaction time in schizophrenia. *Journal of Psychiatry Research, 25(3)*, 131–140.

Seidman, L.J., Yurfelun-Todd, D., Kremen, W.S., Woods, B.T., Goldstein, J.M., Faraone, S.V., & Tsuang, M.T. (1994). Relationships of prefrontal and temporal lobe MRI measures to neuropsychological performance in chronic schizophrenia. *Biological Psychiatry, 35*, 235–246.

Seitz, R.J., Roland, E., Bohm, C., Greitz, T., & Stone-Elander, S. (1990). Motor learning in man: A positron emission tomographic study. *Neuroreport, 1(1)*, 57–60.

Shallice, T. (1982). Specific impairments of planning. *Philosophical Transactions of the Royal Society of London, 298*, 199–209.

Shallice, T. (1988). *From neuropsychology to mental structure*. Cambridge: Cambridge University Press.

Shallice, T., Burgess, P.W., Schon, F., & Baxter, D.M. (1989). The origins of utilization behaviour. *Brain, 112*, 1587–1598.

Sheridan, M.R., Flowers, K.A., & Hurrell, J. (1987). Programming and execution of movement in Parkinson's disease. *Brain, 10*, 1247–1271.

Shibasaki, H., Barrett, G., Halliday, E., & Halliday, A.M. (1980). Components of the movement-related cortical potential and their scalp topograph. *Electroencephalography and Clinical Neurophysiology, 49*, 213–226.

Shibasaki, H., Shima, F., & Kuroiwa, Y. (1978). Clinical studies of the movement-related cortical potential (MP) and the relationship between the dentatorubrothalamic pathway and readiness potential (RP). *Journal of Neuroscience, 219*, 15–25.

Shimizu, N., & Okiyama, R. (1993). Bereitschaftpotential preceding voluntary saccades is abnormal in patients with Parkinson's disease. *Advances in Neurology, 60*, 398–402.

Siegfried, J., & Lippitz, B. (1994). Bilateral chronic electrostimulation of ventroposterolateral pallidum: A new therapeutic approach for alleviating all parkinsonism symptoms. *Neurosurgery, 35*, 1126–1130.

Simpson, J.A., & Khuraibet, A.J. (1987). Readiness potential of cortical area 6 preceding self paced movement in Parkinson's disease. *Journal of Neurology, Neurosurgery and Psychiatry, 50*, 1184–1191.

Singh, J., & Knight, R. (1990). Frontal contribution to voluntary movements in humans. *Brain Research, 531*, 45–54.

Singh, J., Knight, R.T., Rosenlicht, N., Kotun, J.M., Beckley, D.J., & Woods, D.L. (1992). Abnormal premovement brain potentials in schizophrenia. *Schizophrenia Research, 8*, 31–41.

Slade, P.D. (1971). Rate of information processing in a schizophrenic and a control group: The effect of increasing task complexity. *British Journal of Social and Clinical Psychology, 10*, 152–159.

Soliveri, P., Brown, R.G., Jahanshahi, M., & Marsden, C.D. (1992). The effect of practice on the performance of a skilled motor task in patients with Parkinson's disease. *Journal of Neurology, Neurosurgery and Psychiatry, 55*, 454–460.

Spatt, J., & Goldenberg, G. (1993). Components of random generation by normal subjects and patients with dysexecutive syndrome. *Brain Cognition, 23(3)*, 231–242.

Spence, S.A., Brooks, D.J., Hirsch, S.R., Liddle, P.F., Meehan, J., & Grasby, P.M. (1997). A PET study of voluntary movement in schizophrenic patients experiencing passivity phenomena (delusions of alien control). *Brain, 120*, 1997–2011.

Spence, S.A., Hirsch, S.R., Brooks, D.J., & Grasby, P.M. (in press). PET studies of prefrontal activity in schizophrenics and normals: Evidence for remission of "hypofrontality" with recovery from acute schizophrenia. *British Journal of Psychiatry*.

Starkstein, S.E., Mayberg, H.S., Presiosi, T.J., Andrezejewski, P., Leiguarda, R., & Robinson, R.G. (1992). Reliability, validity and clinical correlates of apathy in Parkinson's disease. *Journal of Neuropsychiatry, Clin. Neuroscience, 4(2)*, 134–139.

Stelmach, G.E., & Worringham, C.J. (1988). The control of bimanual aiming movements in Parkinson's disease. *Journal of Neurology, Neurosurgery and Psychiatry, 51(2)*, 223–231.

Stuss, D.T., & Benson, D.F. (1984). Neuropsychological studies of frontal lobes. *Psychological Bulletin, 95*, 3–28.

Tanji, J., & Kurata, K. (1982). Comparison of movement-related activity in two motor cortical areas of primates. *Journal of Neuroscience, 48*, 633–653.

Tanji, J., & Kurata, K. (1985). Contrasting neuronal activity in supplementary and precentral motor cortex of monkeys. I. Responses to instructions determining motor responses to forthcoming signals of different modalities. *Journal of Neuroscience, 53*, 129–141.

Taylor, A.E., Saint-Cyr, J.A., & Lang, A.E. (1986). Frontal lobe dysfunction in Parkinson's disease. *Brain, 109*, 845–883.

Thaler, D., Chen, Y.C., Nixon, P.D., Stern, C.E., & Passingham, R.E. (1995). The functions of the medial premotor cortex. Simple learned movements. *Experimental Brain Research, 102(3)*, 445–460.

Thaler, D.E., Rolls, E.T., & Passingham, R.E. (1988). Neuronal activity of the supplementary motor area (SMA) during internally and externally triggered wrist movements. *Neuroscience Letters, 93(2–3)*, 264–269.

Thickbroom, G.W., & Mastaglia, F.L. (1985). Cerebral events preceding self-paced and visually triggered saccades. A study of presaccadic potentials. *Electroencephalography and Clinical Neurophysiology, 62*, 277–289.

Thompson, P.D., & Marsden, C.D. (1992). Cortico-basal degeneration. Baillieres. *Clin. Neurology, 1*, 677–686.

Touge, T., Werhahn, K.J., Rothwell, J.C., & Marsden, C.D. (1995). Movement-related cortical potentials preceding repetitive and random-choice hand movements in Parkinson's disease. *Archives of Neurology, 37(6)*, 791–799.

Venables, P.H. (1958). Stimulus complexity as a determinant of the reaction time of schizophrenics. *Canadian Journal of Psychology, 12*, 187–190.

Verfaellie, M., & Heilman, K.M. (1987). Response preparation and response inhibition after lesions of the medial frontal lobe. *Neurologie, 44*, 1265–1271.

Vidailhet, M., Stocchi, F., Rothwell, J.C., Thompson, P.D., Day, B.L., Brooks, D.J., & Marsden, C.D. (1993). The Bereitschaftspotential preceding simple foot movement and initiation of gait in Parkinson's disease. *Neurology, 43*, 1784–1788.

Walter, W.G., Copper, R., Aldridge, V.J., McCallum, W.C., & Winter, A.L. (1994). Contingent negative variation: An electrical sign of sensorimotor association and expectancy in the human brain. *Nature, 203*, 380–384.

Weinberger, D.R., Berman, K.F., & Zec, F.R. (1986). Physiological dysfunction of DLPC in schizophrenia. *Archives of General Psychology, 43*, 114–124.

Weinberger, D.R., Mattay, V., Callicott, J., Kotria, K., Santha, A., Van Gelderen, P., Duyn, J., Moonen, C., & Frank, J. (1996). fMRI applications in schizophrenia research. *Neuroimage, 4*, S118–S126.

Weiss, P., Stelmach, G.E., & Hefter, H (1997). Programming of a movement sequence in Parkinson's disease. *Brain, 120*, 91–102.

Westphal, K.P., Grozinger, B., Becker, W., Diekmann, U., & Kornhuber, H.H. (1992). Spectral analysis of EEG during self-paced movements: Differences between untreated schizophrenics and normal controls. *Biological Psychiatry, 31*, 1020–1037.

Westphal, K.P., Grozinger, B., Diekmann, V., Scherb, W., Reess, J., Leibing, U., & Kornhuber, H.H. (1990a). Slower theta activity over the midfrontal cortex in schizophrenic patients. *Acta Psychiatrica Scandinavica, 81(2)*, 132–138.

Westphal, K.P., Grozinger, B., Diekmann, V., Scherb, W., Reess, J., & Kornhuber, H.H. (1990b). EEG-spectra parameters distinguish pathophysiological and pharmacological influences on the EEG in treated schizophrenics. *Arch. Italian Biol. 128(1)*, 55–66.

Wichmann, T., & DeLong, M.R. (1993). Pathophysiology of Parkinsonian motor abnormalities. In H. Narabayashi, T. Nagastu, N. Yanagisawa, & Y. Mizuno (Eds.), *Advances in neurology, Vol. 60*. New York: Raven Press.

Wichmann, T., & DeLong, M.R. (1996). Functional and pathophysiological models of the basal ganglia. *Current Opinions in Neurobiology, 6(6)*, 751–758.

Wiegersma, S., Van der Scheer, E., & Hijman, R. (1990). Subjective ordering, short-term memory, and the frontal lobes. *Neuropsychologia, 28(1)*, 95–98.

Williams, R.M., & Hemsley, D.R. (1986). Choice reaction time performance in hospitalized schizophrenic patients and depressed patients. *European Archives of Psychiatry and Neurological Science, 236*, 169–173.

Wilson, S.A.K., (1925). Disorders of motility and of muscle tone, with special reference to the corpus striatum. *Lancet, 2*, 1–10, 53–62, 169–178, 215–219, 268–276.

Wolkin, A., Sanfilip, M., Wolf, A., Angrist, B., Brodie, J., & Rotrosen, J. (1992). Negative symptoms and hypofrontality in chronic schizophrenia. *Archives of General Psychiatry, 49*, 959–965.

THE VISUAL PERCEPTION OF HUMAN LOCOMOTION

Ian M. Thornton
University of Oregon, Eugene, OR, USA

Jeannine Pinto
Rutgers University, Newark, NJ, USA

Maggie Shiffrar
Rutgers University, Newark, NJ, USA and UMR CNRS, Université de la Méditerranée, Marseille, France

To function adeptly within our environment, we must perceive and interpret the movements of others. What mechanisms underlie our exquisite visual sensitivity to human movement? To address this question, a set of psychophysical studies was conducted to ascertain the temporal characteristics of the visual perception of human locomotion. Subjects viewed a computer-generated point-light walker presented within a mask under conditions of apparent motion. The temporal delay between the display frames as well as the motion characteristics of the mask were varied. With sufficiently long trial durations, performance in a direction discrimination task remained fairly constant across inter-stimulus interval (ISI) when the walker was presented within a random motion mask but increased with ISI when the mask motion duplicated the motion of the walker. This pattern of results suggests that both low-level and high-level visual analyses are involved in the visual perception of human locomotion. These findings are discussed in relation to recent neurophysiological data suggesting that the visual perception of human movement may involve a functional linkage between the visual and motor systems.

INTRODUCTION

Any animal's survival depends upon its ability to identify the movements of both prey and predators. As social animals, humans behave largely in accordance with their interpretations and predictions of the actions of others. If the visual system has evolved so as to be maximally sensitive to those factors upon which an animal's survival depends (Shepard,

---

Requests for reprints should be addressed to Maggie Shiffrar, UMR CNRS: Mouvement et Perception, Université de la Méditerranée, Faculté des Sciences du Sport, 163, avenue de Luminy CP 910, 13288 Marseille, CEDEX 9, France (Tel: (33) 4 91 17 22 71; Fax: (33) 4 91 17 22 52; E-mail: mag@laps.univ-mrs.fr

This work was funded by NIH:NEI grant 099310 to the third author and NATO Collaborative Research Grant CRG970528 (with J. Pailhous and M. Bonnard of the CNRS at the Université de la Méditerranée) to the second and third authors. Some of these results were presented at the 1995 Congress on Perception and Action in Marseille, France and at the 1996 ARVO Conference. We thank James E. Cutting for kindly providing an updated version of his walker code.

1984), then one would expect to find that human observers are particularly sensitive to human movement. Several decades of perceptual research support this prediction. In a classic study of the visual perception of human movement, Johansson demonstrated that human observers can readily recognise extremely simplified deceptions of human locomotion (e.g. Johansson, 1973, 1975; Johansson, von Hofsten, & Jansson, 1980). Extending a technique first devised by Marey (1972) in 1895, Johansson created "point-light walker" displays by filming human actors with small light sources attached to their major joints. By adjusting the lighting, the resultant film showed only a dozen or so moving points of light, as shown in Fig. 1. Nevertheless, observers of these films report a clear and compelling perception of the precise actions performed by the point-light defined actors. Importantly, observers rarely recognise the human form in static displays of these films (Johansson, 1973). Subsequent research has demonstrated that our perception of the human form in such displays is rapid (Johansson, 1976), orientation specific (Bertenthal & Pinto, 1994; Pavlova, 1989; Sumi, 1984), tolerates random contrast variations (Ahlström, Blake, & Ahlström, 1997), and extends to the perception of complex actions (Dittrich, 1993), social dispositions (MacArthur & Baron, 1983), gender (Kozlowski & Cutting, 1977, 1978), and sign language (Poizner, Bellugi, & Lutes-Driscoll, 1981).

What neural mechanisms underlie the visual perception of human movement? Recent neurophysiological research suggests that

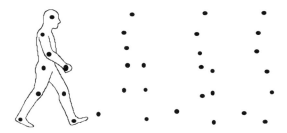

Fig. 1. Four static views of a point-light walker. The outline of the human body, shown in the first frame, is never shown in experimental stimuli. When presented statically, these displays are difficult to interpret. However, when set in motion, observers easily organise the complex patterns of point motion into a coherent perception of human locomotion.

relatively high-level integrative mechanisms may play a fundamental role in the visual analysis of human movement. For example, the superior temporal polysensory area (STP) of the macaque monkey, which receives input from both dorsal and ventral visual pathways (Baizer, Ungerleider, & Desimone, 1991), contains cells that appear to be selectively attuned to precise combinations of primate forms and movements (Perrett, Harries, Mistlin, & Chitty, 1990). Neurons in this area have also been shown to respond to Johansson point-light walker displays (Oram & Perrett, 1994). Furthermore, case studies of patients with extrastriate lesions sparing the temporal lobe demonstrate that individuals can lose their ability to perceive simple motion displays while retaining the perception of point light walker displays (Vaina, Lemay, Bienfang, Choi, & Nakayama, 1990; McLeod, Dittrich, Perrett, & Zihl, 1996).

A behavioural signature of high-level visual processes is their dependence upon global display characteristics. More specifically, most models of the visual system are hierarchical in nature (e.g. Van Essen & DeYoe, 1995; Zeki, 1993). Visual analyses at the lower levels of this hierarchy are thought to occur within brief temporal intervals and small spatial neighbourhoods. The results of these low-level or "local" analyses are then passed onto and processed by higher-level or more "global" mechanisms, which process information across larger spatiotemporal extents. Although local and global are difficult to define as absolute terms, most studies of the visual perception of human movement have defined local analyses as the computations conducted on individual points (joints) or point pairs (limbs). Global analyses are conducted over larger areas and generally involve half to an entire point-light walker. In the temporal domain, local motion processes are thought to be restricted to a window of 50msec or less (Baker & Braddick, 1985), while global motion processes may operate over much longer intervals.

Several psychophysical studies support the hypothesis that the visual perception of human movement depends upon a spatially global mechanism (e.g. Ahlström et al., 1997; Cutting, Moore, & Morrison, 1988). One approach to this issue involves masked point-light displays. In this paradigm, observers view displays containing a point-light walker that is masked by the addition of superimposed moving point lights. This mask can be constructed from multiple point-light walkers that are positionally scrambled so that the spatial location of each point is randomised. The size, luminance, and velocity of the points remain unchanged. Thus, the motion of each point in the mask is identical to the motion of one of the points defining the walker. As a result, only the spatially global configuration of the points distinguishes the walker from the mask. The fact that subjects are able to detect the presence as well as the direction of an upright point-light walker "hidden" within such a scrambled walker mask implies that the mechanism underlying the perception of human movement operates over large spatial scales (Bertenthal & Pinto, 1994). The spatially global analysis of human movement is further supported by studies of the aperture problem. Whenever a moving line is viewed through a relatively small window or aperture, its motion is ambiguous because the component of translation parallel to the line's orientation cannot be measured. As a result, the line's motion is consistent with an infinitely large family of different motion interpretations (Wallach, 1935). The visual system can overcome this measurement ambiguity or aperture problem through local motion analyses (restricted to small spatial regions) or global motion analyses (that link information across disconnected spatial regions). When viewing a walking stick figure through a multiple aperture display, observers readily perceive global human movement. Under identical conditions, however, observers default to local interpretations of moving nonbiological objects and upside-down walkers (Shiffrar, Lichtey, & Heptulla-Chatterjee, 1997). This pattern of results suggests that the visual analysis of human locomotion can extend over a larger or more global spatial area than the visual analysis of other, nonbiological motions.

While the mechanism underlying the visual perception of human locomotion appears to conduct global analyses over space, its temporal characteristics remain unclear. Psychophysical researchers commonly use the phenomenon of apparent motion to investigate the temporal nature of motion processes. In classic demonstrations of apparent motion, two spatially separated objects are sequentially presented within a certain temporal range so that they give rise to the perception of a single moving object. Early studies demonstrated that apparent motion percepts depend critically upon the temporal separation of the displays (Korte, 1915; Wertheimer, 1912). When displays are separated by relatively long inter-stimulus intervals (ISIs), long-range apparent motion processes are thought to integrate information across the displays and to facilitate the perception of motion. On the other hand, when the frames in an apparent motion display are separated by brief temporal intervals (short ISIs), short-range processes are thought to underlie motion percepts (Anstis, 1980; Baker & Braddick, 1985). Long-range processes alone may conserve global cues to image structure such as object orientation (e.g. McBeath & Shepard, 1989), spatial frequency (e.g. Green, 1986), and perceptual grouping principles (e.g. Pantle & Petersik, 1980). Although there has been much debate concerning the precise nature of apparent motion phenomena (Cavanagh, 1991; Cavanagh & Mather, 1989; Petersik, 1989, 1991), the traditional distinction between long- and short-range processes will be adopted here as it provides a useful framework within which to discuss temporal manipulations involving a single class of stimuli.

The perception of human movement in apparent motion displays provides an intriguing demonstration of the difference between short-range (temporally brief) and long-range (temporally extended) motion processes. In all apparent motion displays, the figure(s) shown in each display frame can be connected by an infinite number of possible paths. Under most conditions, however, observers typically report seeing only the shortest possible path of motion (e.g. Burt & Sperling, 1981). Yet, when humans move, their limbs tend to follow curved rather than straight trajectories. Given the visual system's shortest-path bias, will observers of human movement be more likely to perceive apparent motion paths that are consistent with the movement limitations of the human body or paths that traverse the shortest possible distance? This hypothesis has been tested previously with stimuli consisting of photographs of a human model in different positions created so that the biomechanically possible paths of motion conflicted with the shortest possible paths (Shiffrar & Freyd, 1990, 1993). For example, one stimulus consisted of two photographs in which the first displayed a standing woman with her right arm positioned on the right side of her head while the second photograph showed this same arm positioned on the left side of the woman's head. The shortest path connecting these two arm positions would involve the arm moving through the head whereas a biomechanically plausible path would entail the arm moving around the head. When subjects viewed such stimuli, their perceived paths of motion changed with the Stimulus Onset Asynchrony (SOA) or the amount time between the onset of one photograph and the onset of the next photograph. At short SOAs, subjects reported seeing the shortest, physically impossible, motion path. However, with increasing SOAs, observers were increasingly likely to see apparent motion paths consistent with normal human movement (Shiffrar & Freyd, 1990). Conversely, when viewing photographs of inanimate control objects, subjects consistently perceived the same shortest path of apparent motion across increases in SOA. Importantly, when viewing photographs of a human model positioned so that the shortest movement path was a biomechanically plausible path, observers always reported seeing this shortest path (Shiffrar & Freyd, 1993). Thus, subjects do not simply report the perception of longer paths with longer presentation times. Moreover, observers can perceive apparent motion of nonbiological objects in a manner similar to apparent motion of human bodies. However, these objects must contain a global hierarchy of orientation and position cues resembling the entire human form before subjects perceive human-like paths (Heptulla-Chatterjee, Freyd, & Shiffrar, 1996). This pattern of results suggests that human movement is analysed by long-range motion processes that operate over large temporal intervals.

However, this conclusion appears inconsistent with the results of another series of apparent motion experiments (Mather, Radford, & West, 1992). These intriguing studies involved the presentation of synthesised point-light displays depicting the sagittal view of a person walking within a mask of randomly moving point lights. In some of these studies, observers reported whether the animated walker faced leftward or rightward in the picture plane. To create conditions appropriate for both long-range and short-range apparent motion, blank frames were added between the frames containing the masked walker. When the time between successive point-light walker frames (ISI) reached or surpassed 48msec, observers were unable to discriminate the two directions of walker motion. Since subjects could only perform the motion discrimination task under short-range apparent motion conditions, their perception of human movement appears to have depended upon local motion analyses. This finding suggests that the mechanism underlying the visual perception of biological motion analyses information within small temporal windows.

Thus, it is not yet clear whether the visual perception of human locomotion involves temporally local or global processes. Because the temporal studies cited differ significantly in methodology, their apparently conflicting results can not be unambiguously interpreted. Did the difference in results arise from methodological differences in display form, subject task, masking, or display duration? The goal of the following experiments was to resolve this interpretation limitation, and thereby to provide a better understanding of the mechanism underlying this perceptual behaviour. These studies were motivated by the following assumption. If the neural mechanism subserving the visual perception of human locomotion operates over extended temporal windows, then subjects should be able to perform perceptual judgements of human locomotion under long-range apparent motion conditions.

EXPERIMENT 1: TRIAL DURATION

Why were subjects in the experiments of Mather et al. (1992) unable to determine a point-light walker's direction of motion under long-range apparent motion conditions? One possible reason concerns overall display duration. Johansson (1976) found that naive observers could identify a human form and its action from a point-light walker displayed for 200msec. However, the correct identification of a point-light walker presented within a mask requires longer display durations. Specifically performance in a direction discrimination task can fall to chance levels when masked point-light walkers are presented for less than 800msec (Cutting et al., 1988). In the experiments of Mather and his colleagues, the masked point-light walker was visible for as little as 240msec per trial. On the other hand, in the studies by Shiffrar and her colleagues (Heptulla-Chatterjee et al., 1996; Shiffrar & Freyd, 1990, 1993; Shiffrar et al., 1997), human movement displays were usually presented for several seconds. Thus, one possible explanation is that the use of brief display durations may lead to an underestimation of observers' perceptual capacities to interpret human movement. To examine this possibility, a modified replication of one of the studies conducted by Mather et al. (1992) was undertaken. Briefly, subjects performed a two-alternative forced-choice task in which they discriminated between rightward and leftward facing point-light walkers presented within a mask. The experimental modification involved the use of both long-duration and short-duration trials. If poor performance results from the use of excessively brief display durations, then performance in the long-duration trials should be superior to performance in the short-duration trials. Secondly, if above chance levels of performance are found, then the results of this experiment can be used to test whether low-level or high-level motion analyses are involved in the perception of human movement. More specifically, if performance at all ISIs is mediated exclusively by short-range motion processes, then performance should fall to chance levels with ISIs that extend beyond the temporal window for short-range analyses; namely, ISIs greater than approximately 50msec. If, however, the perception of human locomotion involves temporally extended motion analyses, then performance should remain well above chance with increases in ISI.

Method

Subjects

Three experienced psychophysical observers participated in this experiment. All observers had normal or corrected-to-normal vision. One subject was an author whereas the remaining subjects were naive with regard to the purpose of this study.

Apparatus

All stimuli were displayed on a Macintosh 21" (40 × 30cm) RGB monitor with a refresh rate of 75Hz and a 1152 × 870 pixel resolution. Monitor output was controlled by a Macintosh Quadra 950. A chin rest was used to fix the subjects' viewing distance at 90cm from the monitor. The stimuli were presented in a 6.3° by 6.3° window positioned in the centre of the monitor. This window size closely replicated that used by Mather et al. (1992). This apparatus was used in both of the experiments reported here.

Stimuli

The stimuli were generated by modifying, in Think C version 7.0, a classic point-light walker algorithm (Cutting, 1978) together with a simultaneously presented mask of randomly moving dots (Cutting et al., 1988). Each animation frame consisted of 77 identical black dots displayed against a uniform, middle grey background. Eleven of these dots defined the walker while the remaining 66 dots defined the mask. Every dot, whether it belonged to the mask or the walker, was a 5 × 5 pixel square that subtended 6.1 min arc.

The simulated walker was displayed in profile as shown in Fig. 2. The dots that defined the walker were positioned on the simulated head, near shoulder, both elbows, both wrists, near hip, both knees, and both ankles of the walker (Cutting, 1978). As in previous masked point-light walker studies, the walker was always displayed with all 11 dots. That is, dots did not disappear when they would normally be occluded by the walker's torso or limbs. The removal of this natural occlusion cue minimised non-motion related cues to the location of the walker in the mask (Bertenthal & Pinto, 1994; Cutting et al., 1988; Mather et al., 1992). The mask dots themselves were placed randomly around the walker on a frame-by-frame basis. As a result, the dots defining the walker and the mask could only be distinguished from each other by their motion. Mather et al. (1992) nicely described these stimuli, when set in motion, as resembling a "figure striding through a light snowstorm".

The walker figure subtended 4.6° in height (head to ankle) and 2.4° in width at the most extended point of the step cycle. A complete stride cycle (i.e. the sequence of movements that occurs between two consecutive repetitions of a body configuration) was achieved in 40 animation frames. The duration of each frame was fixed at 40msec. As a result, when these frames were presented in immediate succession, a walking speed of 38 strides per minute was simulated. This speed falls within the range of 30–70 strides per minute associated with human walking under normal conditions (Inman, Ralston, & Todd, 1981). The walker figure did not translation across the screen but rather appeared to walk in place as if on a treadmill. On half of the trials, the walker faced and walked to the right while on the other half of the trials, the walker faced and walked to the left. The horizontal and vertical

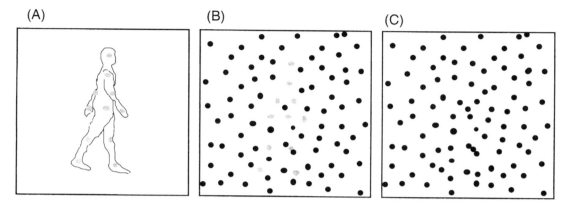

Fig. 2. The creation of a masked point-light walker display. Frame A illustrates a walker with 11 grey points fixed to each of the major body joints and the head. Frame B displays the grey point-light walker within a mask of black points. In the experimental stimuli, the walker points and mask points are identical, as shown in Frame C. The walker can be located within dynamic but not static displays.

position of the walker was randomised within the central display area on a trial-by-trial basis. The walker's position was constrained by the need to ensure that none of the dots defining the walker approached or exceeded the boundary of the display area. The starting position within a stride cycle (e.g. legs far apart or close together) was also randomised on each trial. These display manipulations ensured that subjects would not be able to identify the walker configuration simply by its presentation at a particular location or during a specific animation frame.

To manipulate the ISI, and thereby create long-range and short-range apparent motion, a blank frame was inserted between each of the animation frames. This blank frame contained no dots and was the same uniform grey as the background in the animation frames. Across trials, the duration of these blank frames was varied from 0msec (no blank frame) to 120msec in 15msec increments. This yielded a total of nine different Inter-stimulus Intervals (ISIs) of 0, 15, 30, 45, 60, 75, 90, 105 and 120msec.

There were two types of trials. A short-duration trial consisted of 20 animation frames and corresponded to half of a walker's stride cycle. This short-duration trial condition was selected in order to replicate the findings of Mather et al. (1992). Long-duration trials consisted of an 80-frame sequence and allowed for the presentation of two complete strides. Within each trial duration, the full range of ISIs was used. Trial duration was always equal to or greater than 800msec. More precisely, the overall duration of the 20 frame trials was 800msec when the ISI equalled 0msec and 3.2 sec when the ISI equalled 120msec. The 80 frame trials had durations as brief as 3.2sec and as long as 12.8sec when the ISI was 0 or 120msec, respectively.

Procedure

Subjects were seated in front of the display monitor and were told that they would see a point-light walker within a mask. They were instructed to determine, on each trial, if the walker's direction was to the left or right and then to press one of two buttons on a computer keyboard to indicate their decision. Responses could only be recorded after an animation sequence was completed. Subjects initiated the next trial by pressing another button on the keyboard. No feedback was provided during the practice or experimental sessions.

According to a within-subjects design, each subject completed four blocks of short-duration trials and four blocks of long-duration trials. These eight blocks were intermixed and their order was counterbalanced across subjects. Each block contained 10 trials at 9 different ISIs for a total of 90 trials. On average, subjects completed 1 block of trials in approximately 15 minutes. The order of the trials within each block was randomised independently for each subject. All subjects completed 18 practice trials before beginning each new block of experimental trials.

Results

The results, shown in Fig. 3, are plotted as the mean percentage of trials during which subjects correctly reported the walker's direction at each ISI level in both the short (20 frame) and long (80 frame) trial duration conditions. A 2 (Condition) × 9 (ISI) repeated measures ANOVA was used to analyse these data. A significant main effect of Condition [$F(1,2) = $

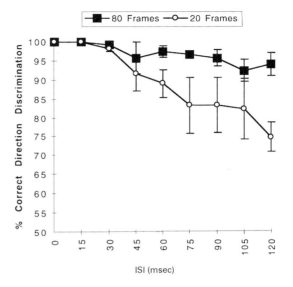

Fig. 3. The results of Experiment 1. The results are collapsed across subjects. Performance in the long-duration trial condition, indicated by the filled squares, remains high across variations in the ISI. Performance in the short-duration trial condition, shown by the empty circles, decreases with increasing temporal delays. The error bars represent the standard error of the mean.

23.07, $MSE = 33.8$, $P < .05$], was identified, with responses to 80 frame trials being more accurate ($M = 96.85$, $SD = 4.01$) than responses to 20 frame trials ($M = 89.26$, $SD = 11.16$). While there was also a significant main effect of ISI [$F(8,16) = 4.7$, $MSE = 40.5$, $P < .01$], this effect should be interpreted in the light of a Condition × ISI interaction [$F(8,16) = 3.12$, $MSE = 22.2$, $P < .05$]. To explore this interaction further, post hoc contrasts were used to compare Condition means at each level of ISI. This analysis revealed a significant divergence in performance by 60 msec [$F(1,16) = 4.7$, MSE

= 22.2, $P < .05$] with the short-duration trials remaining significantly below the long-duration trials for all ISIs beyond this point. Separate repeated measures ANOVAs confirmed this pattern of results with a strong main effect of ISI for the 20-frame condition [$F(8,16) = 4.37$, $MSE = 54.75$, $P < .01$], but only a marginal effect for the 80-frame condition [$F(8,16 = 2.54$, $MSE = 8.0$, $P < .064$]. Finally, it is important to note that even the poorest performance, which occurred in the 20-frame condition when the ISI equalled the 120msec, was still significantly above chance [$t(2) = 6.55$, $P < .01$].

Discussion

The results of this experiment clearly demonstrate that observers can perceive human locomotion under both long-range and short-range apparent motion conditions. More precisely, in the 20-frame condition, ceiling levels of performance were recorded when the temporal delay or ISI between the frames displaying the masked point-light walker was less than 60msec. This value is consistent with the 0 to 50msec temporal window associated with short-range apparent motion processes (Baker & Braddick, 1985). Beyond this point, performance dropped with increasing ISIs. This pattern of results replicates those of Mather et al. (1992, Expt. 2) in which direction discrimination performance dropped with ISIs greater than 48msec. However, in the present experiment, performance in the 80-frame trial duration condition remained relatively flat across increases in ISI. Since the long duration trial condition was constructed by simply increasing the number of walker frames from 20 to 80, the responses of low-level motion detectors should have remained unchanged. Nonetheless, subjects were better able to determine the point-light walker's direction of motion under long-range apparent motion conditions when trial durations were extended beyond those used by Mather et al.

Although the pattern of results from the short trial duration condition is very similar to the pattern reported in Mather et al. (1992), absolute performance differs. Subjects in the current experiment performed the direction discrimination task more accurately than subjects in the direction discrimination experiment of Mather et al. This difference may reflect our use of only trained psychophysical observers. However, we have since replicated this same pattern of results with more than 20 naive observers (Pinto, Thornton, & Shiffrar, 1998). Superior overall performance may have also resulted from differences in frame duration. Each walker frame was displayed for 40msec in the current experiment but for only 24msec in the direction discrimination experiment by Mather and his colleagues. Thus, superior performance with longer frame durations is completely consistent with the hypothesis that subjects perform relatively poor perceptual judgements of masked human locomotion when displays are presented only briefly (Cutting et al., 1988).

Previous investigators of the visual perception of biological motion have used masked point-light walker displays to examine the spatial nature of this perceptual process. The results of their studies suggest that the perception of human movement involves spatially global analyses (Bertenthal & Pinto, 1994; Cutting et al., 1988). In earlier studies of the temporal characteristics of biological motion perception, researchers have varied the delay between photographs of a human model in different positions. The results of these studies support the existence of a temporally global mechanism (Shiffrar & Freyd, 1990, 1993). The current methodology involved a combination of these strategies, since a temporal delay was inserted between frames depicting a masked point-light walker. The current results therefore suggest that subjects can make subtle perceptual judgements about human locomotion even when these judgements require visual analyses that are global across both space and time. This findings is consistent with the hypothesis that a high-level mechanism, rather than low-level motion processes alone, underlies the visual perception of human movement.

However, it is important to note that the results of this experiment can not be convincingly interpreted as exclusively representing a high-level mechanism. That is, if performance in the long trial duration condition were solely the function of a temporally global analysis, then performance should have been independent of ISI. Yet, performance varied with ISI. One possible interpretation of this result is that local motion analyses may be involved in the perception of human movement. The goal of the following experiment was to determine more precisely whether low level motion analyses play a role in the visual perception of human movement.

EXPERIMENT 2: MASK COMPLEXITY

The mask used in Experiment 1 and in Mather et al. (1992) consisted of randomly moving points. Thus, the position of each point in the mask was uncorrelated from frame to frame. Since the walker points had pendular trajectories that simulated normal human locomotion, the position of these points was correlated across frames. As a result, the motions of the individual points of the mask and walker differed. These local differences were therefore available to low-level motion detectors and may have contributed to the detection of the walker in the mask. Therefore, a different type of mask is needed to eliminate the utility of low-level motion processes.

Previous research has shown that subjects can accurately discriminate the direction of a point-light walker in a mask even when the motion of each mask point mimics the motion of a walker point (Bertenthal & Pinto, 1994). These so-called "scrambled walker" masks are constructed by duplicating a point-light walker several times and then scrambling the starting position, but not the motion trajectory, of each point. This process yields a mask which might, for example, consist of points corresponding to seven left wrists plus seven right wrists plus seven left ankles plus seven heads, etc., and each having a randomly determined location within the 2D plane of the mask. Only the configuration of points that define the walker can be used to distinguish the walker from the mask. Thus, such "scrambled walker" masks more thoroughly camouflage human location than "random dot" masks (Cutting et al., 1988). In other words, "scrambled walker" masks can be used to eliminate or drastically reduce the influence of low-level motion processes in the perception of point-light walkers.

If the visual analysis of human locomotion is global across both space and time, then subjects should be able to interpret a point-light walker within a scrambled walker mask even under conditions of long-range apparent motion. To test this prediction, subjects performed a modified replication of Experiment 1 in which the same point-light walker was presented within a scrambled walker mask rather than a random dot mask.

Method

The same three psychophysical observers from Experiment 1 served as subjects in this experiment. As before, two of the subjects were naive to the hypothesis under investigation.

The subjects' task in this experiment was identical to that of the previous experiment. The displays were also identical except for the motion trajectories of the dots making up the mask. In the previous experiment, the mask dots moved randomly. In this experiment, each dot in the mask had a motion trajectory that was identical to the trajectory of one of the dots defining the walker. This "scrambled walker" mask was created by generating six copies of the walker within the display area. The initial vertical and horizontal positions of each dot were then randomised within the display window. As a result, each mask dot had the same velocity as one of the walker dots but bore no predictable spatial relationship to any other dot. As before, the mask dots also had the same size, colour, and luminance as the walker dots. The experimental procedure replicated that of Experiment 1.

Results

The results, shown in Fig. 4 as the mean percentage of trials during which subjects correctly reported the walker's direction at each ISI level, were analysed in a 2 (Condition) × 9 (ISI) repeated measures ANOVA. This yielded a significant main effect of Condition [$F(1,2) = 51.12$, $MSE = 32.1$, $P < .05$], with responses to 80-frame trials being more accurate ($M = 74.72$, $SD = 13.36$) than responses to 20-frame trials

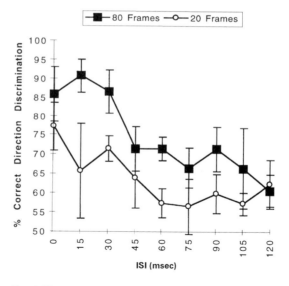

Fig. 4. The results of Experiment 2 collapsed across subjects. Performance in the long-duration trial condition (filled squares) is above chance for ISIs less than 90msec and superior to performance in the short-duration trial condition (empty circles). Error bars represent the standard error of the mean.

($M = 63.7$, $SD = 11.82$). Unlike in Experiment 1, there was no Condition × ISI interaction. Separate analysis of the data from the two conditions revealed only a marginal main effect of ISI for the 20-frame condition [$F(8,16) = 2.46$, $MSE = 61.01$, $P < .06$] and a significant main effect of ISI for the 80-frame condition [$F(8,16) = 6.13$, $MSE = 53.65$, $P < .01$]. Polynomial contrasts revealed that this main effect had a strong linear component [$F(1,16) = 40.13$, $MSE = 53.65$, $P < .001$], reflecting a gradual drop in performance between the 0msec ($M = 85.83$, $SD = 12.58$) and 120msec ($M = 60.83$, $SD = 7.2$) ISI increments. T-tests indicated that performance in the 20-frame condition remained at chance levels (50%) for all ISI increments except 0 and 30msec. In contrast, in the 80-frame condition, performance remained significantly above chance (all $ps < .05$) for all ISIs except those of 105msec ($P = .13$) and 120msec ($P = .06$).

Discussion

Three general conclusions are suggested by the results of this study. First, performance in this direction discrimination task is better at long (80-frame) trial durations than at short (20-frame) trial durations. This finding further supports the hypothesis that poor performance in this task can stem from the use of trials presented over insufficient durations. Second, performance in the long-duration trial condition suggests that subjects can integrate motion correctly over large spatial and temporal extents in the analysis of human locomotion even when masking renders local motion signals uninformative. This finding clearly suggests that high-level or temporally global motion analyses are involved in the visual perception of human movement. Finally, comparison with the results of Experiment 1 demonstrates that the perception of a point-light walker is more difficult when it is presented within a mask of identically moving points than in a mask of randomly moving points. Local differences in motion trajectories are available in random dot masks but not in scrambled walker masks. These local motion differences may account for the performance differences between Experiments 1 and 2. This interpretation is further supported by the results of the long-duration trial condition in this experiment. Although performance was generally above chance, it also dropped with increasing ISI. The influence of low-level motion detectors is thought to decrease as temporal delays increase (e.g. Baker & Braddick, 1985). If so, when considered together, these results suggest that both low-level (Mather et al., 1992) and high-level (Bertenthal & Pinto, 1994; Shiffrar & Freyd, 1990, 1993) visual mechanisms may be involved in the visual perception of human locomotion

GENERAL DISCUSSION

The goal of this behavioural research project was to develop a better understanding of the mechanisms underlying the visual interpretation of human movement by examining the temporal characteristics of locomotion perception. In two experiments, subjects viewed

Johansson-like point-light walkers presented within a mask of moving points and reported the walker's direction of motion. Apparent motion displays were created by inserting blank frames of variable duration (or ISIs) between the walker frames. In Experiment 1, subjects viewed point-light walkers within a mask of randomly moving points over short and long trial durations. When only 20 walker frames were presented, performance dropped with ISIs greater than 60msec. This performance pattern replicates earlier findings (Mather et al., 1992). When the same masked walker was shown for 80 frames per trial, near-ceiling levels of performance were found across variations in ISI. This finding, that longer trial durations can improve performance, supports previous demonstrations that subjects report the perception of human movement under long-range apparent motion conditions (Heptulla-Chatterjee et al., 1996; Shiffrar & Freyd, 1990, 1993). When considered together, the results of this experiment suggest that the perceptual processes tapped by point light walker displays can operate over extended spatiotemporal neighbourhoods. Such global behaviour is generally considered to be a signature of mechanisms resisting within relatively late stages of the visual system.

In Experiment 2, the point-light walker was presented within a "scrambled walker" mask rather than in a "random dot" mask. As a result, the motion trajectories of the points defining the mask were identical to the motion trajectories of the walker points. Under these conditions, subjects generally performed at chance levels in the short trial duration condition. In the long trial duration condition, performance was generally above chance and depended upon ISI. Above-chance performance with ISIs greater than 50msec is thought to reflect high-level motion processes (Anstis, 1980; Baker & Braddick, 1985). Such processes may allow for attentional tracking of the point-light walker over extended temporal intervals (Cavanagh, 1992; Lu & Sperling, 1995; Thornton, Rensink, & Shiffrar, 1998). Interestingly, neural representations of action are influenced by attentional processes (Decety, 1996). However, other aspects of the results of this experiment cast serious doubt on the hypothesis that the visual perception of human movement depends exclusively on high-level neural processes. First, in the long trial duration conditions, performance was at ceiling when random dot masks were used but significantly below ceiling when scrambled walker marks were employed. Since scrambled walker masks effectively eliminate the utility of local motion analyses, suboptimal performance with these masks can be attributed to the loss of input from local analyses. Second, in the long trial duration condition of Experiment 2, subjects could not accurately judge the walker's direction at long ISIs. This finding further supports the importance of temporally restricted, or low-level motion analyses. Thus, the results of these experiments suggest that both local and global processes contribute to our visual interpretation of the movements of others.

Since low-level motion detectors may serve as the gateway to the perception of object motion, it might not be surprising that they play an important role in the visual perception of human movement. Indeed, models involving strictly local computations do capture some aspects of the visual perception of human movement (Hoffman & Flinchbaugh, 1982; Webb & Aggarwal, 1982). However, such approaches cannot explain the orientation specificity (Ahlström et al., 1997; Bertenthal & Pinto, 1994; Pavlova, 1989; Sumi, 1984) nor the spatio-temporal limits within which we can visually identify a moving human. It is also unclear how such models can be extended to account for our ability to visually classify different human actions (Dittrich, 1993; MacArthur & Baron, 1983). Thus, the critical question becomes, what is the nature of the high level mechanism(s) involved in the visual perception of locomotion? Neurophysiological and case studies suggest that area STP may play an important role in the visual perception and/or interpretation of human movement (McLeod et al., 1996; Oram & Perrett, 1994; Perrett et al., 1990; Vaina et al., 1990). Since this region receives convergent input from the dorsal and ventral pathways (Baizer et al., 1991), it may be involved in the integration of form and motion cues (Perrett et al., 1990). This integration may contribute to the visual perception of a moving human form across space and time.

Another line of research suggests that the visual perception of human movement may involve a functional linkage between the perception and production of motor activity (Viviani, Baud-Bovy, & Redolfi, 1997; Viviani & Stucchi, 1992). In other words, the perception of human movement may be constrained by knowledge of human motor limitations (Shiffrar, 1994; Shiffrar & Freyd, 1990, 1993). Given our extensive visual exposure to people in action, it is possible that this implicit knowledge may be derived from visual experience. However, physiological evidence increasingly suggests that motor experience may be crucial to this visual process. For example, "mirror" neurons in monkey premotor cortex respond both when a monkey performs a particular action and when that monkey observes another monkey or a human performing that same action (Rizzolatti, Fadiga, Gallese, & Fogassi, 1996). Recent imaging data clearly suggest that, in the human, the visual perception of human movement involves both visual and motor processes. That is, when subjects are asked to observe the actions of another human so that they can later imitate those actions, PET activity is found in those brain regions involved in motor planning (Decety et al., 1997). Thus, visual observation of another individual's movement can lead to activation within the motor system of the observer.

Interestingly, action observation without the intent to imitate does not consistently engage motor planning areas (Decety et al., 1997). Intentionality is known to play a fundamental role in the production of human movement (Bonnard & Pailhous, 1991, 1993; Laurent & Pailhous, 1986). Indeed, intentionality or the ability to actively modify muscle activity, marks the critical difference between animal and object movement. Since intentionality controls both the motor production and visual analysis of human movement, it may serve to connect the two processes. This proposed linkage is consistent with the hypothesis that the perception of human movement may differ from the perception of other complex but non-intentional, motions. Taken together, these intriguing results suggest that we may understand the actions of others in terms of our own motor system. The high-level visual mechanism suggested by the results of the current behavioural experiments may well reflect this linkage between the visual and motor systems.

REFERENCES

Ahlström, V., Blake, R., & Ahlström, U. (1997). Perception of biological motion. *Perception, 26*, 1539–1548.

Anstis, S.M. (1980). The perception of apparent movement. *Philosophical Transactions of the Royal Society of London, 290*, 153–168.

Baizer, J., Ungerleider, L., & Desimone, R. (1991). Organisation of visual inputs to the inferior temporal and posterior parietal cortex in macaques. *Journal of Neuroscience, 11*, 168–190.

Baker, C., & Braddick, O. (1985). Temporal properties of the short-range process in apparent motion. *Perception, 14*, 181–192.

Bertenthal, B.I., & Pinto, J. (1994). Global processing of biological motions. *Psychological Science, 5*, 221–225.

Bonnard, M., & Pailhous, J. (1991). Intentional compensation for selective loading affecting human gait phases. *Journal of Motor Behaviour, 23*, 4–12.

Bonnard, M., & Pailhous, J. (1993). Intentionality in human gait control: Modifying the frequency-to-amplitude relationship. *Journal of Experimental Psychology: Human Perception and Performance, 19*, 429–443.

Burt, P., & Sperling, G. (1981). Time, distance, and feature trade-offs in visual apparent motion. *Psychological Review, 88*, 171–195.

Cavanagh, P. (1991). Short-range vs. long-range motion: Not a valid distinction. *Spatial Vision, 5*, 303–309.

Cavanagh, P. (1992). Attention-based motion perception, *Science, 257*, 1563–1565.

Cavanagh, P., & Mather, G. (1989). Motion: The long and the short of it. *Spatial Vision, 4*, 103–129.

Cutting, J.E. (1978). A program to generate synthetic walkers as dynamic point-light displays. *Behaviour Research Methods & Instrumentation, 10*, 91–94.

Cutting, J.E., Moore, C., & Morrison, R. (1988). Masking the motions of human gait. *Perception and Psychophysics, 44*, 339–347.

Decety, J. (1996). The neurophysiological basis of motor imagery. *Behavioural Brain Research, 77*, 45–52.

Decety, J., Grezes, J., Costes, N., Perani, D., Jeannerod, M., Procyk, E., Grassi, F., & Fazio, F. (1997). Brain activity during observation of actions: Influence of action content and subject's strategy. *Brain, 120*, 1763–1777.

Dittrich, W.H. (1993). Action categories and the perception of biological motion. *Perception, 22*, 15–22.

Green, M. (1986). What determines correspondence strength in apparent motion? *Vision Research, 26*, 599–607.

Heptulla-Chatterjee, S., Freyd, J., & Shiffrar, M. (1996). Configurational processing in the perception of apparent biological motion. *Journal of Experimental Psychology: Human Perception and Performance, 22*, 916–929.

Hoffman, D.D., & Flinchbaugh, B.E. (1982). The interpretation of biological motion. *Biological Cybernetics, 42*, 195–204.

Inman, V.T., Ralston, H., & Todd, F. (1981). *Human walking*. Baltimore, MD: Williams & Wilkins.

Johansson, G. (1973). Visual perception of biological motion and a model for its analysis. *Perception and Psychophysics, 14*, 201–211.

Johansson, G. (1975). Visual motion perception, *Scientific American, 232*, 76–88.

Johansson, G. (1976). Spatio-temporal differentiation and integration in visual motion perception. *Psychological Review, 38*, 379–393.

hand (Perrett et al., 1989). They also conjectured that Brodmann area 45 in the inferior frontal cortex corresponds to a system for representation of grasping movements, functionally similar to F5 in the monkey, where "mirror" neurons were found.

In a third study, cerebral blood flow was measured during observation of actions where both the cognitive strategy of the subjects during the observation and the semantic content of the actions were manipulated (Decety et al., 1997). Subjects were scanned under four conditions of visually presented meaningful or meaningless actions. In each of the four activation conditions, they were instructed to observe actions carefully with one of two aims: to be able to recognise or to imitate them later. The differences in the meaning of actions, irrespective of the strategy used during observation, led to different patterns of brain activity and clear left/right asymmetries. Meaningful actions were associated with activations located in the left hemisphere, in the inferior frontal gyrus (Ba 45), in the middle temporal gyrus (Ba 21), and in the orbitofrontal cortex. This neural network was interpreted to be implicated in the semantic object processing and in action recognition. In contrast, meaningless actions involved mainly the right occipito-parietal pathway, reaching the premotor cortex when the aim of the observation was to imitate. These activations were interpreted to be processing of the visual properties necessary for generating visuomotor transformation, in order to construct a new representation of actions to be imitated. Observing with the intent to recognise activated memory encoding structures, while observing with the intent to imitate was associated with activations located in the dorsolateral prefrontal cortex, the SMA, and the cerebellum. These regions are known to be implicated in the planning and generation of actions (Decety, 1996).

However, an issue that has not fully been considered in these studies is the respective contribution of vision for perception and vision for action. Indeed, in daily life, actions performed by others are usually perceived without any purpose or without requiring further overt integration. On some occasions, however, perception of actions has a specific goal, such as imitation or whenever a decision has to be taken. These considerations raise the interesting question of whether the actions perceived in such a natural context are processed differently according to their nature or meaning (semantic content). A second related interesting issue is to probe whether the aim of the observation has an effect on the visual information processing.

The aim of the present study was to investigate the respective effect of the perception of meaningful and meaningless actions on the neural pathways during nondirected processing that occurs in the absence of explicit instructions as well as when the subject shifts from a passive observation to a more cognitive mode, i.e. the potential contribution of the aim of observation (top-down effect) on the neural network engaged by the nature of the movements. The neural response to the perception of hand action was assessed by measuring regional cerebral blood flow (rCBF) with PET in normal subjects while they observed meaningful or meaningless actions without any purpose and later while they observed similar actions with the aim of imitating them. Finally, to reveal the whole network engaged during the perception of human movements, a control condition with no movements was performed.

MATERIAL AND METHODS

Subjects

Ten healthy male volunteers (20–28 years) who had given written informed consent participated in the experiment. All were right-handed according to the Edinburgh Inventory test (Oldfield, 1971) and had normal or corrected-to-normal vision. The experiment was performed in accordance with the guidelines from the declaration of Helsinki and with the approval of the local Ethical Committee (Centre Léon Bérard). Subjects were paid for their participation.

Activation Tasks

Subjects were scanned 10 times during the observation of video-filmed scenes. Four experimental conditions and one control condition were used according to the instructions given to the subjects (for definition of the five conditions and their abbreviations, see Table 1). Each condition was repeated once and was separated from the next by a 10min inter-scan period. In all activation tasks, video-films consisted of sequences of five actions executed with the upper limbs; these videos showed an experimenter's upper limbs and trunk only. Each action, which lasted for 4sec, was separated from the next by a 500msec blank screen. The five actions were presented three times in a random order (15 stimuli per condition). The stimuli were presented in the centre of a colour video monitor (36cm), located in front of the subject at 60cm from their eyes. The screen was oriented so as to be perpendicular to the subject's line of sight. The field of view of the subject was 19° and 26° for the vertical and horizontal dimensions respectively. The video apparatus and the subjects were surrounded by a black curtain. Room lights were reduced to a minimum and cooling fans provided low-level background noise.

Table 1. Five Experimental Conditions

Abbreviation	Nature of the Stimulus	Aim of the Observation
MF	Meaningful action	Without any purpose
ML	Meaningless action	Without any purpose
S	Stationary hands	Without any purpose
IMF	Meaningful action	In order to imitate
IML	Meaningless action	In order to imitate

In two conditions (MF & IMF), meaningful actions were presented. They consisted of pantomimes of transitive acts (e.g. opening a bottle, drawing a line, sewing a button, hammering a nail) performed by a right-handed person. These actions mainly involved the right (dominant hand); the left was used to hold the represented object. All meaningful actions were gestures that utilise tools with objects or gestures that utilise objects but no tools. Gestures that utilise no tool and no object such those used in communication were excluded.

In two other conditions (ML & IML), meaningless actions were presented. These actions were derived from American Sign Language (ASL), with the constraint that they be perceptually as close as possible to the actions presented during the meaningful actions (e.g. movements involving mainly the right hand). As the subjects had no knowledge of ASL, the actions bore no overt relation to language nor to symbolic gestures.

In one other condition (S), stationary hands were presented. The stimulus structure was the same as that used in activation tasks with the following characteristic: no movements. Five meaningless spatial positions of the hands and limbs were used and randomly presented throughout the condition. The aim of this condition was to provide a reference level for activation tasks, i.e. to subtract the "low-level" visual analysis of the stimuli (upper limb recognition, colour, texture, etc.).

Experimental Design

The four experimental conditions, duplicated once, were devised in two sessions. In the first, subjects were required to look at video-films without any specific aim. In the second session, subjects were asked to watch the video-films with the aim of imitating the actions presented. The two sessions were performed consecutively using different sets of similar kinds of stimuli.

First session. Subjects were instructed to observe carefully meaningful (MF) or meaningless (ML) actions without any specific purpose. These two conditions (four scans) were presented in a random order in the first session. Subjects were not informed about the nature of the movements presented before the start of the film and thus they were unaware of which category of action would be presented. There was no task performed at the end of the scanning period. Immediately after the end of the PET-data acquisition, subjects were asked to describe what they had seen. Their verbal report were recorded without reinforcement. This question was asked in order to control whether the subjects paid attention to the film and also to see whether they became aware of the nature of the stimuli (i.e. meaningful or meaningless).

Second session. Subjects were asked to observe carefully the meaningful (IMF) or meaningless (IML) actions in order to imitate them after the scanning period. These two conditions (four scans) were presented in a random order. Subjects were instructed that they would have to reproduce accurately the five actions that they had seen during the scan. No mention was made concerning the nature of the actions presented. In these conditions, subjects were specifically asked to avoid verbalisation during the observation period (a word couldn't accurately describe a movement) and during the imitation after the scanning period. Immediately after the scanning acquisition, subjects were asked to reproduce the actions presented. The subjects' performance was recorded on a videotape and then scored by an experimenter on a 2-point scale (1 = correctly reproduced, 0 = unrecognisable or not reproduced).

Prior to the first scanning of the second session, and after general instructions had been given, a practice trial composed of five actions was administrated. The video used in the practice trial was different from those during the scanning period.

The control condition (observation of stationary hands) was presented twice and randomly distributed within the two sessions. In this condition (S), subjects were required to observe stationary hands carefully without any specific task.

Scanning Procedure

Subjects were examined in the supine position on the bed of the PET scanner. Control of the head position throughout the examination was made by laser alignment along with reference points on Reid's line before and after each session. The head was slightly raised above the bed by means of a head holder, which allowed adequate fixation. Subjects could thus look comfortably at the monitor.

PET scans were acquired using a Siemens CTI HR+ (63 slices, 15.2cm axial field of view) PET tomograph with collimating septa retracted operating in high-sensitivity three-dimensional mode. The system has 31 rings which allow acquisition of 63 transaxial images with a slice thickness of 2.42mm without a gap in between. Transmission data were acquired using rotating pin sources filled with ^{68}Ge (9 mCi/pin). A filtered back-projection algorithm was employed for image reconstruction, on a 128×128 matrix (pixel size 2.02mm, Hanning filter with a cut-off frequency of 0.5 cycles/pixel). rCBF was estimated by recording the distribution of radioactivity following the intravenous bolus injection of 333MBq of $_{15}$O-H$_2$O through a forearm cannula placed into the brachial vein. The integrated counts collected for 60sec, starting 20sec after the injection time, were used as an index of rCBF.

At the beginning of the film, specific instructions were given to focus subjects' attention and to tell them what task, if any, was to be performed. The video-film with action sequences was switched on at the same moment as the injection was given. A 10min interval between each condition was necessary for the radioactivity decay.

Each subject also underwent a high-resolution magnetic resonance imaging (MRI) scan (FLASH 3D, T1, 120 slices parallel with the AC-PC plane, 1mm thick) obtained with a Philips 1.5 Tesla magnet.

Data Analysis

Image analysis was performed on Silicon Graphics O2 stations. The data were analysed with statistical parametric mapping (SPM96 software MRC Cyclotron Unit, London; Friston et al., 1995) implemented in Matlab 4.2 (Math Works, Natick, MA). The structural MRI and the realigned PET images were spatially normalised into a standard stereotaxic space (Talairach & Tournoux, 1988) using a reference template image (Friston et al., 1995). The resulting voxel dimensions of each reconstructed scan was $2 \times 2 \times 4$mm in the x, y, and z dimension, respectively. The scan were smoothed using a gaussian filter of 12mm full width at half-maximum. Global differences in cerebral blood flow were covaried out for all voxels and comparisons across conditions were made using t statistics with appropriate linear contrasts, and then converted to Z scores. Only regional activations significant at $P < .001$ uncorrected for multiple comparisons ($Z > 3.09$) were considered. In addition to this standard procedure, and in order to improve the precise anatomical description, the activation foci were superimposed both on a reference MRI from one normal subject, available in SMP96, and on the averaged MRI from the 10 study subjects. Since the comparison between the individual MRI and the averaged MRI was in good correspondence (for the overlapping of the major sulci), we decided to use the individual MRI in order to localise peak sites. Anatomical identification was performed with reference to the atlas of Talairach and Tournoux (1988) as well as the atlas of Duvernoy (1991).

Statistical Analysis

The differences between conditions were assessed by comparisons of specific rCBF maps pertaining to the experimental conditions.

Simple Main Effects

The subtractions of the control condition (S) from the observation without any purpose of meaningful actions (MF–S) and of meaningless actions (ML–S) were performed in order to reveal the whole network engaged by the perception of human movements and also to determine whether meaningful and meaningless actions are processed differently.

The subtractions of the control condition (S) from observation to imitation of meaningful actions (IMF–S) and of meaningless actions (IML–S) allowed the determination of whether the strategy to imitate has an effect on the neural network involved by the perception of actions.

Main Effect

In order to isolate the network engaged by the strategy to imitate during the observation of actions, irrespective of the nature of the actions presented, a factorial subtraction was computed according to the formula: [(IMF + IML) – (MF + ML)].

Task-related Regional Activity

Post hoc analysis, which is a descriptive analysis and not a quantitative one, was occasionally performed, using the stereotactic coordinates of some relevant activation foci found in the categorical comparisons. These profiles of activity, which represent relative rCBF values in each task, were used to demonstrate the differential involvement of a selected brain area in the five experimental conditions.

RESULTS

Subjects' Performances

The subjects achieved 89% correct in the imitation of meaningful actions, and 80% in the imitation of meaningless actions.

PET Results

The results from simple main effect relative to comparisons between experimental conditions and the control condition will first be presented. Then, the main effect relative to the strategy used during the observation (to imitate), irrespective of the nature of the movements will be presented.

Simple Main Effect Related to the Observation of Actions (Meaningful or Meaningless) Relative to the Observation of Stationary Hands

See Tables 2 and 3.

Observation of meaningful movements (MF–S). Observing meaningful movements without any purpose versus stationary hands was associated with activations in the superior occipital gyrus (Ba 19) and in the occipital temporal junction (Ba 19/37) bilaterally. Significant activation was found in the left inferior frontal gyrus, with the peak located in the fundus of the sulcus between Ba 44 and Ba 45. The precentral gyrus (Ba 4), the middle temporal gyrus (Ba 21), and the inferior parietal lobe in its ventrorostral part (Ba 40) were activated in the left hemisphere, whereas the lingual gyrus (19/18) and the superior temporal gyrus were activated in the right hemisphere. rCBF increases were also found in the fusiform gyrus (Ba 20) and in the inferior temporal gyrus (Ba 20/38) in the left hemisphere (Fig. 1). Post hoc exploration, which was a qualitative analysis, performed for the peak site located in the left inferior frontal gyrus (Ba 44/45) showed that this region was predominantly involved during the observation of meaningful actions without any purpose and much less during the other activation conditions (Fig. 2).

Table 2. Cortical Foci (Local Maxima) Demonstrating Significant rCBF Increases During Observation of Meaningful Actions vs. Observation of Stationary Hands

Brain Region	L/R	Brodmann Area	Coordinates			Z Score
			x	y	z	
Precentral gyrus	L	4	−34	−12	48	3.95
Superior occipital gyrus	R	19	24	−86	36	4.15
Superior occipital gyrus	L	19	−28	−86	30	3.91
Inferior parietal lobe ventrorostral	L	40	−64	−26	24	4.75
Inferior frontal gyrus	L	44/45	−48	12	22	3.65
Superior temporal gyrus	R	22	64	−38	16	5.23
Middle temporal gyrus	L	21	−60	−46	10	4.59
Occipital temporal junction	L	37/19	−46	−64	6	5.29
Gyrus lingual inferior	R	19/18	8	−64	0	4.06
Occipital temporal junction	R	37/19	44	−64	0	5.79
Fusiform gyrus	L	20	−50	−46	−24	3.53
Inferior temporal gyrus	L	20/38	−34	−4	−36	4.40

Coordinates are in mm and correspond to the Talairach and Tournoux atlas (1988). L/R: left or right hemisphere. Threshold = 3.09 ($P < .001$).

Table 3. Cortical Foci (Local Maxima) Demonstrating Significant rCBF Increases During Observation of Meaningless Actions vs. Observation of Stationary Hands

Brain Region	L/R	Brodmann Area	Coordinates			Z Score
			x	y	z	
Superior parietal lobule	L	7	−26	−54	70	4.16
Superior parietal lobule	L	7/40	−34	−44	66	5.41
Superior parietal lobule	R	7	32	−46	62	4.19
Precentral gyrus	L	4	−34	−12	48	2.95*
Inferior parietal lobe	R	40	40	−38	46	5.46
Superior occipital gyrus	R	19	24	−84	36	5.60
Superior occipital gyrus	L	19	−18	−88	32	3.62
Inferior parietal lobe ventrorostral	L	40	−62	−28	24	5.69
Superior temporal gyrus	R	22	68	−38	18	6.65
Middle temporal gyrus	L	21	−58	−50	10	3.36
Occipital temporal junction	L	37/19	−46	−64	6	5.94
Occipital temporal junction	R	37/19	44	−62	0	6.70
Cerebellum	R		26	−48	−46	3.67

* $P < .01$ Z = 2.33.

Observation of meaningless movements (ML–S). Activations produced by the observation of meaningless actions without any purpose, as contrasted to stationary hands, were located bilaterally in the superior occipital gyrus (Ba 19) and the occipital temporal junction (Ba 19/37). A large and strong activation was found in the right parietal lobe, whose area covered the superior parietal lobule (Ba 7) following the intraparietal sulcus and extending into the inferior parietal lobe in its upper portion (Ba 40). On the left side, a similar activation was found, but it was weaker and less extensive. This activation spread throughout the superior parietal lobule (Ba 7) and followed the intraparietal sulcus up to its most anterior part. An independent focus of rCBF increase was also located in the left inferior parietal lobe in its ventrorostral part (Ba 40). Activations were also found in the precentral gyrus (Ba 4), in the middle temporal gyrus (Ba 21) in the left hemisphere and in the superior temporal gyrus (Ba 22) and in the cerebellum on the right side (Fig. 1).

Simple Main Effect Related to the Observation of Actions (Meaningful or Meaningless) in Order to Imitate vs. Stationary Hands

See Tables 4 and 5.

Observation of meaningful actions to imitate (IMF–S). Most of the rCBF increases during observation of meaningful actions to imitate versus stationary hands, were located in both hemispheres. The occipital temporal junction (Ba 19/37), the SMA (mesial Ba 6) and the middle frontal gyrus (fundus of Ba 6) were found to be activated. In addition, the superior parietal lobule (Ba 7) extending into the upper part of the intraparietal sulcus (between the

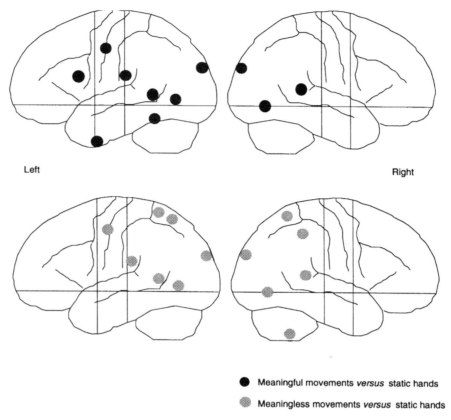

Fig. 1. Lateral view of the left and right hemispheres showing cortical regions of increased rCBF during observation of meaningful and meaningless actions without any purpose, compared to the observation of stationary hands. Results are listed in Tables 2 and 3.

coordinates y = −54 to y = −30, Ba 7/40) and the inferior parietal lobe in its upper part (Ba 40) were found to be activated. Although the centre of gravity of these activations are similar the two hemispheres, the areas belonging to them were broader in the left hemisphere. The inferior parietal lobe in its ventrorostral part (Ba 40) was activated in the left hemisphere. This latter focus was independent of the other foci in the parietal cortex. In the right hemisphere, the superior occipital gyrus (Ba 19), the superior temporal gyrus (Ba 22), and the cerebellum appeared to be involved. Bilateral activations were also found in the orbital gyrus (Ba 11) (Fig. 3).

Observation of meaningless actions to imitate (IML–S). During observation of meaningless actions to imitate versus stationary hands, activations were located in the middle frontal gyrus (fundus of Ba 6), in the occipital temporal junction (Ba 19/37), and in the cerebellum

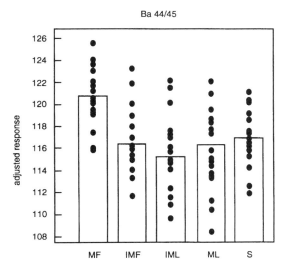

Fig. 2. Results of the post hoc exploration performed from a voxel (local maxima) in the inferior frontal gyrus (Brodmann's areas 44/45: x = –48, y = 12, z = 22). The profile of activity illustrates relative rCBF value in each experimental conditions. The black circles represent the distribution of each scan for the 10 subjects in each experimental condition. Note the prevalent involvement of this region during the observation of meaningful actions without any purpose (MF).

in both hemispheres. Bilateral and comparable activations were found in the parietal cortex, although they were stronger in the left hemisphere. These activations spread out from the superior occipital gyrus (Ba 19) to the superior parietal lobule (Ba 7) extending into the upper section of the intraparietal sulcus (Ba 7/40). Activations were found in the right hemisphere in the superior temporal gyrus (Ba 22) and in the inferior parietal lobe in its upper part (Ba 40). The ventral part of the precentral gyrus (Ba 6) and the inferior parietal lobe in its ventrorostral part (Ba 40) were also activated in the left hemisphere (Fig. 3).

Main Effect Due to the Strategy to Imitate During Observation of Actions vs. Observation without Any Purpose [(IMF + IML) – (MF + ML)]

See Table 6. Observation in order to imitate versus observation without any purpose, irrespective of the nature of the stimuli, was associated with activations located bilaterally in the middle frontal gyrus (Ba 9), in the premotor cortex (Ba 6) and in the SMA. The inferior parietal gyrus (Ba 40), in its upper section, the superior parietal lobule (Ba 7), and the anterior cingulate gyrus (Ba 32) were activated on both sides. At the subcortical level, the medio-dorsal thalamic nucleus was found to be activated in the right hemisphere, whereas the posterior caudate nucleus was activated in both sides. The dorsal frontal gyrus (Ba 10), the cerebellum, and the cuneus (Ba 17/18) were activated in both hemispheres. rCBF increases were also found in the precuneus (Ba 7) and in the middle frontal gyrus (Ba 46) in the left hemisphere (Fig. 4).

Post hoc exploration performed for the activation located in the middle frontal gyrus (Ba 9) showed that this region was involved both during observation to imitate (IMF and IML) and during the control condition (stationary hands S) (Fig. 5).

DISCUSSION

Perception of Human Movements

The subtractions of the baseline condition (stationary hands) from the four target conditions

Table 4. Cortical Foci (Local Maxima) Demonstrating Significant rCBF Increases During Observation of Meaningful Actions in Order to Imitate vs. Stationary Hands

Brain Region	L/R	Brodmann Area	Coordinates			Z Score
			x	y	z	
Intraparietal sulcus	R	7/40	44	−42	64	4.80
Superior parietal lobule	L	7	−18	−70	64	4.73
Mesial frontal gyrus (SMA)	L	6	−6	−4	60	3.44
Intraparietal sulcus	L	7/40	−44	−46	58	4.12
Middle frontal gyrus (fundus)	L	6	−26	−6	56	3.75
Superior parietal lobule	R	7	24	−62	54	3.40
Middle frontal gyrus (fundus)	R	6	32	−6	52	3.77
Mesial frontal gyrus (SMA) anterior	R	6	2	8	52	3.32
Inferior parietal lobe	R	40	40	−38	44	5.40
Inferior parietal lobe	L	40	−58	−32	40	3.94
Superior occipital gyrus	R	19	22	−86	36	5.63
Inferior parietal lobe ventrorostral	L	40	−66	−26	26	4.34
Superior temporal gyrus	R	22	72	−36	16	4.73
Occipital temporal junction	L	37/19	−50	−68	2	4.69
Occipital temporal junction	R	37/19	46	−60	0	6.82
Orbital gyrus	L	11	−22	36	−30	3.69
Orbital gyrus	R	11	28	30	−30	3.68
Cerebellum	R		28	−48	−44	5.39

showed that several cortical areas were involved regardless of either the semantic content of the movements or the purpose of the observation. It is thus natural to interpret these activations as reflecting hand movement analysis.

The activation of the occipito-temporal junction (Ba 19/37) corresponds precisely to the coordinates of V5 given by Watson et al. (1993) from a PET study in humans. This region is known to be specifically engaged by visual motion and would be the homologue of the middle temporal area (MT) in monkey. Vaina, Lemay, Bienfang, Choi, and Nakayama (1990) reported a related case of a patient with bilateral lesions involving the temporal-parietal-occipital junction, presumably damaging the human homologue of MT, sparing the primary visual cortex. This patient was severely impaired on motion tasks but was normal on higher-order tasks, exhibiting no deficits in the Johansson biological motion tasks. Another single-case study, reported by Marcar, Zihl, and Cowey (1997), came to the same conclusion. These findings imply that V5 is not essential for processing biological motion and/or that there is a separate motion pathway specialised for the perception of biological motion (as suggested by Colby, Gattas, Olson, & Gross 1988). Other activations, which pertain to the dorsal stream and are located in the superior occipital gyrus (Ba 19) and in the one could predict that the perception of meaningful actions to imitate would engage the ventral stream in order to identify and recognise the actions observed, whereas meaningless actions, which are unknown and thus impossible to recognise, would only involve the dorsal stream. Yet our results are, to some extent, in contradiction with this hypothesis. Indeed, observation of meaningful and meaningless actions in order to imitate them later were associated with activations located only in the dorsal stream without any participation of the ventral stream. It can be argued that the lack of activation of the ventral stream is due to the fact that our meaningful actions are pantomimes with represented objects, and that the ventral stream would have been activated if meaningful actions involved the utilisation of real objects. Nevertheless, the ventral stream was found to be engaged during the perception of meaningful actions without any purpose as well as during the observation with the aim to recognise in a previous PET-activation study (Decety et al., 1997). It is also conceivable that the design of the experiment in two sessions has influenced the processing of the nature of the stimuli, through a habituation effect, which led in the second session to a pre-eminent involvement of the dorsal pathway.

Thus, it appears that the reproduction of action that involves initial observation, does not require the semantic integration or verbal labelling. Such a result is not so inconsistent with experiments in the field of observational learning that have shown that a facilitatory effect is measured only after eight opportunities to observe the modelled actions (Carrol & Bandura, 1990). In the present experiment, the actions were presented three times. Thus, verbal labelling is no help for reproduction if the movements are observed a few times.

From accumulated findings in clinical neuropsychology on apraxic patients, we could have expected a more clear difference between meaningful and meaningless actions when the intention is to imitate observed actions. To account for the dissociation between meaningful and meaningless, Rothi et al. (1991) proposed a cognitive model that postulates two routes: a lexical (semantic) route and a direct, nonlexical one, which respectively mediate vision and motor control in imitation of actions. The lexical route is composed by long-term memory representations of meaningful and familiar actions. The other route provides a direct link between visual analysis of gestures and motor control. This latter one can be used for imitation of both meaningless and meaningful actions. According to this model, when the lexical route is interrupted, imitation of meaningful and meaningless actions is preserved, yet patients are impaired in action recognition. In contrast, the interruption of the direct route gives rise to the dissociation between impaired imitation of meaningless gestures and preserves performance of meaningful actions. Our results show that observation of meaningful actions without any purpose involved the lexical (semantic) route, whereas observation of meaningful and meaningless actions with the aim to imitate them later engaged only the direct route. This finding is consistent with the discrepancy between gesture comprehension and imitation found in some apraxic patients (Rothi et al., 1986). It is also coherent with the distinction proposed by Milner and Goodale (1995), if one draws a parallel between the lexical route and the ventral pathway and between the direct route and the dorsal pathway. In this case, then, the lexical route is engaged during vision for perception or comprehension, and the direct route is involved during vision for action as well as vision for imitation.

The hypothesis that perception of actions can be directly linked to generative (i.e. motor) processes is supported by several studies. Berger, Carli, Hammersla, Karshmer, and Sanchez (1979) demonstrated by electromyography that observation of body movements is associated with specific innervation in the corresponding muscles. Later, Fadiga, Fogassi, Pavesi, and Rizzolatti (1995) confirmed this result, by applying transcranial magnetic stimulation to subjects' motor cortex during observation of grasping movements without any purpose of imitating them. The results showed that there were an increase in motor evoked potential, recorded from hand muscles, during movement perception. These experiments support the notion of a direct link between perception and action. However, they do not explain how this route translates visual information into motor commands. Goldenberg and Hagmann (1997) recently suggested that imitation of meaningless gestures involves an intermediate step that allows elaboration of a mental representation of the gestures. The fact that cortical areas involved in motor planning are found to be activated during the observation of meaningless actions, when the intention is to imitate later, speaks in favour of an elaboration of motor representations held in working memory.

Observation of meaningless actions in order to imitate them was associated with bilateral rCBF increases in the dorsal stream extending into the dorsal lateral premotor cortex. This latter region is known to be implicated in action planning and externally triggered action (Passingham, 1993) and is connected with the superior parietal lobule (Jackson & Husain, 1996). Thus, this network is used to integrate the actions perceived with the goal to be reproduced. Activations elicited by the observation of meaningful actions in order to imitate them were located in the same regions but additional increases were found in the SMA, in the orbitofrontal cortex, and in the left inferior parietal lobule. Meaningful actions rely on mental representation in long-term memory. The activation of the SMA is coherent with this hypothesis. Indeed, this region is known to participate in the programming and planning of internally guided actions (Passingham, 1996; Tanji & Shima, 1994). This activation is also consistent with the interpretation given by Watson, Fleet, Rothi, and Heilman (1986) concerning its involvement in transitive limb movements. The fact that the primary motor cortex was not found activated when the purpose of the observation was to imitate may sound puzzling. Our interpretation is that, since the task was to reproduce a sequence of five actions, subjects were engaged in action planning at a higher level which didn't require coding of specific motor commands for each action. It could be speculated that if the observation applied only one action and not a sequence, then the primary motor cortex may have been implicated.

The activation located in the orbitofrontal cortex has also been found in the Decety et al. (1997) study. This region may play a role in the inhibition of actions which have to be imitated later. Lhermitte, Pillon, and Serdaru (1986) observed patients with lesions in this part of the frontal cortex who exhibit an exaggerated dependence on environmental cues (imitation behaviour or utilisation behaviour). The authors interpreted these behaviours as a consequence of impaired inhibition on the parietal cortex of automatic actions. Other evidence in support of the inhibitory role of the orbitofrontal cortex has recently been reported by Marshall, Halligan, Fink, Wade, and Frackowiak (1997) from a PET-activation study in a single case of left-sided hysterical paralysis. When asked to attempt to move her paralysed leg, the right orbito-frontal cortex was significantly activated.

Observation of Actions with the Intent to Imitate

The effect of the strategy to imitate, as such (i.e. irrespective of the nature of the stimuli), was associated with activations in several cortical areas (SMA, premotor cortex, prefrontal cortex, anterior cingulate), subcortical nuclei (caudate and thalamus), and in the cerebellum. These regions are similar to those found in several neurophysiological studies during motor preparation tasks as well as during mental simulation of action (Decety et al., 1992, 1994; Stephan et al., 1995). They are usually interpreted as implicating motor representations (Crammond, 1997; Decety, 1996) and, in our study, they implicate motor representations during the stage of observation.

The involvement of the middle frontal gyrus (Ba 9) found in the factorial subtraction disappears when the control condition is subtracted from the target conditions. This could be explained (based on post hoc analysis) by the fact that this region was also involved in observation of stationary hands. According to Shiffrar and Freyd (1993), during the perception of realistic photographs of a human body in different positions, the visual system constructs paths of apparent motion that are consistent with the biomechanical limitations of human body. The perception of different stationary hand positions may have led to such process, although the presentation of each position lasted 4 seconds (which is much longer than the paradigm of Shiffrar & Freyd, 1993). Thus, it cannot be excluded that there is an involvement of the middle frontal cortex in order to keep in mind the stationary position and imagine covertly the movements between two stationary hand positions.

CONCLUSION

Perception of human biological motion led to specific activations depending on the nature of the actions presented when subjects were given no instructed purpose. The activated regions during the observation of meaningful actions lead us to propose that not only fine visual analysis of hand motion takes place but also the semantic integration of their meaning. It can also be proposed that actions are implicitly decoded in terms of motor production. This hypothesis is consistent with motor theories of perception, and it seems to be true for both meaningful and meaningless actions. The latter additionally necessitate a visuospatial analysis of the hand moving in space.

The perceptual system is conceived as being cognitively impenetrable (Fodor & Pylyshyn, 1981). In addition, the direct theory of perception posits that sensory information from behavioural events are organised by an autonomous perceptual system in which inferences, attitudes, and other cognitive influences do not play a major role. To what extent is perception of biological motion encapsulated? We have seen that a common network was activated by the perception of meaningful and meaningless actions, both of which are biological motion. This network may be a cognitive encapsulated part of the perceptual analysis. The specific activations relative to the nature of the biological stimuli, and most notably the influence of the strategy to imitate, clearly demonstrate that there is a modulatory effect on the neural network during the perception phase.

Indeed, when the observation has a goal, namely the intention to imitate, it appears that there is much less specificity of the neural networks involved, relative to the nature of the stimulus. The activations in the ventral pathway are no longer observed when observation of meaningful actions has a goal (to imitate). Only the dorsal pathway was strongly activated, reaching the lateral and the mesial premotor cortex. The same picture holds true for meaningless action and the intention to imitate, with the exception of the mesial premotor cortex. There is thus a clear top-down effect of the strategy (to imitate) upon the information processing, at least in terms of activated regions. These results further demonstrate the pre-eminent role of the dorsal pathway in perception for action, as suggested by Milner and Goodale (1995).

Some of these results are surprising in light of the literature on apraxia (e.g. Heilman et al., 1982; Heilman, Watson, & Rothi, 1997), in which most of the clinical observations have pointed to a role of the supramarginal and angular gyri in movement comprehension and imitation. These two regions were not found to be activated in our study and there are several possible reasons to account for this discrepancy. First, our experimental paradigm was designed to identify the regions engaged during the perception phase which was, in some conditions, followed by imitation. This condition corresponds to deferred imitation and not to immediate imitation as it is often studied in patients. Thus, the differences within the posterior parietal cortex may be explained in terms of task requirement. Indeed, deferred imitation necessitates temporary stored action planning, whereas immediate imitation engages on-line control. These two processes are known to rely on different neural networks. Second, lesion sites can vary in accuracy and location across patients; it is best if the lesions are stable, well demarcated, and referable to a neuroanatomic unit. Finally, our findings are based on group data (i.e. averaged images). All of these remarks may explain the partial overlap between the two sets of data (normal subjects/patients). This overlap illustrates why it is useful to combine empirical work on apraxia with studies performed on normal subjects and neurophysiological recordings to elucidate the neural and cognitive mechanisms involved in the perception, the recognition, and the generation of actions. This also affirmatively answers the provocative question that is often raised as to whether PET is a method capable of generating and testing new hypotheses or merely of confirming old ones.

REFERENCES

Andersen, R.A., & Zipser, D. (1988). The role of the posterior parietal cortex in coordinate transformations for visual-motor integration. *Canadian Journal of Physiology and Pharmacology*, 66, 488–501.

Beardsworth, T., & Buckner, T. (1981). The ability to recognize oneself from a video recording of one's movements without seeing one's body. *Bulletin of the Psychonomic Society*, 18, 19–22.

Berger, S.M., Carli, L.L., Hammersla, K.S., Karshmer, J.F., & Sanchez, M.E. (1979). Motoric and symbolic mediation in observational learning. *Journal of Personality and Social Psychology*, 37, 735–746.

Bonda, E., Petrides, M., Ostry, D., & Evans, A. (1996). Specific involvement of human parietal systems and the amygdala in the perception of biological motion. *Journal of Neurosciences*, 16, 3737–3744.

Carey, D.P., Perrett, D.I., & Oram, M.W. (1997). Recognizing, understanding and reproducing action. In F. Boller & J. Grafman (Eds.), *Handbook of neuropsychology* (pp. 111–129). Amsterdam: Elsevier.

Carroll, W.R., & Bandura, A. (1990). Representational guidance of action, production in observational learning: A causal analysis. *Journal of Motor Behavior*, 20, 85–97.

Colby, C.L., Gattas, R., Olson, C.R., & Gross, C.G. (1988). Topographic organization of cortical afferents to extrastriate visual area PO in the macaque: A dual tracer study. *Journal of Comparative Neurology*, 269, 392–413.

Crammond, D.J. (1997). Motor imagery: Never in your wildest dream. *Trends in Neurosciences*, 20, 54–57.

Cutting, J.R., & Koslowski, L.T. (1977). Recognizing friends by their walk: Gait perception without familiarity cues. *Bulletin of the Psychonomic Society*, 9, 353–356.

Decety, J. (1996). The neurophysiological basis of motor imagery. *Behavioural Brain Research*, 77, 45–52.

Decety, J., Grèzes, J., Costes, N., Perani, D., Jeannerod, M., Procyk, E., Grassi, F., & Fazio, F. (1977). Brain activity during observation of actions. Influence of action content and subject's strategy. *Brain*, 120, 1763–1777.

Decety, J., Kawashima, R., Gulyas, B., & Roland, P.E. (1992). Preparation for reaching: A PET study of the participating structure in the human brain. *NeuroReport*, 3, 761–764.

Decety, J., Perani, D., Jeannerod, M., Bettinardi, V., Tadary, B., Woods, R., Mazziotta, J.C., & Fazio, F. (1994). Mapping motor representations with positron emission tomography. *Nature*, 371, 600–602.

De Renzi, E. (1989). Apraxia. In F. Boller & J. Grafman (Eds.), *Handbook of neuropsychology* (pp. 245–263). Amsterdam: Elsevier.

Di Pellegrino, G., Fadiga, L., Fogassi, L., Gallese, V., & Rizzolatti, G. (1992). Understanding motor events: A neurophysiology study. *Experimental Brain Research*, 91, 176–180.

Dittrich, W.H. (1993). Action categories and the perception of biological motion. *Perception, 22*, 15–22.

Duvernoy, H.M. (1991). *The human brain. Surface, three-dimensional sectional anatomy and MRI.* New York: Springer Verlag.

Fadiga, L., Fogassi, L., Pavesi, G., & Rizzolatti, G. (1995). Motor facilitation during action observation: A magnetic stimulation study. *Journal of Neurophysiology, 73*, 2608–2611.

Fodor, J.A., & Pylyshyn, Z. (1981). How direct is visual perception? Some reflections on Gibson's Ecological approach. *Cognition, 9*, 138–196.

Friston, K.J., Holmes, A.P., Worsley, K.J., Poline, J.B., Frith, C.D., & Frackowiak, R.J.S. (1995). Statistical parametric maps in functional imaging: A general linear approach. *Human Brain Mapping, 3*, 189–210.

Gallese, V., Fadiga, L., Fogassi, L., & Rizzolatti, G. (1996). Action recognition in the premotor cortex. *Brain, 119*, 593–609.

Gibson, J.J. (1979). *The ecological approach to visual perception.* Boston, MA: Houghton Mifflin.

Goldenberg, G., & Hagmann, S. (1977). The meaning of meaningless gestures: A study of visuo-imitative apraxia. *Neuropsychologia, 35*, 333–341.

Goodale, M.A. (1997). Visual routes to perception and action in the cerebral cortex. In F. Boller & J. Grafman (Eds.), *Handbook of neuropsychology* (pp. 91–109). Amsterdam: Elsevier Science.

Goodale, M.A., & Milner, A.D. (1992). Separate visual pathway for perception and action. *Trends in Neurosciences, 15*, 20–25.

Goodale, M.A., & Milner, A.D., Jakobson, L.S., & Carey, D.P. (1991). A neurological dissociation between perceiving objects and grasping them. *Nature, 349*, 154–156.

Heilman, K.M., Rothi, L.G., & Valenstein, E. (1982). Two forms of ideomotor apraxia. *Neurology, 32*, 342–346.

Heilman, K.M., Watson, R.T., & Rothi, L.G. (1977). Disorders of skilled movements: Limb apraxia. In T.E. Feinberg & M.J. Farah (Eds.), *Behavioural neurology and neuropsychology* (pp. 227–235). New York: McGraw-Hill.

Jackson, S.R., & Husain, M. (1996). Visuomotor functions of the lateral pre-motor cortex. *Current Opinion in Neurobiology, 6*, 788–795.

Jeannerod, M., & Decety, J. (1990). The accuracy of visuomotor transformation: An investigation into the mechanisms of visual recognition of objects. In M.A. Goodale (Eds.), *Vision and action: The control of grasping* (pp. 33–48). Norwood, NJ: Ablex.

Jeannerod, M., Decety, J., & Michel, F. (1994). Impairment of grasping movements following bilateral posterior lesion. *Neuropsychologia, 32*, 369–380.

Johansson, G. (1973). Visual perception of biological motion and a model for its analysis. *Perception and Psychophysics, 14*, 201–211.

Koslowski, L.T., & Cutting, J.E. (1977). Recognizing the sex of a walker from a dynamic point-light display. *Perception and Psychophysics, 21*, 575–580.

Kupferman, I. (1991). Localizing of higher cognitive and affective functions: The association cortices. In E.R. Kandel, J.H. Schwartz, & T.M. Jessell (Eds.), *Principles of neural science* (pp. 823–851). New York: Elsevier.

Lhermitte, F., Pillon, B., & Serdaru, M. (1986). Human autonomy and the frontal lobes. Part I: Imitation and utilization behavior: A neuropsychological study of 75 patients. *Annals of Neurology, 19*, 326–334.

Liepmann, H. (1977). The syndrome of apraxia (motor asymboly) based on a case of unilateral apraxia. [Trans. from *Monatsschrift für Psychiatrie und Neurologie*, 1900, 8, 15–44.] In D.A. Rottenberg & F.H. Hockberg (Eds.), *Neurological classics in modern translation.* New York: Macmillan.

Marcar, V.L., Zihl, J., & Cowey, A. (1997). Comparing the visual deficits of a motion blind patient with the visual deficits of monkeys with area MT removed. *Neuropsychologia, 35*, 1459–1465.

Marshall, J.C., Halligan, P.W., Fink, G.R., Wade, D.T., & Frackowiak, R.S.J. (1997). The functional anatomy of hysterical paralysis. *Cognition, 64*, 1–8.

Matelli, M., & Luppino, G. (1997). Functional anatomy of human motor control areas. In F. Boller

& J. Grafman (Eds.), *Handbook of neuropsychology* (pp. 9–26). Amsterdam: Elsevier.

Matelli, M., Rizzolatti, G., Bettinardi, V., Gilardi, M.C., Perani, D., Rizzo, G. & Fazio, F. (1993). Activation of precentral and mesial motor areas during the execution of elementary proximal and distal arm movements: A PET study. *NeuroReport, 4,* 1295–1298.

Melher, M.F. (1987). Visuo-imitative apraxia. *Neurology, 37,* 129.

Meltzoff, A.N., & Gopnik, A. (1993). The role of imitation in understanding persons and developing a theory of mind. In S. Baron-Cohen, H. Tager-Flusberg, & D.J. Cohen (Eds.), *Understanding other minds* (pp. 335–366). New York: Oxford Medical Publications.

Meltzoff, A.N., & Moore, M.K. (1977). Imitation of facial and manual gestures by humane neonates. *Science, 198,* 75–78.

Milner, A.D., & Goodale, M.A. (1995). *The visual brain in action.* Oxford: Oxford University Press.

Oldfield, R.C. (1971). The assessment and analysis of handedness: The Edinburgh inventory. *Neuropsychologia, 9,* 97–113.

Orliaguet, J.P., Kandel, S., & Böe, L.J. (in press). Visual perception of motor anticipation in cursive writing: Influence of spatial and movement information on the prediction of forthcoming letters. *Perception.*

Parsons, L.M., Fox, P.T., Downs, J.H., Glass, T., Hirsch, T.B., Martin, C.C., Jerabek, P.A., & Lancaster, J.L. (1995). Use of implicit motor imagery for visual shape discrimination as revealed by PET. *Nature, 375,* 54–58.

Passingham, R.E. (1993). *The frontal lobes and voluntary action.* New York: Oxford University Press.

Passingham, R.E. (1996). Functional specialisation of the supplementary motor area in monkeys and humans. In H.O. Lüders (Ed.), *Advances in Neurology, Vol. 70* (Supplementary Sensorimotor Area) (pp. 105–116). Philadelphia: Lippincott-Raven.

Perani, D., Cappa, S.F., Bettinardi, V., Bressi, S., Gorno-Tempini, M., Matarrese, M., & Fazio, F. (1995). Different neural systems for the recognition of animal and man-made tools. *NeuroReport, 6,* 1637–1641.

Perrett, D.I., Harries, M.H., Bevan, R., Thomas, S., Benson, P.J., Mistlin, A.J., Chitty, A.J., Hietanen, J.K., & Ortega, J.E. (1989). Frameworks of analysis for the neural representation of animate objects and actions. *Journal of Experimental Biology, 146,* 87–113.

Perrett, D.I., Mistlin, A.J., Harries, M.H., & Chitty, A.J. (1990). Understanding the visual appearance and consequence of actions. In M.A. Goodale (Ed.), *Vision and action* (pp. 163–180). Norwood, NJ: Ablex.

Petrides, M., & Pandya, D. (1994). Comparative architectonic analysis of the human and macaque frontal cortex. In F. Boller & J. Grafman (Eds.), *Handbook of neuropsychology, Vol. 9* (pp. 17–58). Amsterdam: Elsevier Science.

Porro, C.A., Francescato, M.P., Cettolo, V., Diamond, M.E., Baraldi, P., Zuiani, C., Bazzocchi, M., & Prampero, P.E. (1996). Primary motor and sensory cortex activation during motor performance and motor imagery: A functional magnetic resonance imaging study. *Journal of Neuroscience, 16,* 7688–7698.

Rizzolatti, G., Fadiga, L., Matelli, M., Bettinardi, V., Perani, D., Fazio, F. (1996). Localization of cortical areas responsive to the observation of hand grasping movements in humans: A PET study. *Experimental Brain Research, 111,* 246–252.

Roth, M., Decety, J., Raybaudi, M., Massarelli, R., Delon-Martin, C., Segebarth, C., Marand, S., Gemignani, A., Decorps, M., & Jeannerod. M. (1996). Possible imvolvement of primary motor cortex in mentally simulated movement: A functional magnetic resonance imaging study. *NeuroReport, 7,* 1280–1284.

Rothi, L.J.G., Mack, L., & Heilman, K.M. (1986). Pantomime agnosia. *Journal of Neurology, Neurosurgery and Psychiatry, 49,* 451–454.

Rothi, L.J.G., Ochipa, C., & Heilman, K.M. (1991). A cognitive neuropsychological model of limb praxis. *Cognitive Neuropsychology, 8,* 443–458.

Runeson, S., & Frykholm, G. (1983). Kinematic specifications of dynamics as an informational basis for person-and-action perception: Expectation, gender recognition, and deceptive intention. *Journal of Experimental Psychology: General, 112,* 585–615.

Sakata, H., Taira, M., Kusunoki, M., Murata, A., & Tanaka, Y. (1997). The parietal association cortex on depth perception and visual control of hand action. *Trends in Neurosciences, 20*, 8, 350–357.

Shiffrar, M., & Freyd, J.J. (1993). Apparent motion of the human body. *Psychological Science, 1*, 257–264.

Sirigu, A., Cohen, L., Duhamel, J.R., Pillon, B., Dubois, B., & Agid, Y. (1995). A selective impairment of hand posture for object utilization in apraxia. *Cortex, 31*, 41–55.

Sirigu, A., Grafman, J., Bressler, K., & Sunderland, T. (1991). Multiple representations contribute to body knowledge processing. *Brain, 114*, 629–642.

Stephan, K.M., Fink, G.R., Passingham, R.E., Silbersweig, D., Ceballos-Baumann, A.O., Frith, C.D., & Frackowiak, R.S.J. (1995). Functional anatomy of the mental representation of upper extremity movements in healthy subjects. *Journal of Neurophysiology, 73*, 373–386.

Stränger, J., & Hammel, B. (1996). The perception of action and movement. In W. Prinz & B. Bridgeman (Eds.), *Handbook of perception and action, Vol. 1* (pp. 397–451). New York: Academic Press.

Talairach, J., & Tournoux, P. (1988). *Co-planar stereotaxic atlas of the human brain.* Stugggart: Thieme.

Tanji, J., & Shima, K. (1994). Role for supplementary motor area cells in planning several movements ahead. *Nature, 371*, 413–416.

Ungerleider, L.G. (1995). Functional brain imaging studies of cortical mechanisms for memory. *Science, 270*, 769–775.

Ungerleider, L.G., Mishkin, M. (1982). Two cortical visual systems. In D.J. Infle, M.A. Goodale, & R.J.W. Mansfield (Eds.), *Analysis of visual behaviour* (pp. 549–586). Cambridge, MA: MIT Press.

Vaina, L.M., Lemay, M., Bienfang, D.C., Choi, A.Y., & Nakayama, K. (1990). Intact biological motion and structure from motion perception in a patient with impaired motion mechanisms: A case study. *Visual Neuroscience, 5*, 353–369.

Vivani, P., & Stucchi, N. (1992). Biological movements look uniform: Evidence for motor-perceptual interactions. *Journal of Experimental Psychology: Human Perception Performance, 18*, 603–623.

Vogt, S. (1996). Imagery and perception-action mediation in imitative actions. *Cognitive Brain Research, 3*, 79–86.

Watson, R.T., Fleet, W.S., Rothi, L.G., & Heilman, K.M. (1986). Apraxia and the supplementary motor area. *Archives of Neurology, 43*, 787–792.

Watson, J.D.G., Myers, R., Frackowiak, R.S.J., Hajnal, J.V., Woods, R.P., Mazziotta, J.C., Shipp, S., & Zeki, S. (1993). Area V5 of the human brain: Evidence from a combined study using PET and MRI. *Cerebral Cortex, 3*, 79–94.

THE NEURAL BASIS OF IMPLICIT MOVEMENTS USED IN RECOGNISING HAND SHAPE

Lawrence M. Parsons and Peter T. Fox

Research Imaging centre, University of Texas Health Science centre at San Antonio, San Antonio, TX, USA.

Psychophysical studies indicate that observers recognise the laterality of single, randomly oriented, visually presented hands by implicitly moving their left hand into the orientation of any left hand stimulus and their right hand into the orientation of right hand stimuli. Such data imply that lateralised somatic sensorimotor representations are engaged as one's action is mentally simulated to perceive stimulus handedness. This hypothesis was evaluated in a positron emission tomography study in which (mostly) right hands were tachistoscopically presented to the left visual field (right cerebral hemisphere) and vice versa. Exploiting known cerebral lateralisations, the design allowed the observation of a spatial dissociation of visual-field-lateralised and limb-lateralised neural activity. Although the primary somatosensory and motor cortices were not activated, strong regional cerebral blood flow increases (task minus rest) were observed in frontal (motor) and parietal (somatosensory) areas similar to those activated by actual and imagined movement. Limb-contralateral activations were in pre-supplementary motor area (pre-SMA), cerebellum, and opercular and superior frontal sulcal premotor cortex; these areas are implicated in a variety of motor and somatosensory functions. Activations in the dominant left hemisphere regardless of the limb were localised in SMA proper, superior premotor cortex, and inferior parietal cortex, areas involved in planning, guidance, and attention to motor performance. Activations in the nondominant right hemisphere for both limbs were present in dorsal superior premotor cortex, insula, superior parietal cortex, and occipitotemporal cortex; these areas underlie motor planning, high-level somatic representation, evaluation of visuospatial information, and representation of object/action identity. Overall, these data confirm or extend prior understanding of the distributed system of brain areas supporting motor imagery and the left–right judgement of body parts.

Requests for reprints should be addressed to: Lawrence M. Parsons, Research Imaging Center, University of Texas Health Science Center at San Antonio, 7703 Floyd Curl Drive, San Antonio, Texas, 78284-6240, USA. (Tel: (210) 567-8189; Fax: (210) 567-8152; E-mail: parsons@uthscsa.edu).

We are grateful to Michael Martinez, Frank Zamarripa, and Shawn Mikiten for assistance with image processing, data analysis, and graphic production.

INTRODUCTION

Visual recognition of objects requires an active process of mapping visual sensations onto stored mental representations. Humans can recognise or discriminate the shapes of objects seen at different orientations in most instances (Rock, 1973; Biederman, 1987; Tarr, 1995; Ullman, 1996). If the shapes in question are sufficiently similar, such as an object and its mirror image, observers will reorient the objects or themselves by either physical or mental means in order to compare the shapes from corresponding viewpoints (Shepard & Cooper, 1982; Hinton & Parsons, 1988). Hence, this is an instance where perceiving an object's identity (which handed version of a shape it possesses) depends on an action (or at least a spatial transformation), be it imagined or real. Psychophysical studies of observers performing the mental rotation of one object into another's orientation indicate that the time to complete the rotation is an approximately linear function of the size of its angle (Shepard & Metzler, 1971; Shepard & Cooper, 1982; Parsons, 1987c, 1995).

However, psychophysical data also indicate that when the object that is imagined to be spatially transformed is one's own body, there is a much more complicated relationship between completion time and rotation angle. These imagined spatial transformations are elicited in a task in which subjects decide whether a randomly-oriented body part belongs to the left or right side of the body (Cooper & Shepard, 1975; Sekiyama, 1982; Parsons, 1987a). Specifically, psychophysical studies of observers making a left–right judgement of a disoriented hand (Parsons, 1987b, 1994) indicate that observers imagine moving their own hand from its orientation during the task into the orientation of the stimulus for comparison. Typically, subjects imagine their left hand in the orientation of left hand stimuli and their right hand in the orientation of right hand stimuli. A rapid, initial perceptual analysis of hand shape allows subjects to imagine moving first the hand that often turns out to match the stimulus. Subjects making this left–right judgement can be characterised as executing an implicit movement, because actual movements do not occur, imagined movements are not requested, and subjects are often unaware of an imagined movement. Imagining one's body at the orientation of a stimulus may reflect a reliance on imitation, our most fundamental learning and bootstrapping strategy, wherein one system observes another, forms an internal model of its behaviour, and then interprets the model in order to guide performance of the corresponding actions itself.

The property of imagined limb movement that makes it a more complicated phenomenon than the mental rotation of other objects is that the trajectory imagined for the observer's hand (or foot) is strongly influenced by the biomechanical constraints specific to its actual movements. Hence, trajectories imagined for the left hand are constrained by actual left hand joints and trajectories imagined for their right hand reflect actual right hand constraints. Some of the principal data supporting the process model described earlier for the left–right judgement and accompanying imag-

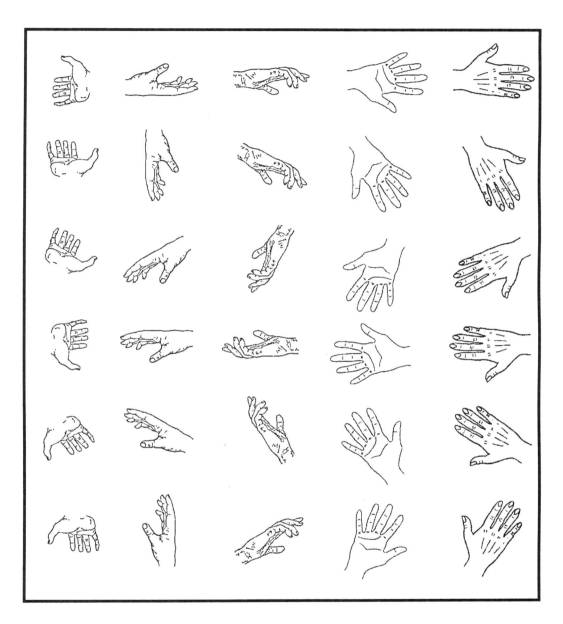

Fig. 2. Right hand versions of the experimental stimuli.

SuperLab (Cedrus, Inc) running on a Macintosh IIci. The elapsed time between the offset of one stimulus and onset of the next was adjusted to allow just barely enough time for task performance, based on prior chronometric pilot studies. In particular, the intertrial time was adjusted on the basis of individual stimuli (varying from 500 to 1600 msec) and on the basis of the number of performed trials (i.e. the extent of learning and practice effects across trials during the PET sessions.) These conditions, allowing no time for introspection, heightened the implicit nature of the performance.

Experimental Procedure

Each subject underwent nine PET blood-flow scans, three in the left visual field task, three in the right visual field task, and three in fixation-point rest. These three conditions were performed in approximately counterbalanced order across subjects such that trial numbers 1–3, 4–6, and 7–9 each contained a left field task, a right field task, and a fixation-point rest task. During the experimental tasks, subjects made a covert decision as to whether a visual stimulus was a left or right hand. Subjects stared at the fixation point as stimuli appeared in either the left or right visual hemifield. Subjects reported at the conclusion of the session that they were very often able to maintain central fixation during each PET trial. Subjects performed the tasks while lying in a supine position in the bed of the tomograph. Head position was controlled by a form-fitting plastic mask; head position was checked prior to each scan with reference to a laser alignment along anatomical marks on the face and mask. Subjects were not informed as to which hemifield the stimuli would appear in, but they were informed that the stimuli could be either a left or right hand. They were unable to see their own hands and were not allowed to make actual movements. During the rest condition, subjects were instructed to fixate the central dot. During the intertrial interval, no stimuli were presented and the subject rested quietly. Prior to the start of the PET session, subjects studied the stimuli and were given 15 minutes of practice in the task. An experiment was performed at the conclusion of the scanning session in which a hand was presented in central fovea until subjects pressed a left or right button to indicate whether it was a left or right hand (Parsons, 1987b, 1994). Subjects' response time by orientation functions and their accuracy during the practice trials and during the post-PET session were comparable to those of prior studies (Parsons, 1987b, 1994) and for sake of brevity will not be included in this report.

PET Method

The PET scans were performed on a GE 4096 camera (General Electric, Inc), which possesses a pixel spacing of 2.6mm; a spatial resolution of 5mm full width half maximum; an interplane, centre-to-centre distance of 6.5mm; 15 scan planes; and a z-axis field of view of 10 cm. Attenutation correction was performed with at

Ge68/Ga68 pin source. Regional cerebral blood flow (rCBF) was measured with $H_2^{15}O$ as the blood-flow tracer; it was administered as an intravenous bolus of 8–10ml of saline containing 60mCi (Fox et al., 1984). A 40sec scan was triggered as the tracer bolus entered the field of view (the brain), by the rise in the coincidence-counting rate. There was a 10min interval between each scan, sufficient for isotope decay (5 half-lives) and the re-establishment of resting-state levels of (rCBF) within activated areas.

PET Data Analysis

All data were assessed for inter-scan, intra-subject head movement. Movement assessment and correction were performed using the Woods algorithm (Woods et al., 1992, 1993). Images were averaged within condition for each subject. Images were spatially normalised into bicommissural coordinate space (Talairach & Tournoux, 1988) using SN^{TM} (Lancaster et al., 1995) which employs a 9-parameter, affine transformation, derived from 14 structural landmarks. Then, the images were averaged within condition among subjects. The grand-mean image for each task was compared to control, forming grand-mean images of the task-induced changes in brain activity. Grand-mean images were converted to z-score images using the population variance (Fox & Mintun, 1989). Activation loci were identified by 3D centre-of-mass address, peak-voxel intensity and z-value, cluster size, and statistical significance.

RESULTS

Significant task-induced increases in rCBF relative to rest ($P < .01$, Table 1) were quite compatible with the model described earlier that we derived from psychophysical studies. Activations were observed in all of the brain areas known to represent somatosensory and motor information, except the primary somatosensory (s1) and motor (m1) cortices (e.g. Ghez, 1991; Passingham, 1993; Porter & Lemon, 1993).

Activation foci were detected in a superior region of the supplementary motor area (SMA) in the dominant (left) hemisphere for both left and right hand trials (Fig. 3, Table 1). In addition, a larger pair of foci were present for both left and right hand trials in the pre-SMA of the hemisphere contralateral to stimulus hand (Fig. 4).

The pattern of activation just described was also observed in inferior premotor cortex (Brodmann Area [BA] 44 and 46), in which small foci were contralateral to stimulus hand (Fig. 5).

Moderate strength foci were detected in the superior frontal sulcal premotor areas (BA 4) (Fig. 6) such that on right hand trials activation was predominantly in the left hemisphere and on left hand trials activation was bilateral, with more in the right hemisphere than the left.

Strong foci were present in bilateral superior premotor cortex (BA 6), with more extensive activation in the dominant (left) hemisphere for both left and right hand trials. In both conditions, these activations appeared in a dorsal bilateral cluster (RH2, LH5, LH6,

Table 1. Maximal Foci of Activation. All focal increases in brain blood flow occurring during either experimental condition (relative to rest) are listed.

Site	Brodman Area	Gyrus	Side	Extent (mm^3)	Z-score	P	Coordinate (mm)		
							x	y	z
LH1	6	GFd	L	432	3.26	***	-4	2	56
LH2	7	LPs	R	488	3.03	**	36	-56	52
LH3	7	LPs	L	496	3.54	****	-28	-58	50
LH4	4	GPrC	R	168	2.75	**	24	-13	50
LH5	6	GFm	R	416	2.75	**	26	-6	50
LH6	6	GPrC	L	688	3.40	****	-32	-4	48
LH7	6	GPrC	L	576	3.46	****	-24	-11	46
LH8	6	GFd	R	592	3.57	****	1	12	46
LH9	40	LPi	L	392	3.09	***	-38	-36	45
LH10	7	PCu	L	336	2.66	**	-16	-64	42
LH11	19	PCu	L	368	2.92	**	-33	-66	40
LH12	40	LPi	L	120	2.49	*	-46	-30	40
LH13	40	LPi	L	848	3.83	*****	-34	-40	38
LH14	40/1	GSm	L	256	2.92	**	-42	-40	32
LH15	19	Cu	L	224	2.46	*	-8	-90	31
LH16	6	GPrC	L	376	3.00	**	-43	0	28
LH17	6	GPrC	L	168	2.89	**	-37	-6	26
LH18	46	GFm	R	128	2.35	*	40	32	24
LH19	39	GTm	L	104	2.58	**	-30	-64	24
LH20	39/19	GO	R	72	2.49	*	28	-70	24
LH21	19	Cu	R	160	2.75	**	12	-88	24
LH22	19	GO	L	840	3.69	*****	-24	-76	22
LH23	18	SPO	L	832	3.46	****	-20	-84	22
LH24	19	GOm	L	680	3.40	****	-28	-86	22
LH25		Nc	L	88	2.38	*	-13	-2	20
LH26	18	GOm	L	584	2.92	**	-26	-86	12
LH27	18	Cu	L	768	4.08	******	-4	-96	12
LH28	17	Cu	R	72	2.40	*	6	-86	8
LH29	19	GOm	L	976	4.51	******	-38	-78	8
LH30	17	GL	R	184	2.69	**	2	-76	6
LH31		INS	R	192	2.63	**	32	16	4
LH32		Nc	R	96	2.35	*	14	16	2
LH33		GP2	R	160	2.77	**	20	1	0
LH34	37	GTi	R	896	3.94	*****	46	-74	0
LH35	17	GL	R	304	2.86	**	18	-82	0
LH36	17	GL	R	384	3.06	**	8	-90	-4
LH37	19	Gh	L	72	2.35	*	-18	-44	-6
LH38	19	GOm	L	760	4.60	******	-46	-74	-6

Table 1. Continued

Site	Brodman Area	Gyrus	Side	Extent (mm^3)	Z-score	P	Coordinate (mm)		
							x	y	z
LH39	18	GL	R	152	2.52	*	18	-81	-8
LH40	18	GOm	R	784	3.51	****	42	-76	-8
LH41	18	GL	R	392	2.72	**	0	-90	-8
LH42	18	GL	L	784	3.94	*****	-10	-80	-8
LH43		lingual	R	144	2.80	**	0	-42	-10
LH44	19	GF	L	496	3.29	****	-18	-68	-10
LH45	19/37	GOm	R	840	4.34	******	44	-66	-10
LH46	19	GF	L	920	4.40	******	-36	-66	-12
LH47		declive	L	232	2.77	**	-22	-56	-12
LH48		declive	R	80	2.40	*	22	-54	-14
LH49	37	GF	R	648	4.20	******	50	-68	-16
LH50		declive	R	120	2.43	*	40	-76	-18
LH51		culmen	L	408	3.29	****	-31	-54	-20
LH52		pyramis	R	120	2.63	**	24	-66	-26
LH53		culmen	R	72	2.43	*	43	-52	-28
RH1	6	GFd	L	240	2.49	*	-2	0	56
RH2	6	GFm	R	320	2.47	*	28	-4	52
RH3	7	LPs	R	128	3.37	****	36	-56	52
RH4	7	LPs	R	672	3.37	****	34	-56	52
RH5	7	LPs	L	488	3.80	*****	-28	-58	50
RH6	4	GPrC	R	160	2.47	*	34	-10	50
RH7	7	PCu	L	336	2.69	**	-16	-62	46
RH8	4	GFm	L	568	3.03	**	-24	-10	46
RH9	6	GFd	L	304	2.75	**	-1	10	44
RH10	7	PCu	R	416	3.32	****	25	-47	44
RH11	6	GPrC	L	56	2.84	**	-34	-3	42
RH12	6	GPrC	L	456	2.84	**	-36	-2	42
RH13	40	LPi	R	320	3.26	***	44	-32	40
RH14	40	LPi	L	832	4.43	******	-34	-40	38
RH15	40	LPi	R	680	3.77	*****	30	-38	38
RH16	31	GC	L	136	2.98	**	-20	-28	34
RH17	19	GO	R	480	2.84	**	30	-76	28
RH18	44	GPrC	L	96	2.35	*	-46	-2	26
RH19	31	GC	L	88	2.47	*	-19	-54	26
RH20	23	GC	R	72	2.38	*	4	-12	26
RH21	19	Cu	R	672	3.18	***	24	-88	26
RH22	39	GOs	L	344	2.44	*	-26	-72	26
RH23	18	Cu	R	744	3.29	****	12	-88	22

Table 1. *Continued*

Site	Brodman Area	Gyrus	Side	Extent (mm³)	Z-score	P	Coordinate (mm)		
							x	y	z
RH24	19	GOm	R	520	2.78	**	38	-80	14
RH25	18	Cu	R	864	3.86	*****	14	-90	12
RH26	18	Cu	R	600	3.29	****	12	-82	12
RH27		Th	R	120	2.64	**	2	-10	10
RH28	32/24	GC	R	256	2.81	**	22	33	10
RH29		INS	R	88	2.35	*	40	14	10
RH30	17	Cu	L	320	3.06	**	-10	-96	9
RH31		Th	L	88	2.47	*	-28	10	8
RH32	19	GOm	L	712	3.94	*****	-44	-80	6
RH33	18	GL	R	224	3.06	**	2	-74	6
RH34	18	Cu	R	672	3.06	**	4	-94	4
RH35		Nc	R	224	2.89	**	16	18	4
RH36	17	GL	R	680	3.60	****	16	-82	2
RH37	37	GTi	L	784	3.40	****	-44	-72	0
RH38	37	GTi	R	1000	4.40	******	46	-72	0
RH39	37	GTi	L	216	2.64	**	-42	-58	-2
RH40	37/19	GOm	R	880	3.83	*****	44	-62	-6
RH41	19	GF	R	216	2.61	**	30	-58	-6
RH42	18	GL	R	728	4.26	******	8	-86	-6
RH43	18	GOm	L	456	3.26	***	-46	-76	-8
RH44	18	GL	R	568	3.43	****	26	-72	-8
RH45		declive	R	728	3.55	****	14	-74	-12
RH46	37/19	GF	L	664	4.14	******	-36	-66	-12
RH47	37	GF	L	128	2.47	*	-36	-44	-12
RH48		declive	R	320	3.57	****	24	-52	-14
RH49	19	GF	R	560	3.15	***	49	-70	-14
RH50		declive	R	176	2.84	**	40	-74	-19
RH51		culmen	L	448	3.12	***	-32	-54	-20
RH52		culmen	R	192	2.81	**	36	-50	-20
RH53		nodule	R	104	2.52	*	4	-59	-24

Site: indexes foci from superior to inferior for each condition. Brodmann Area: indicates probable Brodmann cytoarchitectonic zone. Gyrus: indicates probable gyrus, sulcus or subcortical structure. Side: indicates the hemisphere of activation. Extent: is the total volume in cubic millimeters of all contiguous pixels above statistical threshold ($P < .01$) of each focus. Z-score: intensity divided by image noise. Coordinates: expressed in millimeter distances from the anterior commissure: x = right (+)/ left (-); y = anterior(+)/posterior(-); z = superior(+)/inferior(-). (Coordinates, Gyrus, Brodmann, & Lobe all follow the conventions of Talairach & Tournoux, 1988.)

*$P = .01$; **$P = .005$; ***$P = .001$; ****$P = .0005$; *****$P = .0001$; ******$P = .00005$.

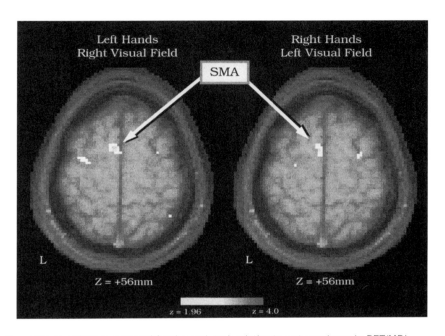

Fig. 3. Examples of activation detected for the each task relative to rest are shown in PET/MRI superpositions. Images were formed by superposition of the averaged (grand mean) PET functional data (colour) onto the averaged (grand mean) MRI anatomical data (gray scale). PET data are Z-scores (Table 1) displayed on a colour scale ranging from 1.96 (white-yellow; $P < .05$) to 4.0 (red; $P < .0001$). Probable Brodmann cytoarchitectonic areas are given in parenthesis. (This description also applied to Figs. 4–9). Foci in a superior region of the supplementary motor area (SMA) (BA 6) are indicated by arrows.

LH7 in Table 1) and in a ventral cluster in dominant left hemisphere (RH11, RH12, LH16, LH17 in Table 1). These right dorsal BA 6 foci were present in the same site for both left and right hand trials (LH5 and RH2), however, the activation in left BA 6, both dorsal and ventral clusters, varied in location depending on stimulus hand. Examples of the dorsal foci are shown in Fig. 7.

Moderate strength activations were also detected in posterior cerebellum and in inferior anterior cerebellum, and more often in the former than the latter. The activation was such that predominantly right activation was present for right hand trials and predominantly left activation was detected for left hand trials (Fig. 8). Once again, this pattern of activation indicates that neural processes supporting motor imagery of one's left and right hand are contralateralised because the left cerebellar hemisphere projects to the right cerebral hemisphere, and vice versa. Three of the cerebellar foci were in the same location for both left and right hand trials (LH48, LH50, LH51, RH48, RH50, RH51 in Table 1) and the others varied in location for left and right hand trials (LH43,

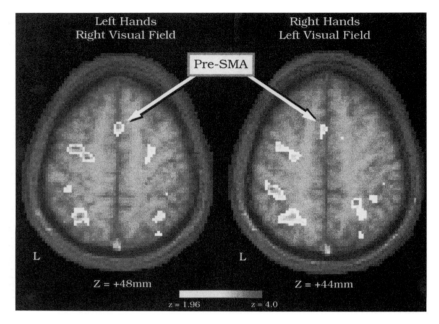

Fig. 4. See description of Fig. 3. Foci in pre-SMA (BA 6) are indicated by arrows.

LH47, LH52, LH53, RH45, RH52, RH53 in Table 1).

For both conditions, superior parietal regions (BA 7) were moderately active bilaterally, with more activation in the hemisphere contralateral to stimulus hemifield. In addition, one moderate focus (LH3, RH5 in Table 1) was present in the right superior parietal (BA 7) on both left and right hand trials.

Inferior parietal cortex (BA 40, including the supramarginal gyrus) was primarily active in the dominant (left) hemisphere during both left and right hand trials. In particular, there was one strong focus of activation (Fig. 9) detected in left inferior parietal cortex for both left and right hand trials (LH13, RH15 in Table 1). The other foci in inferior parietal cortex were contralateral to visual field stimulation.

Visual processing areas (BA 17, 18, and 19) were highly active (Table 1), showing bilateral activation, with greater activation in the hemisphere contralateral to the visual field stimulation. An example of this activation can be seen in Fig. 5.

Inferior occipitotemporal cortex (BA 37/19) was activated bilaterally, with more activation on the right. In particular, there was a strong focus of activity in BA 37 in the right hemisphere (LH34, LH45, RH38, RH40 in Table 1) during both left hand and right hand trials.

A similar pattern of activation was detected in the insula, wherein small foci were present in the right hemisphere during both conditions (LH31, RH29 in Table 1).

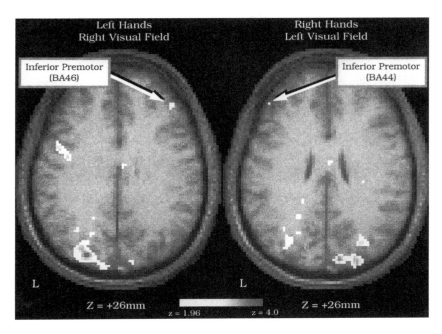

Fig. 5. See description of Fig. 3. Indicated by arrows are foci in an inferior region of premotor cortex (BA 44/46). Also visible (although not indicated) are activations in visual cortex (BA 18/19) contralateral to hemifield of visual stimulation.

Various subcortical structures, including the thalamus, caudate, and globus pallidus (Table 1), were active bilaterally for both conditions.

Finally, there was moderately strong bilateral activation in cingulate (BA 31, 32) but only during right hand trials (Table 1).

DISCUSSION

The task-induced rCBF increases in motor and somatosensory areas observed here are probably caused by the implicit movement that accompanies the discrimination of the shape of left and right visually presented hands (Parsons, 1987a, 1987b, 1994; Parsons, Gabrieli, & Gazzaniga, 1998). The implicit movement performed by subjects performing this task corresponds to a reaching and grasping task targeted to the visual stimulus. As discussed in detail later, the activations we recorded are very similar to neural activations observed for the actual performance of such a movement, with the exception of the primary somatosensory and motor cortices. The lack of activation in the latter areas indicates that motor imagery does not require the primary sensorimotor cortices (see following), perhaps unlike visual mental imagery (Kosslyn et al., 1995).

In addition, our observations confirm that motor imagery is subserved by neural systems

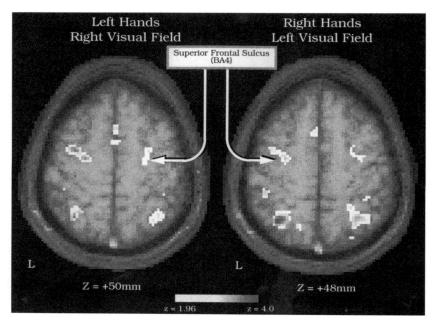

Fig. 6. See description of Fig. 3. Indicated by arrows are foci in premotor cortex of the superior frontal sulcus (BA 4). The foci are adjacent to, but separate from, a foci in BA 6.

of movement (for reviews see, e.g. Jeannerod & Decety, 1995; Crammond, 1997). This conclusion is compatible generally with corresponding conclusions for mental imagery representing other sensory modalities: e.g. that visual and spatial mental imagery are supported by visual and spatial neural systems (e.g. Kosslyn, 1988; Farah, 1994); that auditory mental imagery is supported by auditory neural systems (e.g. Reisberg, 1992; Paulesu et al., 1993; Zatorre et al., 1996); and that cutaneous or tactile mental imagery is supported by the somotosensory neural systems (Hodge et al., 1996). Moreover, the demonstration that a task that requires visual object discrimination, but that disallows explicit motor behaviour, produces strong activation in somotosensory and motor neural systems, indicates that mental imagery is not constrained by the perceptual systems of the presented stimuli.

It is probably due to the fact that motor imagery shares principal neural mechanisms with motor performance that motor imagery spontaneously (implicitly) possesses many of temporal and kinematic properties that motor performance possesses (Parsons, 1994; Sirigu et al., 1995, 1996). Those properties are likely to be intrinsic to the computational and representational structures implemented in motor mechanisms and are thus naturally expressed in the use of those structures for motor imagery. Recent hypotheses about the latter

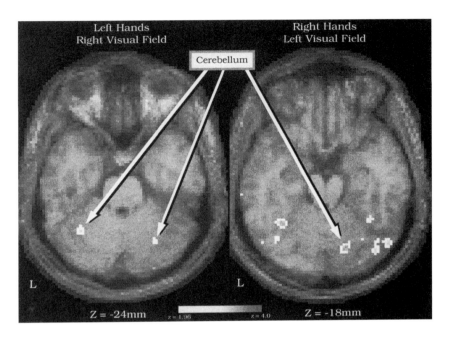

Fig. 7. See description of Fig. 3. Indicated by arrows are examples of foci in superior premotor (BA 6).

observation propose that the operative structures are the efferent motor command signals sent to the brain as a corollary to the motor act (Jeannerod, 1995; Crammond, 1997). The implicit knowledge the brain possesses about movements it can actually generate may also influence our interpretation of observed actions (Johansson, 1973; Viviani & Stucchi, 1992; Decety et al., 1997). At the same time, the implicit verisimilitude of motor imagery, as well as other implicitly represented ecological kinematic constraints, is highly adaptive because that structure serves as a platform enabling varied explicit reasoning and planning which would otherwise be arduous and limited or impossible (e.g. Shepard, 1984). More generally, that a visual shape task is solved via the creation of an action plan supports the view that cognition often employs mental models of physical objects and actions, reasoning by physical analogy (Stevens & Gentner, 1983; Shepard & Cooper, 1982; Johnson-Laird, 1989).

This handedness discrimination task elicits operations on sensory-motor information rather than on higher visual representations alone, probably because the somatomotor system represents, and operates on, body part representations that specify handedness. Humans do not in general appear to represent object shape so as to discriminate a shape from its mirror image (i.e. they do not explicitly code the handedness of shape) (Corballis & Beale, 1976; Hinton & Parsons, 1988).

The power of our experimental design to resolve differences between the neural systems

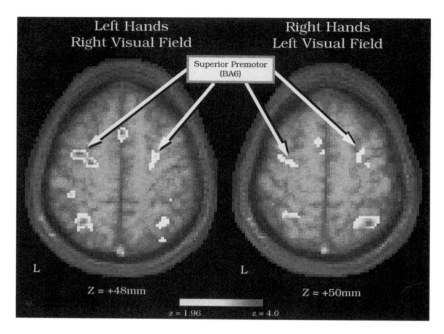

Fig. 8. See description of Fig. 3. Indicated by arrows are examples of foci in cerebellum.

for implicit left hand movement and those for implicit right hand movement is diminished because in left hand PET trials, for example, 20% of stimuli were right hand stimuli. Thus, the time-integrated images for a left hand PET trial contained a minority of activations for processes engaged specifically for right hand stimuli, and vice versa. In addition, the limited field of view of our PET scanner excluded some superior cerebral cortex (e.g. BA 5 and BA 7) and some inferior cerebellum, regions in which limb-specific activations could be expected to occur. Hence, the present data probably underestimate the size and number of neural sites active for limb-specific motor imagery.

Nonetheless, we observed limb-contralateral neural structures activated by limb-specific motor imagery. These structures include pre-SMA, inferior premotor cortex (BA 44/46), superior frontal sulcal premotor cortex (BA 4), and cerebellum. Broadly speaking (see following), these areas are implicated in higher order aspects of motor control, movement preparation/selection, action recognition/copying, spatial working memory, guidance/execution of self-paced movements, and sensory acquisition and control. These observations confirm the hypothesis that mentally simulating one's action and discriminating body part handedness both activate lateralised motor and somatosensory representations. This interpretation is strongly supported by the results of a recent psychophysical study of commissurotomy patients making left–right judgements of a hand

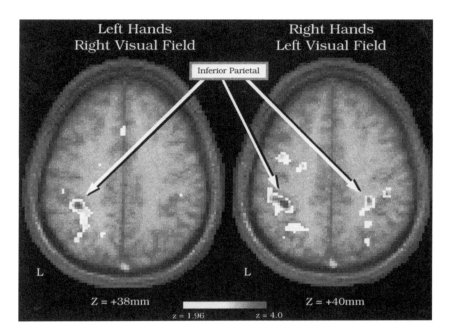

Fig. 9. See description of Fig. 3. Indicated by arrows are examples of foci in inferior parietal cortex (BA 40).

presented selectively to the left or right hemisphere (Parsons, Gabrieli, and Gazzaniga, 1998). The patients' left and right cerebral hemispheres could represent the shape and movement of the contralateral hand for comparison to the stimulus, but could not for the ipsilateral hand. This result suggests that within each cerebral hemisphere there are representations of the contralateral hand and its movement that are accessible for spatial cognition and motor imagery, but that there are no such representations of the ipsilateral hand. Thus, the left hemisphere is sufficient and necessary for the right hand motor imagery and the right hemisphere is sufficient and necessary for the left hand motor imagery.

We observed a number of other lateralised activations. There were neural structures active in the dominant left hemisphere regardless of the limb involved in the motor imagery. These structures include SMA proper, superior premotor (BA 6), and inferior parietal (BA 40). These areas (see following) appear to be involved in planning, guidance, and attention to motor performance. There were also neural structures active in nondominant right hemisphere, regardless of the limb involved in motor imagery. These structures include dorsal superior premotor (BA 6), insula, superior parietal (BA 7), and inferior occipitotemporal (BA 37). These areas have been implicated in motor planning, high-level somatic representation, the evaluation of visuospatial information, and the representation of the identity of objects and actions. Finally, there were neural structures active in the hemisphere contralateral to the visual hemifield of stimulation. These structures include occipital and superior parietal cortex. These areas are generally known to be involved in visual information processing and orienting attention. In the remainder of this report, the activations observed in particular brain areas are discussed in greater detail.

Activation was observed in SMA proper in the dominant (left) hemisphere for both left and right hand trials. This finding confirms earlier indications of SMA participation in motor imagery (Roland et al., 1980b; Fox et al., 1987; Rao et al., 1993; Tyszka et al., 1994; Grafton et al., 1996; Roth et al., 1996). SMA is hypothesised to subserve memory and sequencing of self-generated, internally-guided, composite movements (Goldberg & Bruce, 1990; Deiber et al., 1991; Grafton et al., 1992a, 1994; Passingham, 1993, 1996; Donoghue & Sanes, 1994; Tanji, 1994; Crammond & Kalaska, 1996; Tanji & Shima, 1994). The motor performances mentally simulated by our subjects are internally-guided, composite movements, so SMA activation during its mental simulation is consistent with SMA serving an analogous role during implicit movement.

Our observed pre-SMA activation was larger than that in SMA proper and was contralateral to stimulus handedness, rather than fixed in the dominant left hemisphere. Pre-SMA activation in the left hemisphere has been reported when humans observe right hand motions in order to imitate them (Stephan et al., 1995; Decety et al., 1997). Pre-SMA activation has also been observed to increase during preparation of self-paced executed movements as contrasted with cued ones (Frith et al., 1991; Deiber et al., 1996; Jahanshahi et al., 1995). The exact function of pre-SMA is not yet clear. Across a variety of motor tasks and moved body parts, performances requiring higher order aspects of motor control produce activation in pre-SMA, often as well as in SMA proper (for a review, see Picard & Strick, 1996). Simple motor tasks typically activate SMA proper without also activating pre-SMA. Based on anatomical and physiological data in monkey and human, the pre-SMA seems to have a function more closely related to the selection and preparation of movement and the SMA proper appears to be more closely related to motor execution. The observed strong limb-contralateral activation of the pre-SMA here suggests it has an important role in mental simulation of reaching a specific limb into a specific target posture.

This task activated a variety of ventrolateral and dorsolateral premotor frontal cortex. For both left and right hand trials, dorsolateral premotor regions (BA 6) was active bilaterally but ventrolateral premotor regions (BA 6) were active only in the dominant (left) hemisphere. Other neuroimaging studies of motor imagery have also reported activation in lateral premotor cortex (BA 6) (e.g. Rao et al., 1993; Decety et al., 1994; Leonardo et al., 1995; Stephan et al., 1995; Grafton et al., 1996; Rizzolatti et al., 1996; Roth et al., 1996). In general, lateral premotor cortex (BA 6) has been associated with the planning, programming, initiation, guidance, and execution of simple and skilled motor tasks and of limb movements in extrapersonal space (for a review, see Passingham, 1993). The demonstration here, that these areas are active during the mental simulation of reaching into a stimulus target, is compatible with the idea that motor imagery and preparation for actual movement share common mechanisms (Jeannerod, 1995). Recent research on lateral premotor cortex (BA 6) has indicated that there are functional differences between its ventral and dorsal regions (for a review, see Jackson & Husain, 1996). Studies of nonhuman primates suggest that these two areas support two fairly independent neural circuits, which simultaneously influence or control different properties of limb and grasping movements. The function of these dorsal and ventral circuits has been tied respectively in different hypotheses to the following distinctions: movement planning as opposed to on-line control of movement (Kurata, 1993, 1994); movements executed in visual as opposed to somatic space (Rizzolatti et al., 1988; Rossetti et al., 1994; Graziano et al., 1994; Fox, 1995); and the control of reaching as opposed to the control of grasping (Caminiti, Ferraina, & Johnson, 1996). The different pattern of activation found here in dorsal and ventral lateral premotor cortex is consistent with the possibility that these regions serve different functions during the mental simulation of reaching. However, the present data do not differentially favour any of these three functional dichotomies.

Activations were present in the superior frontal sulcal premotor areas (BA 4). Activation on right hand trials was contralateral to stimulus hand and on left hand trials it was bilateral, with more on the side contralateral to the stimulus hand. One recent neuroimaging study found this area to be active contralateral to the right hand performing a thumb–finger opposition task in which the finger to be touched was determined by an unpredictable external cue (Van Oostende et al., 1997). The superior frontal sulcal area was selectively active when the activation during the cued thumb-finger opposition task was contrasted with that during a self-generated thumb–finger opposition task. In the movements mentally simulated by our subjects, an external cue determined which of the two hands and which of the five digits were to be moved. This similarity in cue-guided motor performance of the present task and that of the Van Oostende et al. (1997) study suggests that our activation in BA 4 superior frontal sulcus was related to the guidance and execution of mentally simulated limb and hand movement. The fact that this activation was limb-contralateral suggests it supports the mental simulation of reaching a specific limb into a specific target posture.

Limb-contralateral activations were also observed in BA 44/46, regions that are also assumed to have premotor functions (e.g. Passingham, 1993). Our finding is consistent with reports of activation in left hemisphere BA 46 both during tasks involving motor imagery of the right hand and during tasks where subjects observed right hand motion for the purpose of recognition or imitation (Decety et al., 1994, 1997; Grafton et al., 1996). This observation is also consistent with activation during a variety of self-initiated actions (Frith et al., 1991). BA 44/45 were also activated by a related motor imagery task (joystick movement) (Stephan et al., 1995). The functions of BA 44 and 46 are not well understood. Four functions for area 44 have been recently proposed. One suggestion is that BA 44 supports movement preparation and motor imagery (Jeannerod, 1995). A second hypothesis is that the area is concerned with action recognition (Rizzolatti et al., 1996a; Carey et al., 1997). A third view is that the area is involved in the planning and selection of movements that possess nonarbitrary ('standard') mappings between the stimulus (hand picture) and the response (hand movement) (Wise et al., 1996). A fourth view is that this region is involved in the preparation of copied movements (Krams et al., in press). Each of these hypothesised functions would seem likely to be engaged in the present task. The fact that these activations are in the limb-contralateral hemisphere suggests that the activated BA 44 site subserves the mental simulation of reaching a specific limb into a specific target posture. Electrophysiological studies in nonhuman primates (Wilson et al., 1993; Goldman-Rakic, 1996) and PET studies of healthy humans (Smith et al., 1995) suggest that BA 46 supports spatial working activation. Activation in spatial working memory areas are consistent with the demands of our task. The mental simulation of reaching into target postures is a staged series of computationally-intense spatial representations sustained over the course of 1–2sec duration (Parsons, 1987b, 1994). In addition, the comparison of imagined and presented object shapes (hands) is likely to require working memory (Shepard & Cooper, 1982). Further research is necessary to clarify the role of these activated areas in motor imagery and left–right hand judgement.

No activation was detected in primary motor cortex (m1). Neuroimaging studies of motor imagery have often either detected no m1 activation (Rao et al., 1993; Sanes et al., 1993; Decety et al., 1994; Stephan et al., 1995) or detected it in only a minority of subjects (Leonardo et al., 1995; Sabbah et al., 1995; Hodge et al., 1996; Roth et al., 1996). Nonetheless, there is evidence for activation of primary somotomotor cortex during motor imagery, even when electromyographic recordings monitor subject body movement and find it negligible or uncorrelated with m1 activation (Porro et al., 1996). Group mean activation in m1 during motor imagery has been reported in recent studies using PET of healthy subjects (Lang et al., 1994); fMRI of healthy subjects (Porro et al., 1996) and of phantom limb patients (Ersland et al., 1996); in magnetoencephalography (Lang et al., 1996; Schnitzler et al., 1997); in topographic EEG (Beisteiner et al., 1995); and in transcortical magnetic stimulation (Abbruzzese et al., 1996). Given the complexity of the motor and premotor systems and that different motor tasks have been used in different studies, it is perhaps not surprising that the strength of activation varies across studies of motor imagery. It seems likely that the strength of activation in primary motor cortices depends in an unknown way on the parameters of the motor behaviour imagined in the task. At present the most appropriate conclusion is that real movement is not required for there to be activation in m1 and that motor imagery can occur without detectable activation of m1.

Our motor imagery tasks activated primarily posterior cerebellum and, to a lesser extent, inferior anterior cerebellum. The activations were primarily ispilateral to the stimulus hand: Hence, in accordance with the contralateral connectivity between cerebellar and cerebral hemispheres, the cerebellar activations were contralateral to hemisphere that is contralateral to the stimulus hand. The presence of cerebellar activation during motor imagery confirms earlier indications (Decety et al., 1990; Ryding et al., 1993; Decety et al., 1994; Bonda et al., 1995). Our result is also fairly consistent with another recent PET study reporting that motor imagery activated primarily posterior cerebellum, whereas motor execution activated primarily anterior cerebellum (Grafton et al., 1996). Classical theory would attribute this activation to a motor control function (Holmes, 1939; Ito, 1984; Thach et al., 1992; Houk & Wise, 1995; Welsh et al., 1995). However, more recent findings suggest that this activation is related to the acquisition and control of somatosensory information rather than motor control per se (Gao et al., 1996; Bower, 1997; Jueptner et al., 1997; Parsons, Bower, Gao, Xiong, Li, & Fox, 1997; Parsons, Bower, Xiong, Saenz, & Fox, 1997; Parsons & Fox, 1997). The activation of these cerebellar regions during imagined movement suggests the cerebellum has a comparable role in controlling the acquisition of the mentally-simulated sensation accompanying the mentally-simulated movement.

Multiple sites in parietal cortex were activated by our tasks. The parietal areas are generally thought to be multimodal association areas important to body schema, egocentric space, and action planning and representation (Andersen, 1987; Wise & Desimone, 1988; Stein, 1991; Heilman, Watson, & Valenstein, 1993; Kalaska & Crammond, 1995; Bonda et al., 1996; Faillenot et al., 1997). Recent neuropsychological studies of patients with lesions in parietal cortex indicate this area is critical for producing motor imagery that accurately corresponds to real movement time (Sirigu et al., 1995, 1996). An hypothesis based on this and other results asserts that evaluations of executed and imagined motor performance take place in parietal cortex based upon an on-line comparison of both corollary discharge (efference copy) and multimodal reafferent (feedback) sensory information against a stored mental representation of the motor plan (Crammond, 1997).

Superior parietal regions (BA 7) were active bilaterally during both left and right hand trials, with more activation in the hemisphere contralateral to stimulus hemifield. There was in addition a focus of activity present in the right superior parietal lobe on both left and right hand trials. Other neuroimaging studies of motor imagery or motor observation have observed strong right superior parietal activation (e.g. Bonda et al., 1995; Stephan et al., 1995; Decety et al., 1997). As hypothesised (Fig. 1), a portion of both the visual hemifield-driven and the right parietal activation may be related to shifts of covert visual attention to the stimulus (Pardo et al., 1991; Corbetta et al., 1993). In addition, superior parietal areas have been shown recently to be active during a shape comparison task that requires the imagined rotation of abstract non-body-part objects (Tagaris et al., 1996). The amount of activation recorded in the superior parietal area was proportional to the number of errors subjects made, suggesting that this area may evaluate visuospatial information involved in the comparison of shape (Crammond, 1997). Given that the latter task has a shape comparison component similar to that in the present left–right hand judgement task and given the evidence suggesting that the right hemisphere is specialised for visuospatial processing (Yin, 1970; Warrington & Taylor, 1973; Hecaen & Albert, 1978; Hamilton & Vermeire, 1988; but see Hellige, 1993; Ivry & Robertson, 1997), it is possible that our observed right superior parietal activation subserves the evaluation of visuospatial information.

Activation was also observed here in inferior parietal cortex (BA 40, including the supramarginal gyrus) primarily in the dominant (left) hemisphere during both left and right hand trials. Left inferior parietal cortex (BA 40) also activates during related motor imagery tasks involving the right hand (Decety et al., 1994; Stephan et al., 1995; Grafton et al., 1996). A variety of clinical and neurophysiological work suggests that action-oriented visual information is synthesised in the posterior parietal area (Milner & Goodale, 1995; Sakata, 1996). In addition, recent PET and neuropsychological studies strongly suggest that an area in left parietal cortex supports motor attention and preparation (Deiber et al., 1996; Rushworth et al., 1997; Krams et al., in press). Patients with left parietal damage were found to be impaired at disengaging attention to limb movement. PET studies of healthy subjects showed activation in the supramarginal gyrus during attention to (preparation for) hand movement. The location of the area supporting this covert motor attention is anterior to the covert orienting area in the more posterior parietal cortex (Posner et al., 1984; Heilman & Rothi, 1993; Corbetta et al., 1993, 1996; Nobre et al., 1997). These somatosensory (motor attention) and oculomotor (orienting) functions may converge in the vicinity of the intraparietal sulcus, which receives strong vestibular projections, at which the location of retinal coordinates of visual stimuli may be transformed into body-centred frames of reference (Krams et al., in press). That the hypothesised motor attention function is lateralised to the left hemisphere is indicated by the fact that the patients with left parietal damage show the impaired motor attention/preparation functions for the (ipsilateral) left hand. The left inferior parietal activation during motor imagery of either the left or right hand here is consistent with this hypothesis, suggesting that this area supports motor attention during the mental simulation of hand and limb movement.

Inferior occipitotemporal cortex (BA 37/19) was activated bilaterally, with much more activation overall on the right. This area has also been reported to be strongly activated during a task in which right hand movement is observed either passively or for the purpose of imitation (Decety et al., 1994, 1997). Physiological studies in monkeys and humans suggest that these regions are involved in representing the features of an object for the purpose of coding object identity per se, independent of spatial orientation (Perrett et al., 1990, 1994; Farah, 1991; Grady et al., 1991; Tanaka, 1996). Indeed, neurons in monkey inferotemporal cortex respond during the observation of hand actions (Perrett et al., 1989). It is seems likely that this area represents hand shape identity information during the perception of the presented stimulus hand and its comparison with the imagined hand of the subject.

Right insula was activated by both left and right hand motor imagery tasks. A related study of a left–right hand judgement task also recorded activation in insula, with more in the right hemisphere (Bonda et al., 1995). Insula is involved in higher levels of somatic function (Showers & Lauer, 1961; Friedman et al., 1986; Schneider et al., 1993), with its role often being compared to that assumed by inferotemporal cortex for visual function (Mishkin et al., 1979). Single cell recordings show that the insula responds to complex somatosensory information, responding even to the whole body (Schneider et al., 1993). In addition, lesions in the insula provoke somatic illusions (Roper et al., 1993) and intra-operative electrical stimulation of the insula induces illusions of shifts in body position and being outside one's body (rather than simpler, more localised sensory experiences) (Penfield, 1955). Note that physically adopting the posture of the presented hand in many instances in the present task would require coordinated activity at more than one arm joint, motion near extremes of joint limits, uncomfortable kinaesthetic sensations, and long trajectories (Parsons, 1987b, 1994). These considerations suggest in combination that the right insula activity we observe is related to higher level somatic representation accompanying mentally-simulated limb movement.

As the present data and discussion indicate, the circuits of somatosensory and motor areas active during the mental representation of motor acts overlap those active during the actual performance of motor acts. When real and simulated actions are compared within a single neuroimaging study, there is activation in both conditions in dorsal premotor cortex (Stephan et al., 1995; Roth et al., 1996), inferior parietal lobule (Stephan et al., 1995), and the primary motor cortex (Porro et al. 1996). In all cases, there is weaker activation in these areas during the imagined than actual movement. Relative to motor execution, motor imagery has been observed to activate in addition ventral premotor cortex, a caudal part of superior parietal cortex (medial BA 7), more rostral regions of cingulate cortex, more rostral regions of pre-SMA, and posterior, rather than anterior, cerebellum (Stephan et al., 1995; Grafton et al, 1996).

Finally, while the focus of the present report has been on motor imagery, a detailed process model of the left–right judgement of a hand is indicated by close analysis of the known relevant psychophysical data (Parsons, 1987b, 1994; Parsons, Gabrieli, & Gazzaniga, 1995). An outline of the model is as follows. Accurate information about stimulus handedness is derived in the early preattentive stages of information processing and later mental simulation of movement and shape-matching operations provide confirmation for conscious decision-making. The task appears to proceed in these stages: (1) analysis of the orientation and handedness of the stimulus; (2) analysis of the current orientation of one's own corresponding hand; (3) planning a path for one's hand to move (within its joint constraints) to the orientation of the stimulus; (4) mental simulation of that planned action; and (5) exact-match confirmation of the shape of imagined and perceived hands. The early implicit knowledge of the handedness of the stimulus (1) is probably the result of analysis utilising associations among object or hand shape and action patterns. The planning of a path (3) may involve using a forward kinematic map of the arm to make small changes in joint angles (i.e. producing hand position changes by changing joint angles within their limits), and then choosing a "satisficing" movement based on an error to the target computed in body-centred (or visual) frames of reference. The integration of further neuroimaging, psychophysics, and neurological studies should allow this model to be evaluated and improved, and to be mapped onto brain structures and their interactions.

REFERENCES

Abbruzzese, G., Trompetto, C., & Schieppati, M. (1996). The excitability of the human motor cortex increases during execution and mental imagination of sequential but not repetitive finger movements. *Experimental Brain Research, 111*, 465–472.

American Academy of Orthopaedic Surgery, Committee for the Study of Joint Motion. (1965). *Joint motion: Method of measuring and recording.* Chicago, IL: American Academy of Orthopaedic Surgery.

Andersen, R.A. (1987). Inferior parietal lobule function in spatial perception and visuomotor integration. In F. Plum & V.B. Mountcastle (Eds.) *Handbook of Physiology* (pp. 483–518). Bethesda, MD: American Physiology Society.

Annet, J. (1995). Motor imagery: Perception or action? *Neuropsychologia, 33*, 1395–1417.

Beisteiner, R., Hollinger, P., Lindinger, G., Lang, W., & Berthoz, A. (1995). Mental representations of movements. Brain potentials associated with imagination of hand movements. *Electroencephalogr. Clin. Neurophysiol. 96*, 183–193.

Biederman, I. (1990). Higher level vision. In D. Osherson, S.M. Kosslyn, J.M. Hollerbach (Eds.), Invitation to cognitive science, Volume II. (pp. 41–72). Cambridge. MA: MIT Press.

Bonda, E., Petrides, M., Ostry, D., & Evans, A. (1995). Neural correlates of mental transformations of the body-in-space. *Proc. Natl. Acad. Sci. USA, 92*, 11181–11184.

Bonda, E., Petrides, M., Ostry, D., & Evans, A. (1996). Specific involvement of human parietal systems and the amygdala in the perception of biological motion. *Journal of Neuroscience 16*, 3737–3744.

Bonnet, M., Decety, J., Requin, J., & Jeannerod, M. (1997). Mental simulation of action modulates the excitability of spinal reflex pathways in man. *Cogn. Brain Res. 5*, 221–228.

Bower, J.M. (1997). Is the cerebellum sensory for motor's sake, or motor for sensory's sake? The view from the whiskers of a rat? *Progress Brain Res., 114*, 483–516.

Brinkman, J., & Kuypers, H.G. (1972). Splitbrain monkeys: Cerebral control of ipsilateral and contralateral arm, hand, and finger movements. *Science, 176*, 536–539.

Caminiti, R., Ferraina, S., & Johnson, P.B. (1996). The sources of visual information to the primate frontal lobe: A novel role for the superior parietal lobe. *Cereb. Cortex, 6*, 319–328.

Table 2. MLAT Study Design

MLAT Condition	Form A	Form B	Form C
Solo-Basic	Present	Toast	Lunchbox
Solo-Distractors	Lunchbox[a]	Present[b]	Toast[c]
Dual-Basic	Toast + Coffee	Lunchbox + Schoolbag	Present + Letter
Dual-Search	Present + Letter	Toast + Coffee	Lunchbox + Schoolbag

[a] Distractors are applesauce, paper towel roll, spatula, hot-dogs, glass, schoolbag.
[b] Distractors are electric tape, stapler, snippers, paper bag.
[c] Distractors are hot-plate, nail file, vegetable shortening, mustard.

condition ends when the subject indicates that she or he is finished.

Subjects performed the MLAT in a single, hour-long session. Conditions were separated by a short break.

Scoring

Performance was simultaneously videotaped from head-on and overhead cameras, and the videotapes were used for scoring. Each condition was scored for errors, and an *Error Score* was computed that expressed the sum of errors on the entire form. Table 4 shows the taxonomy used to code errors. This taxonomy, and the determination of what constitutes an error, emerged from pilot studies with normal adults and neurological patients prone to errors of action (see Schwartz, Buxbaum, et al., 1998; Schwartz, Montgomery, et al., 1998).

The four error types of greatest theoretical interest are Object Substitution, Sequence, Omission, and Action Addition. Object Substitutions are coded when the subject performs an action that is appropriate to the task, but involves a distractor or another semantically or functionally related object. Sequence is a

Table 3. Demographic and Clinical Factors Balanced across Forms

	Form A		Form B		Form C	
Characteristics	M	SD	M	SD	M	SD
Controls						
Age	55.3	10.0	65.5	8.4	67.8	5.1
Years of education	14.7	2.4	13.2	2.7	13.7	1.8
LCVA Patients						
Age	64.6	15.5	57.7	5.3	56.7	21.7
Years of education	13.2	5.3	11.7	0.6	10.7	2.0
FIM	80.4	17.1	80.5	31.6	79.4	21.0

In Forms A, B, and C, there were 6 control subjects each and 5, 4 and 7 LCVA patients, respectively. FIM = Functional Independence Measure.

Table 4. Summary of MLAT Error Taxonomy

Omission
E.g. fail to use stamp on letter; fail to use cream in coffee.

Sequence
- Anticipation/omission: seal thermos before filling; close lunchbox before packed.
- Reversal: stir mug of water, then add grinds.
- Perseveration: make two sandwiches.

Object Substitution
E.g. stir coffee with *fork* (instead of spoon); place bread on *hot-plate* (instead of toaster).

Action Addition
Action not interpretable as step in task; includes "utilisation behaviour" and anomalous actions; e.g. cut gift box, pack extraneous items into schoolbag.

Gesture Substitution
Correct object used with incorrect gesture; e.g. spoon (rather than pour) cream into cup.

Grasp/Spatial Misorientation
Misorientation of the object relative to the hand or to another (reference) object; e.g. grasp wrong end of scissors (misoriented relative to hand); place stamp on envelope sideways (misoriented relative to reference object).

Spatial Misestimation
Spatial relationship between two or more objects incorrect; act otherwise well executed; e.g. cut paper much too small for gift.

Tool Omission
e.g. spread jelly with finger (instead of knife).

Quality
Inappropriate or inexact quantity (spatial or volume); e.g. fill thermos with juice to point of overflow.

composite category that includes performing a later step in advance of an earlier step (Anticipation/Omission errors); transposing two ordered steps (Reversal) and inappropriately repeating a step or subtask (Perseveration).

The Omission code is used for "pure" omissions—those not conditioned by another error. An action impacted by an Object Substitution is not coded as an Omission, nor is the deleted action in an Anticipation/Omission error (e.g. closing the thermos without filling it). Examples of Omissions, and other error types, are shown in Table 4.

The Action Addition code denotes an action that is not clearly in the service of the task; this includes utilisation behaviours (Shallice, Burgess, Schon, & Baxter, 1989) and "double substitutions" (e.g. stamps into paper bag rather than gift into box; lunchmeat into schoolbag instead of sandwich into lunchbox).

Scorers used detailed scoring sheets which enumerate, for each condition of each form, an exhaustive listing of error descriptions and their associated codes. Scorers simply enter a check mark next to the appropriate error description, or, if the observed error is not listed, write it in. Using this system to score the MLAT videotapes of 14 closed head injury patients independently, pairs of trained coders agreed on a mean of 76% of recorded errors (SD 11) (Schwartz, Montgomery, et al., 1998).

Results

Error Scores

Control Subjects. As we have previously reported (Schwartz, Buxbaum, et al., 1998), the

18 control subjects committed a total of 29 errors (Mean 1.6, range 0–4). There was no correlation between Error Score and age ($P = .52$) or education ($P = .83$), and males and females did not differ (Mann-Whitney test, $P > .40$).

LCVA. The LCVA group committed 213 errors (Mean 13.3, SD 11.4, range 0–38). Error Scores did not correlate with age ($r = .009$, $P > .9$) or education ($r = .09$, $P > .7$). Males and females did not differ (Mann-Whitney test, $U = 22.5$, $P > .3$). Error Score did, however, correlate with clinical severity, as measured by both the FIM ($r = -.79$, $P < .01$) and the (Expanded) Stroke Scale ($r = .79$, $P < .01$).

To assess whether the presence of aphasia contributed to the prediction of Error Score in those subjects for whom Stroke Scale data were available ($N = 11$), we recoded the aphasia item from the NIH Stroke Scale as a dichotomous variable (0; > 0) and carried out a univariate regression analysis with aphasia and FIM; Error Score was the dependent variable. The interaction term in the full model was not significant ($F = 0.3$) and so was removed (Kleinbaum & Kupper, 1978). In the reduced model, the FIM variable was significant [$F(1,8) = 12.17$, $P > .01$]; the aphasia variable was not significant [$F(1,8) = 0.30$]. The same result obtained when we added apraxia score (0; > 0) to the model containing FIM. There was no interaction of apraxia and FIM and no independent contribution of apraxia to Error Score.

Effect of Conditions

The effect of the conditions manipulation was assessed by a repeated measures ANOVA which had Condition (Solo-Basic; Solo-Distractors; Dual-Basic; Dual-Search) as the within-subjects factor and Group (LCVA; Controls) and Form (A,B,C) the between-subjects factors. In this, and all other parametric tests reported, the dependent measure (number of errors) was transformed by square root to normalise variance. Effects are significant at $P < .05$, or in the case of post hoc t-tests, at the level required by Bonferroni correction.

Main effects were significant for Group [$F(1,28) = 27.6$] and Condition [$F(3,84) = 6.7$]. Means for LCVA exceeded controls on all four conditions ($t > 3.7$) for all paired comparisons. Figure 1 shows the untransformed means. Inspection of the figure suggests that LCVA had relatively greater difficulty with Dual-Search (and see following); however, the Group X Condition interaction did not approach significance ($F < 1$).

As in our previous work, we explored the Condition effect through a series of planned contrasts. There was not a significant difference between Solo-Basic and Solo-Distractors ($F < 1$) nor between Dual-Basic and Dual-Search ($F < 1$). The difference between Solo vs. Dual conditions was significant ($F = 19.5$), but this is hardly surprising given the greater opportunities for error in the dual conditions.

While Form did not emerge as a main effect, it did qualify the effect of Condition as manifest in a significant Condition X Form interaction [$F(6,84) = 6.7$]. Inspection of the data showed that this was due to the fact that the

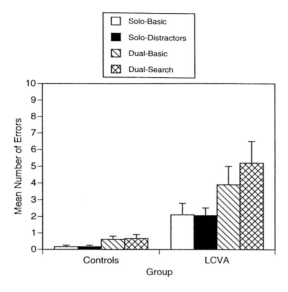

Fig. 1. Mean number of errors (with standard error bars) for the LCVA group (N = 16) and control group (N = 18) on each of the four conditions of the MLAT, collapsed across forms.

toast-making task proved disproportionately easy. In each form, the condition(s) involving toast making, whether solo or dual, yielded fewer errors than expected for that condition. Schwartz, Buxbaum, et al., (1998) reported the same effect in a group of RCVA patients run on the MLAT.

That LCVA patients are seriously compromised in naturalistic action production is confirmed by the finding that they make more errors than controls on even the simplest MLAT condition, in which there is only a single task to perform and no distractors. However, the failure to find a significant Group X Condition interaction raises doubts about the specificity of the deficit. With this in mind, we next compare the performance of LCVA patients to the other patients groups we have studied on the MLAT.

Comparison with RCVA and closed head injury. As noted earlier, the LCVA data were collected as part of a larger study that also looked at closed head injury (see Schwartz, Montgomery, et al., 1998) and right-hemisphere stroke (Schwartz, Buxbaum, et al., 1998). Table 5 shows the summary statistics for the 30 closed head injury (CHI) and 30 RCVA patients reported in previous studies. The subject selection criteria and testing procedures in the present study were identical to those previously used, permitting a comparison of subjects across studies. Both the LCVA and RCVA groups were older than the CHI group and contained a higher proportion of females. Clinical severity, as measured by FIM, was greater for RCVA than CHI ($t = 3.1$), and also tended to be greater for LCVA than CHI ($t = 1.7$, $P = .09$). LCVA and RCVA were of approximately equal clinical severity ($t = 0.7$).

Table 5. Summary Statistics for Closed Head Injury (CHI) and Right CVA (RCVA) Groups

Characteristics	CHI (n = 30)		RCVA (n = 30)	
	M	SD	M	SD
Age	31.3	11.4	59.1	13.1
Years of education	12.7	2.6	11.2	2.7
Days post-onset	97.5	59.4	55.3	106.6
FIM	89.9	17.5	76.1	17.2

The CHI group consisted of 21 males and 9 females. The RCVA group consisted of 19 males and 11 females. FIM = Functional Independence Measure.

To assess whether the groups differed in error rate, and whether they responded differently to the conditions manipulation, we carried out a repeated measures ANOVA in which the within-subjects factor was Condition (Solo-Basic, Solo-Distractors, Dual-Basic, Dual-Search) and the between-subjects factors were Group (LCVA, RCVA, CHI) and Form (A,B,C).

The analysis produced main effects of Group [$F(2,67) = 6.6$] and Condition [$F(3,201) = 16.7$], and most importantly, no suggestion of a Condition X Group interaction ($F < 1$) (see Fig. 2, which shows the untransformed means). Analysis of the Group effect indicated that the number of errors made by the LCVA group was not significantly different than CHI ($t = 1.19$) or RCVA ($t = 1.70$), however, the difference between the RCVA and CHI was significant, with RCVA making more errors ($t = 3.39$).

Planned contrast analysis of the Conditions effect again showed more errors on Dual conditions than Solo conditions ($F = 44.3$), and no difference between Solo-Basic and Solo-Distractors ($F = 1.2$). There was, however, an effect within the Dual conditions ($F = 4.7$). Confirming the trend noted earlier in the LCVA group, there were more errors committed in Dual-Search than Dual-Basic (see Fig. 2).

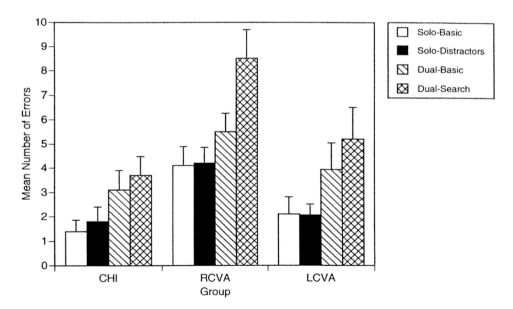

Fig. 2. Mean number of errors (with standard error bars), across the four conditions of the MLAT, of closed head injury (CHI) patients reported in Schwartz, Montgomery, et al. (1998) (N = 30), RCVA patients reported in Schwartz, Buxbaum, et al. (1998) (N = 30), and LCVA patients from the present study (N = 16).

Error Patterns

Previously, we reported a similar ranking of error types in CHI and RCVA groups. The modal type in both groups was Omission, followed by Sequence, Action Addition, and Object Substitution. As Table 6 shows, the ranking is the same for LCVA and the error proportions are strikingly similar across all three groups.

As in earlier studies, we standardised error rates for differences in the number of opportunities afforded for the different error types across conditions and forms. We did this by taking the total number of errors of a given type committed by the subject, dividing it by the number of opportunities that exist for that error type in that condition, and multiplying it by 100. We estimated the error opportunities by counting the number of exemplars listed on the coding sheet, restricting ourselves to those error types for which the coding sheets provide an exhaustive listing of unique exemplars (Omissions, Object Substitutions, and a subset of Sequence errors—Anticipation/Omission and Reversals, hereafter denoted as "Sequence".

Table 7 shows the LCVA group's mean standardised rates (rates per 100 opportunities) for these three error types, broken down by condition. As was previously found with RCVA and CHI, Omission errors continue to predominate over other error types even when opportunities are controlled. Object Substitution and "Sequence" errors are, by comparison, infrequent occurrences.

Although the relative proportions of each error type are similar across groups, it remains possible that the groups differ in the rates of errors of each type. In fact, as Fig. 3 shows, Omission and "Sequence" rates do appear to differ across the groups. In our previous studies, however, we found that standardised error rates for Omissions and "Sequence" tended to correlate more strongly with severity than did Substitutions (Schwartz, Buxbaum, et al., 1998; Schwartz, Montgomery, et al., 1998). Since the three patients groups were not completely matched for severity, this could be the explanation for the apparent differences between them.

To assess this, we carried out an analysis of covariance using FIM score as the covariate. In addition to the group factor (LCVA, RCVA, CHI), there were two within-subjects factors: Condition (Solo-Basic; Solo-Distractors; Dual-Basic; Dual-Search) and Error Type (Omission, Substitution, "Sequence"). Standardised error rate was the dependent variable.

Table 6. Error Frequencies, Expressed as Proportions of All Errors[a]

	Controls[b]	CHI	RCVA	LCVA
	(n = 18)	(n = 30)	(n = 30)	(n = 16)
Omission	.21	.38	.47	.44
Sequence	.10	.20	.19	.27
Object Substitution	.24	.09	.07	.08
Action Addition	.24	.12	.08	.06
Gesture Substitution	.03	.01	.01	.01
Spatial Misorientation	.07	.07	.06	.04
Spatial Misestimation	.00	.01	.04	.01
Tool Omission	.03	.06	.01	.01
Quality	.00	.08	.07	.08

[a]Total Errors for Controls = 29, CHI = 299, RCVA = 625, LCVA = 213.

[b]Controls reported in Schwartz, Buxbaum, et al.,1998: Mean age 62.9, SD 9.5 yrs.

Table 7. Standardised Error Rates (Errors per 100 Opportunities) for the Left CVA (LCVA) Group

Type of Error	Condition 1 (Solo-Basic)		Condition 2 (Solo-Distractors)		Condition 3 (Dual-Basic)		Condition 4 (Dual-Search)		Average Conditions 1–4	
	M	SE	M	SE	M	SE	M	SE	M	SE
Omission	7.8	4.8	11.1	4.9	11.8	4.4	15.8	6.6	11.6	5.2
"Sequence"	4.1	1.8	4.5	1.2	5.3	2.0	6.4	2.4	5.1	1.9
Object Substitution	0.7	0.5	1.6	0.9	0.8	0.6	3.4	2.0	1.6	1.0
Average	4.2	2.4	5.7	2.3	6.0	2.3	8.5	3.7		

We first examined whether the interaction of the covariate and the between-subjects factor was significant. Finding that it was not, we deleted the interaction term from the model (Kleinbaum & Kupper, 1978). In the reduced model, there was a highly significant main effect of FIM [$F(1,70) = 36.7$] but no Group effect [$F(2,70) = 2.5$] and, most important, no Group × Error Type interaction [$F(4.70) = 2.0$]. There was also a significant Condition × Error Type interaction [$F(6,420) = 4.3$]. As we found in our prior studies, there is a Conditions effect for Omissions [$F1,350 = 15.8$], but not for "Sequence" [$F(1,350) < 1$] or Substitution [$F(1,350) = 1.2$]. Standardised Omission rates are higher in the Dual as compared to Solo tasks ($F = 28.3$); and within the dual tasks, higher in Dual-Search compared to Dual basic ($F = 8.0$). These effects are present in all three patient groups, as indicated by the absence of a three-way interaction with Group[2]

These analyses indicate that the apparent differences in error pattern across the groups (see Fig. 3) can be reduced to differences in clinical severity.

Discussion

The evidence from Study 1 is that LCVA patients as a group are no more vulnerable to errors in naturalistic action production than similarly selected groups of RCVA and CHI patients. Nor do they show a specific pattern of breakdown. There was no evidence that the LCVA patients were differentially sensitive to particular conditions of the MLAT, relative to normal controls (Fig. 1) or to the other patient groups (Fig. 2). In addition, there was striking uniformity in the error proportions across the three patient groups. Omissions are the predominant error types, whereas Substitutions and Sequence errors are relatively rare events.

---

[2] There was, however, a three-way interaction of Condition, Error Type, and FIM. As in our previous work, the interaction with FIM is likely to be due in part to the relatively strong relationships of Omissions and "Sequence" (but not Substitutions) to clinical severity. In support of this, when we regressed standardised error rates on FIM severity, the adjusted R-square was .36 for Omissions, .28 for "Sequence", and only .11 for Substitutions. (Similarly, when we regressed standardised error rates on severity as measured by overall Error Scores, the adjusted R-square was .80 for Omissions, .41 for "Sequence", and .15 for Substitutions).

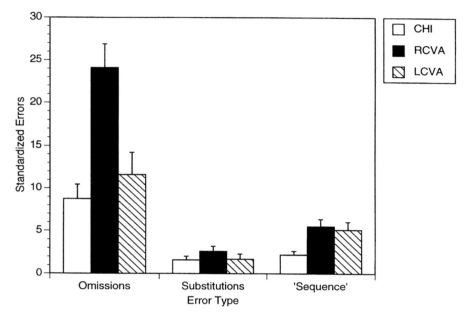

Fig. 3. Means (with standard error bars) of the standardised error rate (errors per 100 opportunities) for Omission, Substitution, and "Sequence" errors on the MLAT. The data are from closed head injury (CHI) patients reported in Schwartz, Montgomery, et al. (1998) (N = 30), RCVA patients reported in Schwartz, Buxbaum, et al. (1998) (N = 30), and LCVA patients from the present study (N = 16). "Sequence" errors comprise Anticipation, Anticipation/Omission, and Reversal errors.

Moreover, it is primarily Omission rate that varies across conditions; and this, too, obtains for all three patient groups.

The role of severity looms large in this study, as it has in our previous group studies. FIM is a strong predictor of overall Error Score, and different error types are more strongly or less strongly associated with FIM. Severity differences across patient groups or across individuals with a group may give rise to differential error patterns. This factor must be considered when evaluating the specificity of an action disorder. Study 2 illustrates and elaborates this point.

STUDY 2

In this study, we examine the performance of GL, a closed head injury patient who exhibits a severe action impairment in the context of left-hemisphere syndrome. Based on site(s) of lesion and neuropsychological profile, this patient would qualify as an ideational apraxic. The question posed here is whether GL's behaviour on the MLAT is explained entirely by his level of clinical severity or, alternatively, whether there are aspects of his action performance that can be attributed to his particular neuropsychological deficits.

Method

Subject

GL was a 48-year-old right-handed professional with a Master's degree who sustained a head injury in an automobile accident 5 months prior to the evaluations reported here. He was intubated at the scene of the accident and Glasgow Coma score was not determined. CT scan performed 6 days after the accident revealed depressed skull fracture of the left temporal and frontal bones, multiple haemorrhagic contusions of the left hemisphere, including the left inferior frontal, temporal, parietal, and occipital cortices, left frontoparietal and temporal subdural haematoma, intraventricular haemorrhage, and midline shift to the right. There was also a small right posterior parietal haemorrhage. T1- and T2-weighted MRI scans performed 9 days after the injury demonstrated moderate oedema associated with compression of the left lateral ventricle in addition to the haemorrhages and midline shift noted on the CT (see Fig. 4).

Neurologic examination at the time of the present investigations revealed a flaccid hemiplegia on the right and an inferior right quadrantanopsia. There was no visual, auditory, or tactile neglect or extinction. Proprioception of the left upper and lower extremities was intact; there was a mild proprioceptive deficit on the right. GL exhibited a nonfluent aphasia with mildly impaired comprehension and relatively spared repetition, consistent with a clinical diagnosis of transcortical aphasia. FIM score was 61 (Mean 3.4), indicating that he needed minimal to moderate assistance with daily activities.

As an inpatient at MossRehab, GL was observed to make errors in everyday activities. Therapists indicated that although he was able to follow several-step commands, he required supervision with daily activities due to object misuse (e.g. using toothbrush for comb) and omission of task steps (e.g. failure to use toothpaste on toothbrush). He was referred to our group for further evaluation of this action impairment. GL was not a subject in any of our previous studies.

Results of neuropsychological testing are summarised in Table 8. It is evident that in addition to aphasia, GL exhibits other signs frequently attributed to left-hemisphere pathology, including severe deficits in limb praxis manifested in gesturing to command and imitation (ideomotor apraxia) and with tool in hand (ideational apraxia; see DeRenzi et al., 1968). He also exhibits severe deficits in semantic knowledge of objects, a factor that has been associated with ideational apraxia (De Renzi & Lucchelli, 1988; Ochipa et al., 1989).

There are also areas of remarkable cognitive sparing. On executive function tests of planning and problem solving (e.g. Tower of London), interference control (Dual Tasks), and working memory (Self-Ordered Pointing), GL's performance is frequently superior to that of normal controls. In addition, his visuospatial functioning is intact. His unimpaired performance on these tasks is also evidence that he is able to comprehend complex instructions and to remain "on task" for extended periods of time.

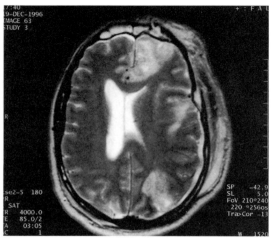 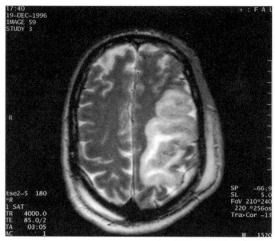

Fig. 4. (Left): Midventricular section of T-2 weighted MRI scan demonstrating inferior frontal, lateral occipital, and temporal lobe contusions on the left, and effacement of the left lateral ventricle. (Right): Supraventricular section of T-2 weighted MRI scan demonstrating extensive left fronto-parietal contusion.

Procedures

GL performed all three forms of the MLAT in the order C, A, B according to the methods described in Study 1. Each form was performed on a separate day. The order of conditions within each form was random. In addition to Error Scores, an Accomplishment Score was calculated for each condition, based only on the steps accomplished and without penalising for commission errors (see Appendix). The four Accomplishment Scores from each form were then summed, divided by the maximum Accomplishment Score for that form, and expressed as a percentage.

Results

GL's performance on the MLAT was highly impaired. To convey the quality of his performance, we describe two segments. The first is from the Solo-Distractors condition of Form A, involving lunchbox preparation; the second is from the Dual-Search condition of Form B, involving toast and coffee preparation. Recall that the two forms were performed on separate days.

(Lunchbox) GL began the task appropriately by removing two pieces of bread from the bread bag and placing them beside one another on the table. He then opened the juice bottle, picked up a spatula (distractor object for the knife which was also present), and attempted to insert it into the bottle, earning an Action Addition error. Failing, he closed the juice. Nearly immediately, he re-opened the juice, poured it on the bread with apparent care (another Action Addition), and closed the bottle. He next opened the mustard jar and attempted to insert the spatula (an Object Substitution: spatula for knife). Unsuccessful again, he closed the jar. He next made a sandwich of the two pieces of moistened bread. He did not add juice to the thermos (an Omission error) and sealed the thermos incorrectly with a cap, but not lid (an Anticipation/Omission). He

Table 8. Results of Neuropsychological Testing

Neuropsychological Test	Score	Comment
Language		
Western Aphasia Battery (WAB)[a]		
Fluency	2/10 (20%)	Severe impairment
Naming	1/60 (2%)	Severe impairment
Auditory verbal comprehension	48/60 (80%)	Mild impairment
Repetition	85/100 (85%)	Mild Impairment
Limb Praxis		
Gesture to sight of tool	6/24 (25%)	Severe impairment
Gesture with tool in hand	7/23 (30%)	Severe impairment
Gesture to imitation	1/24 (4%)	
Semantic Knowledge		
Pyramids and palm trees[b]	35/52 (67%)	Below normal range (94%–100%)
Object classification test[c]		
Variant	23/25 (92%)	Below normal (100%)
Coordinate	23/24 (96%)	Normal performance
Associate	12/24 (50%)	Below normal range (88%–100%)
Function Matching Test[c]	21/36 (58%)	Below normal range (80%–100%)
Serial Action Photographs[c]	3/7 (43%)	Moderate impairment
Memory		
Self-ordered Pointing-Figure Version[d]	10/17	Control mean 10, SD 1.2
Wechsler Memory Scale, revised		
Visual memory span forward	5	34th percentile
Visual memory span reversed	4	18th percentile
Executive Function		
Tower of London[e]		
No. correct < 60sec	2/4	Control mean 2.1, SD 1.3
No. correct irrespective of time	4.4	Control mean 3.5, SD 0.5
Tinker Toy Test[f]	10/12 (83%)	Control mean 65%, SD 16%
Dual Task (Decrement)[d]		
RT decrement: RT with digit span	247msec	Control mean 248msec, SD 190msec
RT decrement: RT with counting	30msec	Control mean 40msec, SD 61msec
Visual/Spatial Processing		
Visual Obj. & Space Percept. Battery[g]		
Object percept. screening test	19/20	Normal Performance
Object decision	15/20	Normal Performance
Position discrimination	19/20	Normal Performance
Rivermead Behavioural Inattention[h]		
Line cancellation	40/40	Normal Performance

[a]Kertesz, 1982; [b]Howard & Patterson, 1992; [c]Buxbaum, Schwartz, & Carew, 1997; [d]D'Esposito & McDowell, unpublished data; [e]Shallice, 1982; [f]Lezak, 1983; [g]Warrington & James, 1991; [h]Wilson, Cockburn, & Halligan, 1987.

concluded the task by packing the thermos in the lunchbox, wrapping his bread and juice "sandwich" neatly in foil, packing it into the lunchbox, and closing the lid.

(Toast & Coffee) Although he was instructed to prepare a single slice of toast with butter and jelly, GL removed two slices of bread from the bread bag in succession, earning a Perseveration error. Using a spoon, he spread jelly on one of the pieces of untoasted bread (Object Substitution error for Spoon; Omission error for not toasting bread). Appropriately using a knife, he then spread butter on the remaining piece of bread (Sequence error for jelly before butter), and placed the two pieces of bread together to form a sandwich. He cut the sandwich in half with a fork, earning an Action Addition. He then turned to coffee preparation. He poured, rather than spooned, instant coffee grinds and sugar into the cup of hot water, with spoon nearby (two instances of Tool Omission). Finally, he stirred the coffee with a fork (Object Substitution).

Both episodes convey the impression that much of the general outline or "plan" of action is intact. Nevertheless, GL's willingness to accept inappropriate objects, and the sense that such objects are interchangeable (spatula for knife, juice for mustard, spoon for knife, fork for knife, fork for spoon) suggests inadequate specification of the attributes of objects needed to meet task goals, and/or a failure to verify whether the objects selected meet such specifications. On some occasions, his actions are so deviant as to violate the affordances of the objects involved, for example, when he pours juice onto the bread slice (a non-container). Along with GL's errors of commission are numerous instances in which essential task steps are omitted, even when the requisite items are highly salient (e.g. omission of toasting, with toaster present in the array). Below we will examine whether the quantity and quality of these various errors are predictable from his level of severity.

Table 9 provides a quantitative summary of GL's MLAT performance (along with that of two other patients, see following). There is substantial improvement in Accomplishment Score across the three forms (days), due primarily to reduction in the number of Omission errors. It is not clear how this improvement should be interpreted, as practice and form effects (e.g. on error opportunities) are confounded[3].

GL's mean Error Score across the three forms is 26.0. A simple regression model derived from the three patient groups (LCVA, RCVA, CHI) predicts, for a FIM score equal to GL's, an Error Score of 25.03 (90% prediction interval [PI] is 7.33–42.72).

Across the 3 forms, GL produced a total of 78 errors. Forty-nine per cent of these were Omissions; the percentage of Sequence, Substitutions, Action Additions, and miscellaneous others was, respectively, 22%, 6%, 6%, and 17%. This closely approximates the group error patterns reported in Table 6.

Summing across the three forms, we calculated GL's standardised error rates for Omission (28.9), "Sequence" (6.7), and Object Substitutions (3.5). The simple regression models derived from the three patient groups predict, for an Error Score equal to GL's, a

---

[3] For example, the opportunity for Omissions is greater in Form C than in Forms A or B (49, 39, and 38 opportunities, respectively); on the other hand, the opportunities for "Sequence" errors is approximately equal across the three forms (61, 56, and 61 opportunities, respectively, for Forms C, A, and B).

Table 9. Performance of Patients, GL, AM, and RC on the MLAT

Form/ Patient	Accomplishment Score	Errors (Inclusive)		
		Type	Number	
Form C			GL	
GL	8%	Omission	21	
		Sequence	7	
		Action Addition	1	
		Spatial Misorientation	1	
		Total	30	
Form A			GL	AM
GL	35%	Omission	14	11
AM	43%	Sequence	4	10
		Object Substitution	3	0
		Action Addition	2	2
		Tool Omission	2	0
		Gesture Substitution	1	2
		Spatial Misestimation	1	4
		Spatial Misorientation	1	1
		Quality	1	1
		Total	29	31
Form B			GL	RC
GL	71%	Sequence	6	2
RC	54%	Omission	3	10
		Tool Omission	3	0
		Object Substitution	2	4
		Action Addition	2	4
		Gesture Substitution	1	0
		Spatial Misorientation	1	0
		Quality	1	0
		Total	19	20

standardised Omission rate of 30.9 (PI 14.05–46.76) and a standardised "Sequence" rate of 6.29 (PI 0.5–12.08). These point estimates, like the estimate of his overall Error Score, are remarkably close to the actual values. (Standardised Substitutions were not predicted due to the weakness of the model; see footnote 2.)

The significance of this result is underscored in Fig. 5. These are histograms of standardised error rates for all patients (LCVA, RCVA, CHI and GL) ordered left to right by total Error Score. (For GL, we took the average across the three forms.) It is evident that GL's standardised error rate for all three error types is essentially indistinguishable

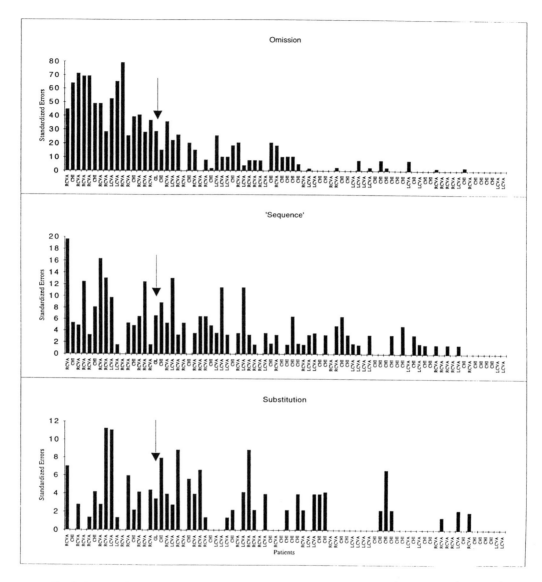

Fig. 5. Standardised error rate for Omission (top), "Sequence" (centre), and Substitution (bottom) errors for individual patients in the MLAT studies. Patients are shown ranked by total MLAT Error Score from left (high Error Scores) to right (low Error Scores). Data are from closed head injury (CHI) patients reported in Schwartz, Montgomery, et al. (1998) (N = 30), RCVA patients reported in Schwartz, Buxbaum, et al. (1998) (N = 30), LCVA patients from the present study (N = 16), and patient GL. Data of patient GL are marked by an arrow. "Sequence" comprises Anticipation, Anticipation/Omission, and Reversal errors.

from that of other patients with similar Error Scores.

Figure 5 highlights a more general point that is also in keeping with the evidence from Study 1: LCVA patients are not disproportionately represented among those showing high standardised rates for "Sequence" or "Substitution", as the literature on ideational apraxia would lead us to expect. If there is a group that appears disproportionately affected, it is the RCVA group (which, of course, is also the most severe).

Admittedly, quantitative analyses of this type may miss important qualitative differences among patients. For example, as we stated in the context of the descriptive segments, GL on occasion makes object-related errors that are more flagrant than the expected substitution of related arguments (object or place). The question is whether these sorts of errors, too, are reflective of general severity, as opposed to aetiology or neuropsychological profile. To address this question, we first identified two patients from the RCVA group who were close to GL in FIM score and overall Error Score.

Patient AM was a 74-year-old retired skilled labourer with a 9th-grade education who had sustained a right temporal-parietal stroke 3 months prior to MLAT testing (Form A). Neurological examination demonstrated a moderate left hemiparesis, left-sided sensory loss for all modalities, left homonymous hemianopia, and consistent extinction to bilateral simultaneous visual and tactile stimulation. He exhibited left neglect in daily activities such as reading, grooming, and feeding. In general, however, he was responsive to verbal cueing to attend to the left. AM's FIM score was 63 (Mean 3.5), indicating that he required minimal to moderate assistance with daily activities. Details of AM's case history are reported in Buxbaum, Coslett, Montgomery, and Farah (1996).

Patient RC was a 64-year-old right handed former nurse who suffered a right basal ganglia haemorrhagic stroke approximately 3 months prior to administration of the MLAT (Form B). Neurological examination revealed a mild left hemiparesis. Visual fields were full, and there was no evidence of extinction or neglect on neurologic exam or in daily activities. RC's FIM score was 64 (Mean 3.5), indicating that she, too, required minimal to moderate assistance.

Table 9 summarises these patients' performance on the MLAT in terms of Accomplishment and Error Score. Note that their overall error rates, as well as the types of errors produced, are similar to those of GL. Additionally, a review of their performance tapes indicated that both patients did occasionally commit errors with objects that flagrantly violated conventional usage. The following are examples:

During the Dual-Search condition involving the Present & Letter tasks, Pt. AM packed the gift box with numerous items from the drawer—thermos lid, scotch tape, knife, fork, food coupons, spatula. Later, he extracted some of these household objects from the box and wrapped them, singly, in tissue paper.

In the Solo-Distractors task (Lunchbox with Distractors), AM placed a large jar containing applesauce into the lid of the lunchbox.

Pt. RC, in the context of the Toast & Coffee, Dual-Search condition, removed the thermos lid and cup from the drawer. She placed the thermos lid onto the cup of hot water intended for instant coffee preparation, then placed the thermos cup over the lid.

Discussion

The results of Study 2 give no support to the suggestion that GL's naturalistic action disorder is different quantitatively or qualitatively from those exhibited by comparably severe patients with other aetiologies and neuropsychological profiles. This evidence does not, however, rule out the possibility that differences might be found. For example, Table 10 shows that GL, but not AM, made Object Substitution errors on Form A, and that GL, but nor RC, made Tool Omission errors on Form C. If more extensive and more focused testing were to show these differences to be reliable, it would challenge the conclusion that all (reliable) differences among patients are attributable to severity.

Even without this evidence, it could be argued that since action errors are multiply determined (Reason, 1990), similar error profiles could arise from drastically different underlying causes. For example, object misuse errors could be due to semantic underspecification in some patients and visual attentional deficits in others. Indeed, we showed a connection between object use errors and hemispatial neglect in the RCVA study (Schwartz, Buxbaum, et al., 1998). While we took care to choose excerpts from AM and RC that were not subject to this interpretation, the general point still holds: Different deficits may map causally onto the same overt behaviours.

We acknowledge the validity of these rhetorical counter-arguments, but maintain that the evidence presented in the present studies, and our earlier ones, is sufficient to shift the burden of proof onto those who would argue for the specificity of the naturalistic action breakdown in ideational apraxia. In the absence of proof, it is worth exploring alternative, nonspecific accounts. We will do so in the General Discussion.

GENERAL DISCUSSION

The data presented here and in our previous group studies (Schwartz, Buxbaum, et al., submitted) indicate that groups of patients with different aetiologies and anatomic loci of damage behave quite similarly in the performance of naturalistic action. Patterns of errors were not obviously affected by the presence of left-hemisphere damage. Even a patient for whom the left-hemisphere syndrome included semantic deficits and ideational apraxia failed to show the distinctive predilections to mis-sequencing and object misuse that are predicted on most accounts of ideational apraxia. This does not mean that the left-hemisphere specialisation for objects and gestures is irrelevant to naturalistic action production (see Buxbaum et al, 1997), only that deficits in these areas do not compromise action in a unique or transparent manner.

In Schwartz and Buxbaum (1997), we proposed that these and other neurocognitive deficits impact naturalistic action production chiefly by undermining automaticity, thereby placing a greater demand upon executive processes like supervisory attention and working memory. These executive processes take advantage of the redundant information sources available to support naturalistic action (e.g. object semantics, script knowledge, sensorimotor associations, gesture representations) to bring about functional reorganisation and compensation. As a result of this reorganisation, the neurocognitive deficit does not manifest transparently in errors. Additionally, if executive deficits accompany the primary neurocognitive deficit, compensation is imperfect and errors arise.

In more recent work, our emphasis has shifted from executive processes to the resources they manipulate. This move was necessitated by several findings. First, we found that differences between patients and controls is one of degree rather than kind (Study 1; Schwartz, Montgomery, et al., 1998). Control subjects made errors on the MLAT at low rates, and their errors were primarily commissions. Mild closed head injury patients made similar errors, and at similar rates. On a more challenging extension of the MLAT, however, the mild closed head injury patients did make more errors than controls, and tended to make more omissions. More severe closed head injury patients made many omission errors even on the MLAT, and their commission errors were more frequent and often more flagrant than controls or those with mild closed head injury. The presence of a continuum of behaviour encompassing intact subject at one end is not consistent with a specific deficit underlying the performance of action-disordered patients. In addition, as noted in the Introduction, the preponderance of omissions was not predicted on any number of deficit-based accounts of action disorder, including those implicating impairments in supervisory attention or working memory. These lines of evidence suggested that the errors in CHI have a nonspecific basis, and pointed to a general resource limitation. In our subsequent work, this was supported by the evidence that right-hemisphere stroke patients—known to have diminished resources—make more errors on the MLAT than the closed head injury patients even when differences in severity were statistically controlled (Schwartz, Buxbaum, et al, 1998).

The fact that GL performed normally on tasks tapping executive processes and working memory constitutes further evidence that a defective executive system is not a necessary feature of naturalistic action breakdown. GL's normal performance under conditions of dual task load, for example (see Table 8), suggests that the central executive of working memory is unimpaired (Baddeley, 1986; Baddeley & Hitch, 1994). In light of this, and in the context of findings from our previous work, we suggest the following explanation for GL's action impairment.

We posit that GL's deficits in semantic memory force a shift from an automatic processing mode to one requiring greater effort, attention, and arousal; that is, greater resource allocation. We suggest, further, that due to the general severity of his neurologic impairment, there are limitations in the capacity of the resource[4]. The dependence on limited-capacity resources has a homogenising effect that obscures any manifestations of underlying neuropsychological deficit, and results in a pattern of naturalistic action performance much like the one seen across patient groups.

This pattern has the following characteristics: First, the burden on limited-capacity resources results in frequent failure to activate any of the candidate object or action schemas sufficiently to surpass a threshold for selection, and this in turn causes a high rate of omissions. The capacity limitation (and resultant selection failure) is most pronounced in complex tasks in which numerous schemas compete for selection. Second, there is an absence of a selective vulnerability to any of the error types associated with specific neuropsychological systems dysfunction. Third, other error types have an "opportunistic" quality, meaning that particular conditions afford different opportunities for different error types, and error rates largely track those opportunities. Fourth, the overall rate of error, and the rates of specific error types, are largely a function of the severity of the resource capacity limitation.

REFERENCES

Allport, A., (1993). Attention and control: Have we been asking the wrong questions? *Attention and performance XIV: A silver jubilee*. Cambridge, MA: MIT Press.

Baddeley, A.D. (1986). *Working memory*. Oxford: Clarendon Press.

Baddeley, A.D., & Hitch, G.J. (1994). Developments in the concepts of working memory. *Neuropsychology, 8*, 485–493.

Brott, T., Adams, J., Olinger, C.P., Marler, J.R., Barsan, W.G., Biller, J., Spilker, J., Holleran, R., Eberle, R., Hertzberg, V., Rorick, M., Moomaw, C.J., & Walker, M. (1989). Measurements of acute cerebral infarction: A clinical examination scale. *Stroke, 20*, 864–870.

Brown, J.W. (1988). *Agnosia and apraxia: Selected papers of Liepmann, Lange and Potzl*. Hillsdale, NJ: Lawrence Erlbaum Associates Inc.

Buxbaum, L.J., Coslett, H.B., Montgomery, M.W., & Farah, M.J. (1996). Mental rotation may underlie apparent object-based neglect. *Neuropsychologia, 34*, 113–126.

Buxbaum, L.J., Schwartz, M.F., & Carew, T.G. (1997). The role of semantic memory in object use. *Cognitive Neuropsychology, 14*, 219–254.

Coslett, H.B., Bowers, D., & Heilman, K. (1987). Reduction in cerebral activation after right hemisphere stroke. *Neurology, 37*, 957–962.

DeRenzi, E., & Lucchelli, F. (1988). Ideational apraxia. *Brain, 111*, 1173–1185.

DeRenzi, E., Pieczuro, A., & Vignolo, L.A. (1968). Ideational apraxia. A quantitative study. *Neuropsychologia, 6*, 41–52.

FIM (1993). *FIM: Guide for the use of the uniform data set for medical rehabilitation, Version 4.0*. Buffalo, NY: State University of New York at Buffalo.

Haaland, K.Y., & Flaherty, D. (1984). The different types of limb apraxia errors made by patient with

---

[4] The notion that cognitive operations may be enhanced by limited-capacity "resources" or attention has a long history (see Allport, 1993, for a critical review).

left vs. right hemisphere damage. *Brain and Cognition, 3*, 370–384.

Harrington, D.L., & Haaland, K.Y. (1991). Hemispheric specialization for motor sequencing: Abnormalities in levels of programming. *Neuropsychologia, 29*, 147–163.

Hecaen, H. (1981). Apraxias. In S.B. Filskov & T.J. Boll (Eds.), *Handbook of clinical neuropsychology*. New York: John Wiley & Sons.

Heilman, K.M., Schwartz, H.D., & Watson, K.T. (1978). Hypoarousal in patients with the neglect syndrome and emotional indifference. *Neurology, 28*, 229–232.

Howard, D., & Patterson, K. (1992). *The Pyramids and Palm Trees Test*. Bury St. Edmunds, UK: Thames Valley Test.

Kertesz, A. (1982). *The Western Aphasia Battery*. New York: Grune & Stratton.

Kimura, D. (1982). Left-hemisphere control of oral and brachial movements and their relation to communication. *Philosophical Transactions of the Royal Society, London, B, 298*, 135–149.

Kimura, D., & Archibald, Y. (1974). Motor functions of the left hemisphere. *Brain, 97*, 337–350.

Kleinbaum, D.G., & Kupper, L.L. (1978). *Applied regression analysis and other multivariate methods*. Belmond, CA: Wadsworth.

Lehmkuhl, G., & Poeck, K. (1981). A disturbance in the conceptual organisation of action in patients with ideational apraxia. *Cortex, 17*, 153–158.

Lezak, M. (1983). *Neuropsychological assessment* (2nd edn.) New York: Oxford University Press.

Liepmann, H. (1920). Apraxie. *Ergebnisse der Gesamten Medizin, 1*, 516–543.

Lucchelli, F., Lopez, O.L., Faglioni, P., & Boller, F. (1993). Ideomotor and ideational apraxia in Alzheimer's disease. *International Journal of Geriatric Psychiatry, 8*, 413–417.

Luria, A.R. (1966). *Higher cortical functions in man*. New York: Basic Books.

Morlaas, J. (1928). *Contribution a l'etude de l'apraxie*. Paris: Legrand.

Norman, D.A., & Shallice. T. (1986). Attention to action: Willed and automatic control of behaviour. In R.J. Davidson, G.E. Schwartz, & D. Shapiro (Eds.), *Consciousness and self-regulation*, Vol. 4. New York: Plenum.

Ochipa, C., Rothi, L.J.G., & Heilman, K.M. (1989). Ideational apraxia: A deficit in tool selection and use. *Annals of Neurology, 25*, 190–193.

Pocock, S.J. (1983). *Clinical trials: A practical approach*. New York: John Wiley.

Poeck. K. (1983). Ideational apraxia. *Journal of Neurology, 230*, 1–5.

Rapcsak, S.Z., Croswell, S.C., & Rubens, A.B. (1989). Apraxia in Alzheimer's disease. *Neurology, 39*, 664–668.

Reason, J.T. (1990). *Human error*. London: Cambridge University Press.

Robertson, I.H. (1993). The relationship between lateralised and non-lateralised attentional deficits in unilateral neglect. In I.H. Robertson & J.C Marshall (Eds.), *Unilateral neglect: Clinical and experimental studies* (pp. 257–278). Hove, UK: Lawrence Erlbaum Associates Ltd.

Robertson. I.H., & Frasca, R. (1992). Attentional load and visual neglect. *International Journal of Neuroscience, 62*, 45–56.

Schwartz, M.F., & Buxbaum, L.J. (1997). Naturalistic action. In L.J.G. Rothi & K.M. Heilman (Eds.), *Apraxia: The neuropsychology of action* (pp. 269–290). Hove, UK: Psychology Press.

Schwartz, M.F., Buxbaum, L.J., Montgomery, M.W., Fitzpatrick-DeSalme, E., Hart, T., Ferraro, M., Lee, S.L., & Coslett, H.B. (1998). *Naturalistic action production following right hemisphere stroke*. Manuscript submitted for publication.

Schwartz, M.F., Montgomery, M.W., Buxbaum, L.J., Lee, S.S., Carew, T.G., Coslett, R.H., Ferraro, M., Fitzpatrick-DeSalme, E., Hart, T., & Mayer, N. (1998). Naturalistic action impairment in closed head injury. *Neuropsychology, 12*, 13–27.

Schwartz, M.F., Reed, E.S., Montgomery, M.W., Palmer, C., & Mayer, N.H. (1991). The quantitative description of action disorganisation after brain damage: A case study. *Cognitive Neuropsychology, 8*, 381–414.

Shallice, T. (1982). Specific impairments of planning. *Philosophical Transactions of the Royal Society of London, B298*, 199–209.

Shallice, T., Burgess, P.W., Schon, F., & Baxter, D.M. (1989). The origins of utilization behavior. *Brain, 112*, 1587–1598.

Warrington, E.K., & James, M. (1991). *The visual object and space perception battery*. Titchfield, UK: Thames Valley Test Company.

Wilson, B., Cockburn, J., & Halligan, P. (1987). *Behavioural Inattention Test*. Titchfield, UK: Thames Valley Test Company.

minute for each category (scores of below 15 in each category are considered to be impaired; see Lezak, 1983).

Although ES presented with a range of cognitive deficits, the most severe were with visuospatial perception and with motor performance. Recall of recent events was good and she had no clinical deficits in semantic memory. Over the test period there was no evidence of increasing dementia. The deficits in spatial perception and the symptoms of simultanagnosia did not impinge on the tests of manual reaching and grasping that we performed with ES. For all the critical tests stimulus presentation times were unlimited and, as we discuss, ES showed clear evidence of having perceived the stimuli correctly. In addition, her good performance in some conditions showed that she was able to localise stimuli appropriately. During the course of testing we did not observe any florid manifestations of anarchic hand activity; however, one hand would frequently attempt to interfere with a required task. For example, in the assessment of dyspraxia, ES was asked to gesture the use of visually presented objects (either by a pantomime, or by actually using the object). The performance of each hand was assessed separately. It was frequently the case that the left hand would attempt the task when the right hand was the focus of testing and vice versa. The interference was of such a magnitude that ES attempted to sit on one hand while the other was tested for gesturing ability. The interfering effect of one hand on the performance of the other was tested more formally in a series of experimental investigations.

EXPERIMENTAL INVESTIGATIONS

In this section we describe the experimental investigations into the factors that elicit manual interference in ES. We define manual interference as occurring when one limb takes control of a required response in a way that is contrary to the verbal instructions given in the task. In Experiment 1, ES had to point to a light, and the hand for the response was determined by the location of the light; a left hand response was to be made to a light appearing on the left of ES's body, a right hand response was to be made to a light on the right of her body. We examined performance when the light cueing the hand for response was blocked over trials, and when it occurred randomly (so that a response from either hand could occur at random). In the subsequent experiments, responses were again determined by the location of the stimulus, but we varied the nature of the stimulus involved and the required response. The stimulus was a cup or a cup-like object, which could be oriented in various directions (e.g. handle to the left or the right). The response was to point to or pick up the object. We show that manual interference is determined both by the nature of the stimulus (a familiar cup rather than a cup-like object, in a particular orientation) and the required action (picking up rather than pointing).

Experiment 1: Pointing to Lights

This experiment consisted of five separate conditions. The methodology for each condition was similar. The first three conditions were blocked (e.g. Condition 1: bilateral lights; Condition 2: unilateral left-side lights; Condition 3: unilateral right-side lights) and in the final two conditions lights were presented randomly to the left or right or both sides of space (Condition 4: lights occurring on left or right sides of space; Condition 5: lights occurring on left or right or both sides of space). The blocked conditions were run before the random conditions. ES was asked to point to the light (or lights) that had brightened, the hand for the response being determined by the location of the light; a left hand response was to be made to a light appearing on the left of ES's body, whereas a right hand response was to be made to a light on the right of her body.

Method

On each trial, ES was required to place her index fingers on two starting positions placed in front of her (one on either side of midline). She faced two light-emitting diodes (LEDs) placed 40cms away from her. Each LED was located 20cms from the midline. When a LED (or LEDs) lit, she was asked to place the index finger onto the LED (left index finger for a left-side LED, right index finger for a right-side LED). ES was asked to perform accurately and told that there were no time limits on responding. She was given the specific instruction for each condition prior to each trial block, and she was reminded of the instruction at intervals throughout. No feedback was given during a trial block. Performance was recorded using a video camera.

Results

The results are given in Table 1. In Condition 1 (bilateral blocked presentation) ES never initiated movements to both left and right LEDs simultaneously (although she had been asked to do so). She would first touch the left-sided LED with her left hand before touching the right-side LED with her right hand. On three trials she failed to use the right hand at all, instead touching both left-side and right-side LEDs with her left hand (going first to the left LED). On one trial she failed to use her left hand, touching both left- and right-side LEDs with her right hand (going first to the right LED).

In Condition 3, ES made one error (using her left hand to point to the right-side LED). The error was made on the first trial of the block after she had just completed Condition 2.

Table 1. Results of Experiment 1

Condition	Total No. of Trials per Condition	Per Cent Correct
Condition 1: Bilateral blocked presentations	32	87.5
Condition 2: Unilateral blocked left presentations	24	100.0
Condition 3: Unilateral blocked right presentations	24	95.8
Condition 4: Random unilateral left or right presentations	64	96.9 left[a] 56.3 right[b]
Condition 5: Random bilateral and unilateral left and right presentations	71	95.8 left[a] 58.3 right[b] 60.9 bilateral[c]

[a]Left-side LEDS.
[b]Right-side LEDs.
[c]Bilateral LEDs.

In Condition 4 (random unilateral left or right presentations) ES scored particularly poorly when responding to the right-side light (18/32 correct). On 14 occasions she reached for the right light with her left hand. In this condition, left hand responses were significantly better than right hand responses ($\chi^2(1) = 14.7, P < .0001$).

In Condition 5 (random bilateral and unilateral left and right presentations) ES again performed poorly when responding to right-side LEDs (14/24) correct. On 10 occasions she reached for the right-side LED with the left hand. When lights appeared randomly either on the left- or on the right-side, ES performed significantly better on left- than on right-side responses [$\chi^2(1) = 9.6, P < .002$]. With bilateral presentations, ES scored 14/23 correct (on 9 occasions she touched both left and right LEDs with the left index finger, going first to the left-side LED). Performance in the bilateral trials condition differed significantly from that in the unilateral left condition [$\chi^2(1) = 8.6, P < .003$], but not from that in the unilateral right condition [$\chi^2(1) = 0.03$, n.s.].

Error Analysis

A total of 40 errors were made in Experiment 1, representing 16% of all trials. Of these errors, 93% (37) were caused by the left hand going to the incorrect location; and 7% (3) were due to the right hand going to the incorrect location.

Discussion

ES performed relatively well when the light positions and the required motor responses were blocked; however, when the task conditions changed and when both the positions of the LED and the hand for response had to change at random across trials she began to make errors (but only to the right-side and not to the left-side LED). The data demonstrate that ES manifested manual interference behaviour with her left hand, but that this behaviour was elicited under certain circumstances: Either when there was locational uncertainty, or when there was a response uncertainty, or both (with stimuli appearing randomly on left or right). For instance, it may be that ES was able to inhibit her involuntary left hand responses when only the right hand had to be used across a block of trials. Alternatively, when the right target was predictable, ES may have preprogrammed a right hand movement to it, giving dominance to this response; when the target's location was unpredictable, the right hand response may not have been preprogrammed and instead the left hand response was directly invoked by the stimulus. The present results are not simply due to the order of presentation of the conditions in the test session and have been replicated with the orders reversed. Experiment 2 sought to tease apart whether locational uncertainty or response uncertainty was crucial for ES's manual interference. First we assessed her performance when the stimulus location was predictable but the hand of response varied. Second, we assessed performance when the hand of response was predictable but the location of the stimulus varied.

Experiment 2a: Pointing to a Constant Location

This experiment consisted of three separate conditions. In response to a centrally-located written cue (either "Left" or "Right"), ES was required to point with the appropriate hand to the position of the cue. The first two conditions were blocked (e.g. "Left" cue for Condition 1 and "Right" cue for Condition 2). In the third condition the written cue varied randomly from "Left" to "Right" (there being equal numbers of left and right trials in total). The experiment was then repeated but the ordering of the first and second conditions was reversed. The hand for the response was determined by the nature of the cue; a left hand response was to be made to the word "Left", a right hand response was to be made to the word "Right".

Method

At the start of each trial, ES was required to place each index finger on two starting positions placed in front of her (one on either side of midline). She faced a VDU placed 40cms away from her and sunk into the table top (so that it faced upwards). SuperLab software was used to set up the experiment. Each trial was initiated by a keypress, which triggered the removal of a white, central fixation cross on a blank screen with a blank for 1000msec followed by a centrally positioned target word (either "Left" or "Right" written in Times font, size 36, in white). The word was presented for 5000msec, its offset triggered another blank for 3000msec, followed by the fixation cross once more. ES was asked to reach forward with the appropriate hand (left in response to the word "Left", right in response to the word "Right") as soon as the word appeared on the VDU. Once ES had touched the VDU screen, the experimenter initiated the next trial. There were 20 trials in Condition 1 and 2, and 40 in Condition 3 (the words "Left" and "Right" appeared randomly, but with equal numbers of trials for each). On the first run through of the experiment, Condition 1 preceded Condition 2, which preceded Condition 3. On the second run through, the ordering of Conditions 1 and 2 was reversed. On the second run through there was a total of 22 trials in Condition 3.

Results

Performance for the different presentations of the same condition did not differ and the data were amalgamated. The results are presented in Table 2. There was no significant difference in performance in Condition 1 and 2 [$\chi^2(1) = 0.39, P < .05$]; nor was there any difference in performance between Condition 1 and the responses to the "Left" cue in Condition 3 [$\chi^2(1) = 1.26, P < .05$]. Performance in response to the "Left" cue and the "Right" cue in Condition 3

Table 2. Results of Experiment 2a

Condition	Total No. of Trials per Condition	Per Cent Correct
1. Cue word: "Left"	40	87.5
2. Cue word: "Right"	40	82.5
3. Random presentation of cue words: "Left" and "Right"	66	77.4 left[a] 35.5 right[b]

[a] Left hand to "Left" cue.
[b] Right hand to "Right" cue.

differed significantly [$\chi^2(1) = 11.09$, $P < .0007$], as did performance in Condition 2 and responses to the "Right" cue in Condition 3 [$\chi^2(1) = 16.38$, $P < .0001$]. Performance with the "Right" cue in Condition 3 (random presentations) was worse than in any other condition.

Error Analysis

A total of 39 errors were made in Experiment 2a, representing 14.7% of all trials. In Conditions 1 and 2, errors always consisted of a response by the incorrect hand. In Condition 3, 19 errors (70.4%) were also of this form; however, 8 errors (29.6%) consisted of no movement being made at all. ES said that she felt unable to move on these occasions. Of the total errors, 8 (20.5%) were caused by incorrect right hand responding, and 23 (59.0%) by incorrect left hand responding.

Experiment 2b: Pointing with a Constant Hand

Experiment 2b consisted of three separate conditions. The first two conditions were blocked. In Condition 1, ES was asked to point to LEDs, which lit randomly in left and right locations, with the left hand. Condition 2 was the same as Condition 1 but ES was asked to respond with the right hand. In Condition 3 the LEDs lit randomly in left and right locations, as before, but now ES was asked to respond to a left-side LED with the left hand and a right-side LED with the right hand.

Method

As in Experiment 1, ES was required on each trial to place her index finger on two starting positions placed in front of her (one on either side of midline). She faced two LEDs placed 40cms away in depth. Each LED was located 20cms from the midline. When an LED (or LEDs) lit, she was asked to place an index finger onto the LED. In two blocked conditions, ES was to respond consistently with the same hand (the left hand in Condition 1, the right hand in Condition 2). In Condition 3, a mixed condition, she was asked to respond to a left-side light with the left hand, and a right-side light with the right hand. Left and right LEDs lit randomly. ES was asked to perform accurately and told that there were no time limits on responding. She was given the specific instruction for each condition prior to each trial block, and she was reminded of the instruction at intervals throughout. No feedback was given during a trial block. Performance was recorded using a video camera. She performed half the blocked location trials first, followed by the random location trials, followed by the remaining blocked location trials.

Results

The results are displayed in Table 3. Performance in the two blocked conditions was not consistent; ES performed significantly better in Condition 1 with her left hand than in Condition 2 with her right hand [$\chi^2(1) = 11.6$, $P < .0007$]. In the mixed condition (Condition 3), she also performed better with her left than with her right hand [$\chi^2(1) = 8.5$, $P < .004$]. There was no significant difference between performance with the right hand in Condition 2 and Condition 3 [$\chi^2(1) = 0.5$, $P < .05$].

Table 3. Results of Experiment 2b

Condition	No. of Trials	Per Cent Correct
1. Left hand to both left and right LEDs	20	100
2. Right hand to both left and right LEDs	20	55
3. Left hand to left LED, right hand to right LED	20[a]	100 left[b] 65 right[c]

[a]Trials for each location.
[b]Left hand to left LED.
[c]Right hand to right LED.

Error Analysis

A total of 16 errors were made in Experiment 2b, representing 20% of all trials. No errors were made in Condition 1; in Conditions 2 and 3 the errors consisted of left hand responses when ES should have been responding with her right hand.

Discussion

In Experiments 2a and 2b we sought to determine whether response uncertainty (Experiment 2a) or locational uncertainty (Experiment 2b) was the significant factor in eliciting manual interference with ES. The results showed that both of these factors are significant; but interestingly, similar results did not obtain in the two experiments. In Experiment 2b, as in Experiment 1, the predominant error type was left manual interference (i.e. the left hand responding when a right hand response was required). In Experiment 2a, some right hand errors occurred (ES made a right hand instead of a left hand reaching response). These data suggest that under conditions of *response* uncertainty, both left and right hands may be preprogrammed to initiate the action and sometimes responses are invoked by the wrong signal. However, in conditions of *locational* uncertainty, left manual interference responses dominate. The data suggests that ES finds it difficult to inhibit hand responses when the processes that initiate the actions are activated under conditions of response uncertainty. Furthermore, processes in the right hemisphere may dominate coding of the spatial parameters of actions under conditions of locational uncertainty, with the result that left manual interference responses are most prevalent under those conditions. In the subsequent experiments we explore more closely how the nature of the required response affects ES's performance. In Experiments 1 and 2 the conditions requiring a response were somewhat artificial (pointing to the location of a target light); in Experiment 3 we explored whether a familiar action (that of grasping and picking up a cup) would also elicit manual interference.

Experiment 3: Picking Up Cups

In Experiment 3 we changed the motor task from pointing to grasping and picking up. Pointing is not a response associated with a particular stimulus; in contrast, actions such as grasping and picking up are, since there are some stimuli to which this response is specifically associated (e.g. a cup). In addition, pointing and grasping may be mediated by different brain areas (see Jeannerod, 1997). It is possible that manual interference responses reflect overlearned (inhibited) asso-

Experiment 4: Pointing to a Cup Handle

Method

As in the previous experiments, ES was asked to place both index fingers on starting positions in front of her. Either a single cup was placed on the left or the right, or two cups were placed on left and right locations (the locations were identical to those in Experiment 3). There were eight conditions, four requiring unimanual responses and four requiring bimanual responses ES was asked to *point* to left-side cups with the left hand and *point* to right-side cups with the right hand regardless of whether the handle was positioned on the left or the right side of the cup. ES was asked to perform accurately and told that there were no time limits on responding. No feedback was given, and the instructions were repeated at frequent intervals. Performance was recorded using a video camera. There were equal numbers of trials (20) in each condition, and conditions were randomised over trials.

Results

Unimanual conditions. Accuracy data for the unimanual conditions are presented in Table 4. The cross indicates the position of the cup relative to midline. As in Experiment 3, ES performed well in the two unimanual conditions where the position of the cup handle was congruent with the side of the responding hand (Conditions 1 and 4). There was no significant difference in the scores in these two conditions (Fisher Exact Probability, $P > 1.0$). Unlike Experiment 3, ES performed poorly in only one of the two conditions where the position of the handle was not congruent with the side of the responding hand, thus performance was poor in Condition 3 but was relatively good in Condition 2 [this difference was significant, $\chi^2(1) = 25.6$, $P < .0001$]. Significantly fewer errors were made in Condition 2 in Experiment 4 relative to the same condition in Experiment 3 [$\chi^2(1) = 31.9$, $P < .0001$]. There was no significant difference in the patterns of performance shown in Condition 3 in Experiment 3 and 4 (Fisher Exact Probability, $P > 1.0$). ES made a total of 21 errors in the unimanual conditions. Of these, 90.5% (17) consisted of the left hand pointing to the right-side cup (the majority of these errors, were made in Condition 3, and 1 such error was made in Condition 4). Only 9.5% (2) errors consisted of the right hand pointing to the left-side cup (these errors were made in Condition 2 when the handle of the cup was located on the right of the cup). Error data are shown in Table 7.

Bimanual conditions. The accuracy results are presented in Table 6. As in Experiment 3, performance was good in Condition 7, where the positions of the cup handles were congruent with the side of the responding hands. Unlike Experiment 3, performance was also good in Condition 6. There was no significant difference in performance between Conditions 6 and 7 in Experiment 3 (Fisher Exact Probability, $P > 1.0$), but there was a significant difference between performance in Condition 6 in Experiments 3 and 4 [$\chi^2(1) = 28.4$, $P < .0001$]. ES did not respond to the left-side cup with her right hand in Experiment 4 as she had done in Experiment 3. In general, performance was better

Table 7. Interference Responses to Handles as a Function of the Conditions with Unimanual and Bimanual Presentations: Experiment 4 (Point to Cup Handle)

a: Unimanual Trials: Total Errors 21/80

Hand	Condition 1	2	3	4
Left hand			18	1
Right hand		2		

b: Bimanual Trials: Total Errors 24/80

Hand	Condition 5	6	7	8
Left hand to both cups	10			10
Right hand to both cups			1	1
Mixed (left hand to right cup, right hand to left cup)				2

in the bimanual conditions in Experiment 4 relative to Experiment 3 [i.e. fewer errors were made in Conditions 5 and 8, $\chi^2(1) = 5.0, P < .03$, and $\chi^2(1) = 9.0, P < .003$, respectively]. Error types are presented in Table 7.

Error Analysis

Overall, a total of 45 errors were made in Experiment 4 (representing 28% of all trials). Significantly fewer errors were made in Experiment 4 relative to Experiment 3 [$\chi^2(1) = 28.1, P < .0001$]. The *nature* of the errors also differed across Experiment 3 and 4. The majority of errors resulted from left-hand pointing (either to a single right-side cup or to both left and right-side cups) (87% or all errors, $N = 39$). The converse error (right hand pointing) occurred relatively infrequently (9% of all errors, $N = 4$) as compared to 39% of total errors in Experiment 3. Two errors were of the "crossed" type (i.e. left hand to right location, right hand to left location).

Discussion

The nature of the required response differed in Experiments 3 and 4 (point rather than pick up). This had the effect of reducing the overall errors in Experiment 4; interestingly, right hand activity was also rarely apparent (as it had been in Experiment 3). For left-side stimuli, performance resembled that in Experiment 1, in that ES rarely made right hand responses to the left side. This suggests that these responses are modulated by activation of the required response (grasping, as in Experiment 3, relative to pointing, as here). For right-side stimuli, however, a different pattern of performance emerged. Now performance was affected by the orientation of the cup (or its handle). Left manual interference responses were more likely when the cup (or the handle) was in a familiar orientation. For the left hand, then, the response associated to the stimulus (the grasp response) did not need to be activated for manual interference to occur. This confirms our findings from Experiment 1, where manual interference responses occurred when ES made left hand pointing actions to right-side stimuli. However, the data also show that the mere presence of a right-side stimulus was not sufficient, and that left hand responses were modulated by the orientation of the stimulus. Left hand responses did not occur here with right-oriented cups, for example:

Thus the left hand, like the right hand, was affected by the stimulus, but the left hand was less affected by the response (errors occurring when ES had to grasp *and* when she had to point). Note that, since stimulus orientation affected left hand performance, it is unlikely that ES's left manual interference simply reflected misjudgement of the spatial location of right-side stimuli.

Experiment 5: Picking Up Nonobjects

Method

As in the previous experiments, ES was asked to place both index fingers on starting positions in front of her. Either a single nonobject was placed on the left or the right, or two nonobjects were placed on left and right locations (the locations were identical to those in Experiments 3 and 4). The nonobjects were constructed to be cup-like in that there was a side-positioned handle. The blocks were 6cm × 6cm × 10cm with a smaller block 2cm × 2cm × 10cm, positioned centrally on one of the long sides. As in Experiments 3 and 4, there were eight conditions, four requiring unimanual responses and four requiring bimanual responses. ES was asked to pick up left-side nonobjects with the left hand and right-side nonobjects with the right hand regardless of whether the "handle" was positioned on the left or the right side of the nonobject. ES was asked to perform accurately and told that there were no time limits on responding. No feedback was given, and the instructions were repeated at frequent intervals. Performance was recorded using a video camera. There were equal numbers of trials (20) in each condition (except for Condition 8 where there were only 18 trials), and conditions were randomised over trials.

Results

Unimanual conditions. Accuracy data for the unimanual conditions are presented in Table 5. The cross indicates the position of the nonobject relative to midline. As in Experiment 3, ES performed well in the two unimanual conditions where the position of the cup handle was congruent with the side of the responding hand (Conditions 1 and 4). ES performed poorly in the two conditions where the position of the handle was not congruent with the side of the responding hand (Conditions 2 and 3). However, Experiments 3 and 5 differed in the nature of the errors performed. For instance, performance in Condition 2 was poor, and did not differ from that in the same condition in Experiment 3 (Fisher Exact Probability, $P = .99$, n.s.). However, while ES used her right (incorrect) hand to pick up the cup in Experiment 3, this error type was performed only twice in Experiment 5. The majority of errors (18) consisted of ES grasping the blocks from above, rather than from the side using the "handle" ("grasp errors"). Since she had been instructed (and was frequently reminded) that she should perform the pick-up by the handle, picking up from the top was classified as an error even though the correct (left) hand was used. Performance was poor in Condition 3 as it had been for the same condition in Experiment 3 (there was no significant difference in numbers of correct trials in this condition in the two experiments, Fisher Exact Probability, $P = .51$, n.s.). However, whereas in Experiment 3 ES typically picked up the cup using her left hand, the same error occurred on only 50% of the error trials here. The other errors consisted of picking the blocks up from above (with the right hand on 9 occasions and with the left hand on 1 occasion). Error responses to handles are presented in Table 8. "Grasp errors" are given in Table 9.

Bimanual conditions. The results are shown in Table 6. The cross indicates the position of the nonobject relative to midline. In general, performance was better in the bimanual conditions in Experiment 5 than it had been in Experiment 3. In Condition 5, ES made a number of errors but performance was better than that in the same condition in Experiment 3 $[\chi^2(1) = 8.1, P < .004]$. ES used her left hand to pick up both left and right blocks on eight occasions (the pick-up was always correct, i.e., by the "handle"). On one occasion ES picked up the right-side block with the correct (right) hand but from above rather than by the "handle". In Condition 6, performance was good and was significantly different from performance in the same condition in Experiment 3 $[\chi^2(1) = 21.3, P < .0001]$. On two occasions ES picked both sets of blocks up (consecutively) with the left hand; the correct grip was used for the left-side blocks, but she picked up the right-side blocks from above with fingers griping the small block. These were classed as "grasp errors". Performance in Condition 7 was good and did not differ significantly from the same condition in Experiment 3 $[\chi^2(1) = 3.7, P >$

Table 8. Interference Responses to Handles as a Function of the Conditions with Unimanual and Bimanual Presentations: Experiment 5 (Pick Up Cup-like Non Objects)

a: Unimanual Trials: Total Handle Errors 12/80

	Condition			
Hand	1	2	3	4
Left hand			10	
Right hand		2		

b: Bimanual Trials: Total Errors 15/78

	Condition			
Hand	5	6	7	8
Left hand to both cups	8			5
Right hand to both cups				2
Mixed (left hand to right cup, right hand to left cup)				

Table 9. Grasp Responses to Cup-like Nonobjects as a Function of the Conditions with Unimanual and Bimanual Presentations: Experiment 5 (Pick Up Cup-like Non Objects)

a: Unimanual Trials: Total Grasp Errors 30/80 (28 with the Correct Hand, 2 with the Incorrect Hand)

Hand	Condition			
	1 □*	2 □*	3 *□	4 *□
Left hand 🖐	1	18	1	1
Right hand 🖐			9	

b: Bimanual Trials: Total Errors 4/78 (All with the Incorrect Hand)

Hand	Condition			
	5 □*□	6 □*□	7 □*□	8 □*□
Left hand to both objects 🖐🖐		2		
Right hand to both objects 🖐🖐	1			1
Mixed (left hand to right object, right hand to left object) 🖐/🖐				

.054]. Performance in Condition 8 was better than in the same condition in Experiment 3 [$\chi^2(1) = 14.9, P < .0001$]. Most errors were of the left hand picking up both left and right blocks (5/9; in all instances the correct grip was used); on two occasions the converse error was performed with the right hand picking up both sets of blocks. On one occasion the left-side block was neglected, and on occasion ES picked up the right-side blocks with the correct hand but grasped the big rather than the little block from the side. Interference responses to handles are given in Table 8. "Grasp errors" are presented in Table 9.

Error Analysis

Overall, a total of 62 errors were made in Experiment 5 (representing 38.8% of all trials); fewer errors were made in Experiment 5 than in Experiment 3 [$\chi^2(1) = 10.1, P < .001$]. The nature of the errors also differed between Experiment 5 and 3 (see Table 10) in that "grasp errors" emerged in Experiment 5. The task required that the blocks were picked up by the "handle". However, ES frequently failed to perform the task in this way, and tended to pick up the blocks from the top rather than by the "handle"; 54.8% (34) of all errors were of this "grasp" form (of these, 67.6% [23] and 32.4% [11] were made by the left and right hands respectively). Interestingly, grasp errors were predominantly made with the correct hand for the side, unlike the manual interference responses to the handle. Only 6/34 grasp errors were with the wrong hand, and 28/34 were with the correct hand (see Table 10). All incorrect responses to the handle were made with the wrong hand.

If responses are scored simply in terms of the hand used (the nature of the grasp being disregarded), the differences between Experiment 3 and 5 become apparent in that there are a greater proportion of left hand to right hand errors in Experiment 5 (see Table 10).

Table 10. Differences in Errors between Experiments 3 and 5

Grasp	Errors	Expt. 3	Expt. 5
Correct grasp (by handle); incorrect target	% with the left hand	47.0	37.0
	% with the right hand	39.0	6.5
Incorrect grasp (by top of object); incorrect target	% with the left hand	0.0	37.0
	% with the right hand	0.0	17.8

Discussion

The overall pattern of correct performance was similar to Experiment 3. Performance was very good when the position of the handle was congruent with the hand of response:

☐ * and * ☐

and it was poor when the hand and the effector were incongruent

☐ * and * ☐

However, unlike Experiment 3, manual interference responses were not predominant in the incongruent conditions. With a left-side, incongruent stimulus, the majority of errors involved the left hand grasping the blocks from above rather than the side. With a right-side, incongruent stimulus, ES made an equal number of top-grasp responses with her right hand and manual interference grasps with her left hand. Hence manual interference behaviour was modulated by the familiarity of the stimulus. For both hands such behaviour was more likely to be invoked by a familiar cup (in Experiment 3) than by an unfamiliar cup-like nonobject (Experiment 5; although, as in Experiment 4, the manipulation affected right hand responses more than left hand responses). In Experiment 6 we tested the effects of stimulus familiarity further by having ES make grasping responses to inverted cups.

Experiment 6: Picking Up Upside-down Cups

Experiment 6 was identical to Experiment 3 except that ES was asked to pick up cups that had been inverted onto the table top.

Method

As in Experiment 1, ES was asked to place both index fingers on starting positions in front of her. Either a single cup was placed on the left or the right, or two cups were placed on left and right locations (the distance from ES to the location markers was 30cm, and each location was positioned 20cm from the midline). There were eight conditions, four requiring unimanual responses and four requiring bimanual responses. There were equal numbers of trials (20) in each condition, and conditions were randomised over trials. ES was asked to pick up left-side cups with the left hand and right-side cups with the right hand *regardless* of whether the handle was positioned on the left or the right side of the cup. She was asked to pick up the cups by the handle. As in all previous experiments, ES was asked to perform accurately and told that there were no time limits on responding. No feedback was given, and the instructions were repeated at frequent intervals. Performance was recorded using a video camera.

Results

The accuracy data are shown in Table 5. The cross indicates the position of the nonobject relative to midline.

Unimanual conditions. Of the unimanual conditions only one was performed well—Condition 1. Performance here did not differ from performance in the same condition in Experiments 3 and 5 [$\chi^2(2) = 0.9$, $P < .05$]. Performance in Conditions 2 and 3 was poor and did not differ significantly from performance in the same conditions in Experiments 3 and 5 [$\chi^2(2) = 5.6$, $P < .05$, and $\chi^2(2) = 2.7$, $P < .05$ for comparisons of Conditions 2 and 3 respectively across Experiments 3, 5, and 6].

The numbers of manual interference responses to handles are given in Table 11, and the numbers of grasp responses are shown in Table 12. In Condition 2 many errors were the same type as those shown in the same condition in Experiment 3; that is, on 8/15 occasions ES picked up the cup with her right hand (i.e. they were right manual interference errors). On seven occasions ES failed to pick up the cup by the handle and instead made a grasp error, lifting the cup by the base. On six of these last occasions she picked up the cup with the left (correct) hand and on one occasion with the right (incorrect) hand. These grasp errors replicate those shown in Experiment 5. In Condition 3, the error patterns were consistent—ES made left manual interference responses to the handle with her left hand on 19 occasions (similar to Experiment 3). On one occasion she made a grasp error to the base using her left hand.

Table 11. Interference Responses to Handles as a Function of the Conditions with Unimanual and Bimanual Presentations: Experiment 6 (Pick Up Upside-down Cups)

a: Unimanual Trials: Total Handle Errors 17/80

Hand	Condition 1	Condition 2	Condition 3	Condition 4
Left hand			19	
Right hand		8		

b: Bimanual Trials: Total Handle Errors 45/80

Hand	Condition 5	Condition 6	Condition 7	Condition 8
Left hand to both objects	17		1	6
Right hand to both objects		9		3
Mixed (left hand to right cup, right hand to left cup)		1		8

Unlike Experiments 3 and 5, Condition 4 was performed poorly [performance was significantly poorer than performance in the same conditions in Experiments 3 and 5; $\chi^2(1) = 8.5$, $P < .004$, and $\chi^2(1) = 5.6$, $P < .01$, for Experiments 3 and 5 respectively). All seven errors consisted of ES making a grasp error by picking the cup up from the base (six times with the left hand, once with the right hand).

Table 12. Grasp Responses to Cups as a Function of the Conditions with Unimanual and Bimanual Presentations: Experiment 6 (Pick Up Upside-down Cups)
a: Unimanual Trials: Total "Grasp" Errors 15/80 (7 with the Correct Hand, 8 with the Incorrect Hand)

	Condition			
Hand	1 ⌒ *	2 ⌒ *	3 * ⌒	4 * ⌒
Left hand		6	1	6
Right hand			1	1

b: Bimanual Trials: Total "Grasp" Errors 12/80 (3 with the Correct Hands)

	Condition			
Hand	5 ⌒*⌒	6 ⌒*⌒	7 ⌒*⌒	8 ⌒*⌒
Left hand to both objects	1	3	1	3
Right hand to both objects		1		
Left hand to left cup, right hand to right cup	1	2		

Bimanual conditions. The accuracy results are shown in Table 7, and the errors in Tables 11 and 12. The cross indicates the position of the nonobject relative to midline. Performance in the bimanual conditions was generally poor except for Condition 7; here performance did not differ significantly from the performance in the same condition in Experiments 3 and 5 $[\chi^2(2) = 2.6, P > .05]$. Two errors were made; on one occasion ES reached for both cups with the left hand, and on the other she picked up both cups from the base using the left hand. Performance in Conditions 5, 6, and 8 did not differ significantly from performance in the same conditions in Experiment 3 [Fisher's Exact Probability, $P = .22$, $\chi^2(1) = 9.2$, $P < .002$, and Fisher's Exact Probability $P > .99$ for comparisons of performance in conditions 5, 6, and 8 respectively in Experiments 3 and 6]. However, performance in these conditions was significantly worse than performance in the same conditions in Experiment 5 $[\chi^2(1) = 11.9$, $P < .0006$; 19.8, $P < .0001$ and 13.1, $P < .0003$ for comparisons of performance in conditions 5, 6, and 8 respectively in Experiments 3 and 5). Predominantly errors resembled those made in Experiment 3 (i.e. the orientation of the handle dictated the hand of response), although a few errors resembled those made in Experiment 5 (i.e. the cup was picked up by the base). In Condition 5, 19 errors were made; of these 17 were made by the left hand picking up first the left and then the right cup (the latter being a manual interference error). On two occasions she picked up the cups from the base, once with the appropriate hand and once with the left hand for both cups. In Condition 6, most of the errors were of the manual interference type and consisted of the right hand picking up both cups (9/16). On one occasion she performed a "cross-error", with the left hand picking up the right-side cup and the right hand picking up the left-side cup. On six occasions she picked up both cups from the bases (using the correct hands twice, the left hand only three times, and the right hand only once). In Condition 8 there were a number of different error types. On six occasions she used the left hand to pick up both cups, on three occasions she used the right hand to pick up both cups, on eight occasions she used a cross response (left hand to right cup and vice versa), and on three occasions she picked up both cups from their bases using the left hand.

Error Analysis

Summing across the unimanual and bimanual conditions, a total of 99 errors were made in Experiment 6 (representing 61.9% of all trials). However, there was a difference in error type between Experiments 3, 5, and 6. In Experiment 3 (with upright cups) the majority of errors consisted of ES responding to the orientation of the cup handle; these occurred with both left and right hands, although there were more left than right hand errors in a ratio of approximately 6:5. Similar errors were made in Experiment 6 (with inverted cups), although the proportions of the left to right hand errors differed (an approximate ratio of 3:1). However, in Experiment 5 (with cup-like nonobjects), ES mainly made left hand errors (the ratio of left:right being approximately 15:1). In Experiments 5 and 6 a substantial proportion of the errors also consisted of ES picking the cup up from the base rather than the handle; 54.8% and 27.2% of total errors in Experiments 5 and 6 respectively. Half of these were made with the correct hand and half with the incorrect hand.

Discussion

ES's performance in Experiment 6, as in the previous experiments, showed that responses were contingent on the familiarity of the object and/or the familiarity of the response. In general, performance fell midway between that in Experiments 3 and 5 in terms of both the overall level of accuracy and the number of manual interference responses that occurred. For instance, the target object(s), though familiar (cups), were presented in an unfamiliar orientation (upside-down). This had an effect of reducing the number of manual interference responses in Experiment 6 relative to those produced in Experiment 3 (although the type of response—grasping—was the same in Experiments 3 and 6). However, the familiarity of the object did have an effect on performance, since right hand performance was not as good in Experiment 6 as it had been in Experiment 5 when nonobjects had been used. Thus, the familiarity of the target object has a role in triggering response associations even when that object is presented in an unfamiliar orientation. Experiment 5 differed from Experiment 3 in that a number of errors were characterised as grasp responses; such errors never occurred in Experiment 3. A possible cause of grasp errors is that stimuli in Experiment 5 had a closed rather than an open top. In Experiment 6 the base of the cup was uppermost, and again grasp responses were seen; however, this error was not performed as frequently in Experiment 6 as it had been in Experiment 5, suggesting that learned stimulus-response associations, which are related to familiar objects, can have significant effects on performance even when the target stimulus is in the incorrect orientation.

Of particular interest was the finding that a substantial number of right hand manual interference errors were observed in Experiment 6. Similar errors had been observed in Experiment 3, leading to the suggestion that ES's right hand responses were evoked by a familiar stimulus, and in particular, were related to the orientation of the handle of the cup, rather than the location of the cup itself. The presence of right hand responses in Experiment 6 supports these conclusions; however, fewer right hand responses were produced here with inverted cups relative to Experiment 3 with upright cups; thus the familiarity of the orientation of the target influenced performance.

GENERAL DISCUSSION

We have reported an experimental study of manual interference responses in a patient with CBD. This has demonstrated that interference responses can be modulated by both the stimulus and the response context. The main findings were that:

1. Interference responses in pointing were predominantly made by the left hand and arose under conditions of spatial and response uncertainty (Experiments 1, 2, and 4).

2. When ES was required to pick objects up, interference responses were found with both hands and these were affected by the position of the relevant part of the object and the effector (e.g. the handle of the cup, Experiment 3).

3. Interference responses when picking up objects were also influenced by stimulus familiarity; fewer interference responses were generated when a cup was replaced by a structurally similar nonobject (Experiment 5) and when the cup was inverted (Experiment 6). The reduction in interference responses with less familiar objects was particularly noticeable with the right hand.

4. With cup-like nonobjects and inverted cups, some errors occurred because ES grasped the top-most part of the stimulus rather than the handle. Grasp errors were typically made with the correct hand (i.e. the hand on the same side as the object).

Factors Eliciting Right vs. Left Manual Interference Responses

ES demonstrates manual interference behaviour with both left and right hands; however, the factors eliciting the manual interference behaviour differ for the two hands. For instance, right manual interference was elicited when a familiar response was made to a familiar stimulus.

Stimulus Effects

When a familiar stimulus (a cup) had to be picked up, right hand interference responses were elicited by left-side cups providing their handle fell to the right. These were eliminated when a cup-like nonobject was used (as in Experiment 5). The orientation of the familiar stimulus was also a significant factor—fewer right manual interference responses were shown in Experiment 6 (with an inverted cup) than in Experiment 3 (with an upright cup).

Response Effects

In addition to the nature of the stimulus, the nature of the response was also important. Cups were used as stimuli in both Experiments 3 and 4 but the response differed from grasping the handle and picking up the cup in Experiment 3 to pointing to the handle in Experiment 4. Relatively few right manual interference responses occurred in Experiment 4 relative to Experiment 3. Right manual interference responses also occurred in Experiment 2a (where the task was to respond to a centrally located written cue); here it appeared that under conditions of response uncertainty, both left and right hands may be preprogrammed to initiate the required action. The manual interference shown by ES seems related to the degree of activation of response initiation processes in the right and left hemispheres. For the left hemisphere (initiating right hand movements), activation is determined by the preprogramming of responses (under conditions of response uncertainty), by the position of the relevant part of the object relative to the effector, and by familiar stimulus-response couplings.

Overall, ES was far more likely to demonstrate left manual interference rather than right, and left manual interference responses were triggered by stimuli under conditions of locational uncertainty as well as response uncertainty (Experiment 2). Left hand responses were also in general less strongly modulated than right hand responses by stimulus-response familiarity and stimulus-effector compatibility. For instance, although ES was more likely to pick up a right-side positioned cup with the left hand when its handle was on the left than when its handle was on the right (Condition 3 in Experiments 3, 5, and 6), the familiarity of the stimulus (cup vs. nonobject) or its orientation (upright vs. inverted) had little effect on performance. Furthermore, there was no difference in the number of left manual interference responses according to the nature of the response (e.g. pointing vs. grasping, Experiments 3 and 4, Condition 3). Nevertheless, the fact that the left hand performance was affected to some degree by the position of the handle of the cup does indicate that left hand responses were not simply determined by poor spatial judgements concerning stimuli on the right of ES's body. Rather it seems that the right hemisphere is strongly activated under conditions of locational and/or response uncertainty, leading to left hand responses being evoked incorrectly. Possibly due to impaired transmission across the corpus callosum, ES failed to inhibit these inappropriate responses when they were activated.

The tendency for the left hand to respond to right-side targets was so striking that we attempted to run a further experiment where ES was asked to respond to lights with the *contralateral* hand (thus, she was asked to respond to a left light with her right hand and a right light with her left hand). Interestingly, the experiment proved impossible to run. ES was completely unable to respond to the crossed location where the action was volitional, even though she made exactly these responses inappropriately in the uncrossed reaching conditions used in the present study. Hence, we conclude that her actions in our experiments were completely *involuntary*. The actions also appeared to operate outside conscious awarenesss, since ES was typically unaware when an incorrect response was made and thought that she had responded correctly. This was the case for all the current experiments. We return to this last point when we discuss the relations between ES's performance and neuropsychological categories of anarchic and alien hand syndromes.

Action Selection

The differential effects of the stimulus and response factors on left and right hand performance in ES are important for understanding how actions to visual stimulus are evoked and selected. Distinctions have been made previously between the dorsal and ventral streams of visual processing in the brain. The dorsal stream is largely thought to subserve spatial vision and action whereas the ventral stream is more implicated in object recognition (Ungerleider & Mishkin, 1982). Neuropsychological data have supported this distinction. For example, Milner, Goodale, and associates have reported an agnosic patient who, after a ventral lesion, was unable to perform simple perceptual matches (e.g. requiring same/different matching of object size) but who was nevertheless able to make the appropriate grasp aperture with the same stimuli, presumably due to the integrity of the dorsal route to action (Goodale, Milner, Jakobson, & Carey, 1991; Milner & Goodale, 1995; Milner et al., 1991). The dorsal stream appears to play a critical role in the visuomotor transformations necessary for skilled visuomotor acts, but to be relatively indifferent to the nature of the stimuli involved (e.g. whether they are familiar or unfamiliar objects (Milner & Goodale, 1995). However, action can also be affected by the familiarity of objects. Jeannerod, Decety, and Michel (1994) report data from a patient with bilateral parietal lesions who showed a contrast between reaching and grasping familiar and unfamiliar objects. With familiar objects, grasp apertures were scaled to the size of the stimuli; with unfamiliar objects, grasp apertures were scaled inaccurately to object size.

Other evidence suggests that actions evoked by familiar objects may be dependent either on the activation of semantic knowledge concerning an object's use, or directly from stored visual knowledge concerning the object (e.g. a direct route from a stored structural description of an object to an action). Direct links between stored visual knowledge and actions can be apparent in patients with optic aphasia. For instance, Riddoch and Humphreys (1987) reported one such patient who was impaired on tests assessing visual access to semantic knowledge, but who nevertheless performed very specific gestures when objects were visually presented (see also Hillis & Caramazza, 1995). Such data suggest that specific learned gestures can be evoked directly via access to stored visual knowledge, even when access to semantic knowledge is impaired. An opposite pattern of results arises from "visual apraxia". Here patients may very well be able to make actions semantically, for instance, from the object's name. However, such patients are impaired at making actions when objects are visually presented. Here there seems to be a "blocking" of action from directly accessed visual information (see Pilgrim & Humphreys, 1991; Riddoch, Humphreys, & Price, 1989). Blocking of the response may not only come from stored visual knowledge but also from several possible responses that may be "afforded" by the parts of objects (e.g. that are not modulated by stored visual knowledge).

The data we have obtained from ES supports the view that actions with the right hand are influenced by the familiarity of the stimulus and its associated response. The results also extend prior findings by demonstrating effects of the familiarity of the object's orientation and the orientation of the object with respect to the relevant effector (e.g. whether the handle of the cup was on the left or right). Visual object recognition itself seems relatively indifferent to the left–right orientation of the object (Biederman & Cooper, 1991), and the semantic information retrieved from objects will be the same irrespective of the way the objects face. Nevertheless, ES's performance was strongly affected by the left–right orientation of the cups. This suggests that right manual interference responses were not generated on the basis of semantic information accessed by the objects, but rather this activity was contingent on direct links between vision and action. The fact that right manual interference responses were sensitive to the particular object (being evoked by cups but not by nonobjects) further suggests that these actions were contingent on stored visual knowledge being activated, although the specificity of the stored knowledge is less clear. It may be that the presence of a handle on the right side of a cylindrical container, when the task is to grasp the object, may be sufficient to trigger a right hand response. There need not be access to stored knowledge specific to the particular cup. Whatever the case, the results show that the visual representations involved are sensitive to object orientation.

In addition to the effects of familiarity, ES's right manual interference responses were influenced also by (a) the spatial compatibility between the parts of objects used for action and the effector, and (b) the goal of the action (reaching or pointing). Right manual interference responses occurred primarily when the cup's handle was facing right, and they arose when the task was to pick up rather than point to the object. These effects, of object part–hand compatibility and goal-state, demonstrate that visual affordances affect action. The ecological psychologist J. J. Gibson (1979) argued that visually guided behaviour depends on direct links (or affordances) between perception and action. The affordance of an environmental stimulus may be defined in terms of the relation between the visual information present and the goal of the organism; for instance, if a surface is flat, reasonably substantial and of an appropriate height, it will "afford" sitting on to the tired observer whether it has been specifically designed for this (e.g. a stool), or "happens" in the environment (e.g. a tree stump). In the experiments reported here, the position of the handle on the cup will more directly afford lifting by one hand than the other (i.e. a right-side handle for the right hand), leading to stronger activation of the response system for that hand. This activation seems particularly strong for the right hand. In the absence of inhibitory signals to the response system, incorrect hand responses are generated. Incorrect responses were not generated, though, when the task goal changed from picking up the cup by the handle to pointing to the cup handle. Thus, the critical object part did not "afford" action when the task did not require picking the object up. We suggest that, for ES, the inhibition of inappropriately activated responses is disrupted by reduced callosal transmission and/or that the frontal motor systems which generate action in response to volitional cues are deficient in some way (either in the nature of the inputs to them, or the systems themselves are damaged).

The left manual interference responses in ES were most affected by locational and response uncertainty, and less affected by stimulus similarity. This dissociation between the right and left hand suggests that the computational processes involved in programming the spatial parameters of actions reside within the right hemisphere and are separated to some degree from processes sensitive to stimulus-response familiarity in action, within the left hemisphere.

Grasp Responses

Other evidence supportive of the role of affordances in directing action comes from the grasp errors found in Experiments 5 and 6. In these experiments, ES was presented with relatively unfamiliar stimuli: cup-like nonobjects and inverted cups. Along with manual interference responses to the handles of these stimuli, ES also made some grasp errors, where she picked the object up (incorrectly) by their topmost part rather than the handle. Interestingly, these responses were often made with the correct hand (on the same side as the object). Since the objects were less familiar in these experiments, responses might tend to be based on affordances rather than learned object–action associations. With cup-like nonobjects and upside-down cups, affordances may be as strong from the top of the object as from the handle. Note, however, that the top has no spatial component linked to the hand of response (unlike the handle). In this case, the strongest affordance should be to the hand nearest the object rather than the opposite hand. As a result, grasp responses are made using the correct hand.

Neurology

On the basis of studies of the movement disorders (apraxias) resulting from left hemisphere lesions, Liepmann (1920) argued that the left hemisphere in right-handed individuals is dominant in the control of some aspects of purposeful skilled movement. In particular, the left hemisphere controlled movement that had to be performed in response to a verbal command, or where imitation of an examiners movement was required. For such actions, the left hemisphere was proposed to control the right hand directly and the left hand via transcallosal links. However, the right hemisphere may house some abilities to generate learned movements to visually presented objects or to objects in naturalistic contexts. Support for this last hypothesis comes from a number of case reports of patients with left hemisphere or callosal lesions who have retained some of the practical functions of the left hand (see Rapcsak, Ochipa, Beeson, & Rubens, 1993). Rapcsak et al. report the practic functions of a man whose left hemisphere was almost completely destroyed as a result of a massive left-hemisphere stroke. Although the patient was severely impaired in gesture imitation and gesture to command, his ability to perform overlearned habitual actions with real objects and intransitive gestures (such as waving goodbye, or shaking a fist) was relatively unimpaired. A somewhat similar proposal has been put forward by Buxbaum, Schwartz, Coslett, and Carew (1995), who also argue that different mechanisms may underlie gesture and naturalistic action; however, they propose that naturalistic action requires the specialised abilities of each hemisphere, integrated across callosal structures. The left hemisphere may be dominant for learned hand gestures especially when performed out of context (with single objects); the right hemisphere may be dominant for control of the spatial parameters of movements. Buxbaum et al.'s proposal is supported by data from a patient with callosal disconnection syndrome following a closed head injury. Although the left hand was particularly impaired on standard gesture tests (suggesting disconnection from the left-hemisphere control systems) both hands performed abnormally in everyday action tasks. The right hand then frequently made spatial errors, whilst the left hand frequently misused objects. ES has a bilateral disturbance of manual performance. It is possible that this may be associated with callosal dysfunction (signal abnormalities are shown on MRI in the region of the posterior centrum-semiovale and the corpus callosum on the left). Impairment in this area would disrupt the axons mediating interhemispheric integration.

Our data show that it is not sufficient to attribute the manual interference behaviour in ES simply to a deficit in fine motor control with the nondominant hand; rather, ES has a problem in modulating activation from stimulus-driven responses to stimuli in response to task demands. This is consistent with a failure to inhibit activation in brain regions in the contralateral hemisphere responsive to stimulus-drive action, possibly as a result of impaired links across the corpus callosum. The left hemisphere may be critical for "abstract" aspects of action (i.e. the ability to gesture the use of an object in its absence, or to imitate an examiner pantomiming an action) (Liepmann, 1920; Rapcsak et al., 1993). ES was shown to be particularly poor at imitating gestures produced by an examiner (see Fig. 1); gesturing the use of an object to command, although not so impaired as gesture imitation, was still performed very poorly. These aspects of ES's performance suggest an impoverished ability for her left hemisphere to generate action. Nonetheless, an influence of left-hemisphere activation on performance was apparent because right hand actions, in particular, were affected both by the familiarity of the stimulus and its associated response. Inappropriate right hand responses were generated when learned responses had to be made to familiar stimuli ("pick up the cup"). This may be because the left hemisphere may be dominant for such responses, although the right hemisphere may be able to initiate them to some degree. In contrast, inappropriate left hand responses may be generated under conditions of spatial uncertainty because the right hemisphere dominates control of the spatial parameters of movement, and is the more strongly activated hemisphere under these conditions. ES appears to be unable to integrate activity across the hemispheres, and there seems to be little inhibition from one hemisphere of inappropriate responses activated in the other—manual interference behaviour results. Overall, more manual interference behaviour was observed with the left rather than the right hand, which probably reflects the greater degeneration of the left hemisphere. In both cases, however, the stimulus driven nature of the deficits suggests some disconnection of the system controlling voluntary action (e.g. SMA) from areas responsive to environmental stimuli (e.g. PMC)—(see the Introduction). This may be a consequence of subcortical damage in ES's case.

Anarchic Hand Syndrome and Utilisation Behaviour

There were manual interference effects and inappropriate actions during everyday life that ES was aware of. The question arises, then, whether the manual interference responses that we have studied experimentally form part of this syndrome. This situation is not clear. One of the critical defining features of anarchic hand syndrome, namely awareness of the incorrect action, was not apparent in our study. ES showed no awareness of making incorrect manual interference responses. This suggests that the interference responses we elicited may arise from a source separate to the source of her action errors in everyday life. On the other hand, the consequences of some of the action errors that befell ES in everyday life could be severe (as when her left hand hit her aunt!). In our experiments, though, there were no adverse consequences for making a manual interference response, and in some respects the goal of the task was fulfilled: ES pointed to or picked up the target object. Speculatively, we might suggest awareness of inappropriate actions in anarchic hand syndrome actually reflects the *consequences* of actions rather than observations of the inappropriate actions per se. In this last case the present manual interference responses may in fact be part of the anarchic hand syndrome in this patient. The "awareness" shown by patients labelled as having anarchic hand syndrome may apply only to consequential acts noted in the clinic. We do not know whether more subtle motor deficits of the type we have examined could be detected in all such patients.

Are the manual interference affects described here with ES the same as utilisation behaviour associated with frontal lobe damage (see Shallice, Burgess, Schon, & Baxter, 1989)? In their formal investigations of a patient with an acute behavioural disturbance[1] as a result of lesions in the distribution of branches of both anterior cerebral arteries, Shallice et al. describe three different forms of utilisation behaviour. Toying (a single action in which an object is manipulated but not in a purposeful way, or in the way it was originally tended to be used; e.g. picking up a pencil but not using it for anything); complex toying (two objects used together in a linked way but in an incomplete fashion or not for the purpose for which they were both designed); and coherent activity (a set of actions integrated in a typical fashion with respect to the objects involved such as picking up a pen and paper and writing). Utilisation behaviour occurred not only when the patient was in conversation with the examiner, but also when he was in the middle of performing both verbal and nonverbal tasks (Shallice et al., 1989). It seems to us that the manual interference effects we have described and utilisation behaviour are not the same thing. The laboratory circumstances for ES and for Shallice et al.'s patient were not comparable (for our experiments only the test items were present); however, ES was never observed to perform utilisation behaviours with other items in situations where other objects were available but inappropriate to the task in hand (such objects present on the table whilst we were performing the neuropsychological assessments). Indeed, in formal tests, ES can reject distractor objects and responds only to targets (even if the response is then inappropriate (Riddoch, Humphreys, & Edwards, in press). Futhermore, even in the present study, ES was able to maintain task goals at some level. For instance, she was required to grasp mugs in Experiment 3, but to point to the handle of the mug in Experiment 4. Here she responded appropriately according to the task set, unlike patients showing utilisation behaviour.

The manual interference effects we have reported here thus may be part of the anarchic hand syndrome in this patient but they dissociate from utilisation behaviour. Manual interference effects can occur under conditions in which patients maintain appropriate task goals, and they are influenced by (a) spatial uncertainty concerning the response (particularly of the left hand) and (b) stimulus familiarity and orientation, when the stimulus orientation is relevant to the task goals (particularly for the right hand). This last property demonstrates that visual affordances play a part in the selection of action to objects.

Manuscript received 16 December 1997
Manuscript accepted 1 April 1998

---

[1] On 17 September, 1987, his son reported that the patient was found early in the morning wearing someone else's shoes, not apparently talking or responding to simple commands but putting coins into his mouth and grabbing imagineary objects. He went round the house, moving furniture, opening cupboards and turning light switches on and off.

REFERENCES

Alexander, G.E., & Crutcher, M.D. (1990). Functional architecture of basal ganglia circuits: Neural substrates of parallel processing. *Trends in the Neurosciences, 13,* 266–271.

Alexander, G.E., Crutcher, M.D., & DeLong, M.R. (1990). Basal ganglia-thalamocortical circuits: Parallel substrates for motor, oculomotor, "prefrontal" and "limbic" functions. In J.M. Uylings, C.G. Van Eden, J.P.C. De Brun, M.A. Corner, & M.G.P. Feenstra (Eds.), *Progress in brain research.* Amsterdam: Elsevier Science Publishers.

Alexander, G.E., DeLong, M.R., & Strick, P.L. (1986). Parallel organisation of functionally segregated circuits linking basal ganglia and cortex. *Annual Review of Neuroscience, 9,* 357–381.

Biederman, I., & Cooper, E.E. (1991). Object recognition and laterality: Null effects. *Neuropsychologia, 29,* 685–694.

Brust, J.C.M. (1986). Lesions of the supplementary motor area. In H.O. Lüders (Eds.), *Advances in neurology: Supplementary motor area.* Philadelphia, PA: Lippincott-Raven Publishers.

Buxbaum, L.J., Schwartz, M.F., Coslett, H.B., & Carew, T.G. (1995). Naturalistic action and praxis in callosal apraxia. *Neurocase, 1,* 3–17.

Della Sala, S., Marchetti, C., & Spinnler, H. (1991). Right-sided anarchic (alien) hand: A longitudinal study. *Neuropsychologia, 29,* 1113–1127.

Della Sala, S., Marchetti, C., & Spinnler, H. (1994). The anarchic hand: A fronto-mesial sign. In F. Boller & J. Grafman (Eds.), *Handbook of neuropsychology.* Amsterdam: Elsevier.

Doody, R.S., & Jankovic, J. (1992). The alien hand and related signs. *Journal of Neurology, Neurosurgery and Psychiatry, 55,* 806–810.

Enns, J.T., Ochs, E.P., & Rensink, R.A. (1990). VSearch: Mackintosh Software for Experiments in Visual Search. *Behavioural Research Methods, Instruments and Computers, 22,* 469–479.

Freund, H.-J. (1996). Historical overview. In H.O. Lüders (Eds.), *Advances in neurology: The supplementary sensorimotor area.* Philadelphia, PA: Lippincott-Raven Publishers.

Geschwind, D.H., Iacoboni, M., Mega, M.S., Zaidel, D.W., Cloughesy, T., & Zaidel, E. (1995). Alien hand syndrome: Interhemispheric motor disconnection due to a lesion in the midbody of the corpus callosum. *Neurology, 45,* 802–808.

Gibb, W.R.G., Luther, P.J., & Marsden, C.D. (1989). Corticobasal degeneration. *Brain, 112,* 1171–1192.

Gibson, J.J. (1979). *The ecological approach to visual perception.* Boston, MA: Houghton Mifflin.

Goldberg, G., Mayer, N.H., & Toglia, J.U. (1981). Medical frontal cortex infarctions and the alien hand sign. *Archives of Neurology, 38,* 683–686.

Goldstein, K. (1908). Zür lehre der motischen Apraxie. *Journal für Psychologie und Neurologie, 11,* 169–187.

Goodale, M.A., Milner, A.D., Jakobsen, L.S., & Carey, D.P. (1991). A neurological dissociation between perceiving objects and grasping them. *Nature, 349,* 154–156.

Hillis, A.E., & Caramazza, A. (1995). Cognitive and neural mechanisms underlying visual and semantic processing: Implications from "optic aphasia". *Journal of Cognitive Neuroscience, 7,* 457–478.

Humphreys, G.W., Riddoch, M.J., & Quinlan, P.T. (1988). Cascade processes in picture identification. *Cognitive Neuropsychology, 5,* 67–103.

Jeannerod, M. (1997). *The cognitive neuroscience of action.* Oxford: Blackwell Publishers.

Jeannerod, M., & Decety, J. (1994). From motor images to motor programmes. In M.J. Riddoch & G.W. Humphreys (Eds.), *Cognitive neuropsychology and cognitive rehabilitation.* Hove, UK: Lawrence Erlbaum Associates Ltd.

Jeannerod, M., Decety, J., & Michel, F. (1994). Impairment of grasping movements following a bilateral posterior parietal lesion. *Neuropsychologia, 32,* 369–380.

Kawashima, R., Yamada, K., & Kinomura, S. (1993). Regional cerebral blood flow changes of cortical motor areas and prefrontal areas in humans related to ipsilateral and contralateral hand movement. *Brain Research, 623,* 33–40.

Kay, J., Lesser, R., & Coltheart, M. (1992). *PALPA: Psycholinguistic assessments of language processing in aphasia.* Hove, UK: Lawrence Erlbaum Associates Ltd.

Kim, S.G., Ashe, J., & Hendrich, K. (1993). Functional magnetic resonance imaging of motor cortex: Interhemispheric asymmetry and handedness. *Science, 261,* 615–616.

Kinsbourne, M., & Warrington, E.K. (1962). A disorder of simultaneous form perception. *Brain, 85,* 461–486.

Leiguarda, R., Starkstein, S., & Berthier, M. (1989). Anterior callosal haemorrhage: A partial interhemispheric disconnection syndrome. *Brain, 112,* 1019–1037.

Lezak, M.D. (1983). *Neuropsychological assessment.* Oxford: Oxford University Press.

Liepmann, H. (1920). Apraxie. *Ergebnisse der Gesamten Medizin, 1,* 516–543.

Marchetti, C., & Della Sala, S. (1998). Disentangling alien and anarchic hand. *Cognitive Neuropsychiatry, 3,* 191–207.

Milner, A.D., & Goodale, M.A. (1995). *The visual brain in action.* Oxford: Oxford University Press.

Milner, A.D., Perrett, D.I., Johnston, R.S., Benson, P.J., Jordand, T.R., Heeley, D.W., Bettucci, D., Mortara, F., Mutani, R., Terazzi, E., & Davidson, D.L.W. (1991). Perception and action in "visual form agnosia". *Brain, 114,* 405–428.

Pilgrim, E., & Humphreys, G.W. (1991). Impairment of action to visual objects in a case of ideomotor apraxia. *Cognitive Neuropsychology, 8,* 459–473.

Rapcsak, S.Z., Ochipa, C., Beeson, P.M., & Rubens, A.B. (1993). Praxis and the right hemisphere. *Brain and Cognition, 23,* 181–202.

Riddoch, M.J., & Humphreys, G.W. (1987). Visual object processing in optic aphasia: A case of semantic access agnosia. *Cognitive Neuropsychology, 4,* 131–185.

Riddoch, M.J., & Humphreys, G.W. (1993). BORB: *The Birmingham Object Recognition Battery.* Hove, UK: Lawrence Erlbaum Associates Ltd.

Riddoch, M.J., Humphreys, G.W., & Edwards, M.G. (in press). Visual affordances and object selection. In S. Monsell & J. Driver (Eds.), *Attention and performance XVIII.* New York: MIT Press.

Riddoch, M.J., Humphreys, G.W., & Price, C.J. (1989). Routes to action: Evidence from apraxia. *Cognitive Neuropsychology, 6,* 437–454.

Rinne, J.O., Lee, M.S., Thompson, P.D., & Marsden, C.D. (1994). Corticobasal degeneration: A clinical study of 36 cases. *Brain, 117,* 1183–1196.

Shallice, T., Burgess, P.A., Schon, F., & Baxter, D.M. (1989). The origins of utilisation behaviour. *Brain, 112,* 1587–1598.

Snodgras, J.G., & Vanderwart, M.A. (1980). A standardised set of 260 pictures: Norms for name agreement, familiarity and name complexity. *Journal of Experimental Psychology: Human Learning and Memory, 6,* 174–215.

Tanaka, Y., Yoshida, A., Kawahata, N., Hashimoto, R., & Obayashi, T. (1996). Diagnostic dyspraxia: Clinical characteristics, responsible lesion and possible underlying mechanisms. *Brain, 119,* 859–873.

Ungerleider, L.G., & Mishkin, M. (1982). Two cortical visual systems. In J. Ingle, M.A. Goodale, & R.J.W. Mansfield (Eds.), *Analysis of visual behaviour* (pp. 549–586). Cambridge, MA: MIT Press.

Warrington, E.K., & James, M. (Eds.). (1991). *VOSP: The Visual Object and Space Perception Battery.* Bury St. Edmunds, UK: Thames Valley Test Company.

DYSSYNCHRONOUS APRAXIA: FAILURE TO COMBINE SIMULTANEOUS PREPROGRAMMED MOVEMENTS

Anna M. Barrett
University of Florida and Veterans Administration Medical Center, Gainesville, FL, USA

Ronald L. Schwartz
University of Florida Health Science Center, Jacksonville, USA

Anastasia L. Raymer
Old Dominion University, Norfolk, VA, USA

Gregory P. Crucian
University of Florida, Gainesville, USA

Leslie Gonzalez Rothi and Kenneth M. Heilman
University of Florida and Veterans Administration Medical Center, Gainesville, USA

Limb apraxia, a defect in skilled, learned purposive movement, may be related to impairment of either representational or innervatory components of praxis processing. Innervatory motor patterns, in turn, may involve on-line motor programs (visual feedback-controlled) or prepared movement programs (independent of continuous visual feedback). We evaluated movement abilities of the innervatory pattern system in TB, a 26-year-old patient with apraxia from a left dorsolateral frontal stroke. TB and four controls performed nonmeaningful single- and multi-joint movements to command, with multi-joint movements combined sequentially (e.g. "open and close your hand and then bend your elbow") or simultaneously (e.g. "open and close your hand; keep doing that while bending your elbow"). TB showed no difference between single-joint (71.5% correct) and multi-joint movements in sequential combinations (68% correct), but she was significantly worse at simultaneous movement combinations (28.6% correct; $P < .02$). Controls performed consistently at > 90% mean accuracy. TB and four normals also performed the Fitts (1954) task, in which they alternately tapped with a pen between two target circles of varying size. TB was proportionately slower than controls on the larger Fitts circles, which call predominantly on prepared movement programs; her performance on the smaller circles (involving more on-line programs) was comparable to

---

Requests for reprints should be addressed to Anna M. Barrett, MD, Department of Neurology, PO Box 100236, University of Florida Health Science Center, Gainesville, FL 32610-0236, USA (E-mail: barrett@medicine.ufl.edu).

Presented before the American Academy of Neurology, San Francisco, California, April 1995, and was published in abstract form in 1996 in *Neurology, 46(2)*: A383. Supported by VA Merit Review #114-30-4813 and the National Institutes of Health (R01 NS28665-05). We are grateful to the patient, TB, for her participation and thank Michelle D. Miller, MA for her help in aphasia testing. We are also grateful for the suggestions of two anonymous reviewers who read this manuscript in its original submitted form.

normals. We conclude that functional synchrony of one innervatory pattern subtype, prepared movement programs, may require late-level frontal processing, and that failure at this level can result in both apraxia and defective programming of nonmeaningful movements.

INTRODUCTION

Limb apraxia is defined as a disturbance of skilled, learned purposive movement, and is still partly defined by exclusion (weakness, akinesia, differentiation, abnormal tone or posture, movement disorder, intellectual impairment, poor comprehension, or uncooperativeness; see Rothi, Ochipa, & Heilman, 1991, 1997 for a review). Multiple, separate representational and motor components supporting praxis have been proposed since the classic work of Liepmann (1905, 1920), who described "time-space-form pictures" from which skilled purposive movements of each hand were generated by activating "innervatory patterns" and motor cortex. He also defined a third module, the "kinaesthetic memory", that stores highly practised movement components. Although Liepmann suggested the space-time representations may be located in the caudal portions of the left hemisphere, Heilman, Rothi, and Valenstein (1982) and Rothi, Heilman, and Watson (1985) specifically proposed and provided evidence that it is the dominant parietal lobe that is crucial for the timing, sequencing, and spatial organisation of skilled purposive movements. Comprehending gestures and pantomimes made by others also depends on parietal-stored, representational praxis knowledge (Heilman et al., 1982).

Gesturing, pantomiming, and working with tools and objects, however, requires that these movement representations be transcoded into a "motor plan of action". This motor plan of action controls the patterns of neuronal firing in motor cortex. Whereas posterior (e.g. parietal) left-hemisphere lesions that destroy movement representations lead to production, comprehension, and discrimination deficits, Heilman et al. (1982), Rothi et al. (1985), and Watson, Fleet, Rothi, and Heilman (1986) demonstrated that more anterior lesions impair production but not comprehension or discrimination of gestures and pantomimes.

There may be two types of innervatory patterns: (1) prepared programs (off-line programs), which are primarily determined before they are executed, and (2) on-line programs requiring continuous sensory (e.g. visual) feedback and monitoring. Gestures performed to verbal command must rely heavily on prepared movement programs, as no visual or tactile stimuli are available to help guide the movement. Prepared movement programs may correspond to the kinaesthetic memories Liepmann proposed (1905, 1920). Haaland and her co-workers (Haaland & Harrington, 1994; Haaland, Harrington, & Yeo, 1987) studied patients with hemispheric dysfunction performing nonpurposive skilled limb movements. They compared patients with left-hemisphere brain damage to patients with right-hemisphere lesions, and demonstrated convincingly that the left-hemisphere-damaged patients are impaired on motor tasks requiring prepared programs. This implies left-hemisphere dominance for prepared movement programs. Additionally, a consistent right-hand superiority is reported in normals for visual feedback-dependent movements using on-line programs (e.g., Annett, Annett, Hudson, & Turner, 1979; Todor & Cisneros, 1985), suggesting that the left hemisphere may be dominant for both types of innervatory patterns.

The medial premotor cortex, or supplementary motor area, may be crucial for translating representations to movement programs and for the development of innervatory patterns. A number of studies report supplementary motor area (SMA) activation in complex movements but not simple, repetitive movements. SMA also activates in imagined complex movement without actual movement execution (Lauritzen, Henriksen, & Lassen, 1981; Orgogozo & Larsen, 1979; Penfield & Welch, 1951; Rao et al., 1993; Roland, Larsen, Lassen, & Skinhøj, 1980; Roland, Skinhøj, Lassen, & Larsen, 1980). Damage to the SMA has been associated with several motor disturbances (Brust, 1996), some of which could be classified as ideomotor apraxia. For example, Watson et al. (1986) reported two patients with left-sided medial frontal lesions. Both of these patients had bilateral ideomotor apraxia sparing gesture comprehension. Dick, Beneke, Rothwell, Day, and Marsden (1986) reported a patient with right SMA infarction and impairment in the production of multiple-joint sequential and simultaneous movements (the patient was reported to be right-handed but it is not noted what functions were assessed). Alternating serial movements of the hands were impaired after SMA lesions in three patients examined by Laplane, Talairach, Meininger, Bancaud, and Orgogozo (1977).

Although Geschwind (1965) proposed that the convexity premotor cortex was important for motor aspects of praxis, he did not describe any patient with apraxia whose lesion was restricted to this area and did not involve deep structures. In this paper, we report a patient with the abrupt onset of limb apraxia following a dorsolateral frontal lesion involving convexity premotor cortex. Based on this lesion location, we first posited that her stroke might have injured praxis production areas that develop innervatory patterns, but not portions of the brain that store movement representations. Therefore, we tested both production and representations of skilled purposive movements. Lastly, the convexity premotor cortex receives strong projections from the visual association cortex. Goldberg (1985) has proposed, based on its connectivity, that convexity cortex may be important in mediating externally controlled movements. Goldberg's hypothesis predicts when convexity premotor cortex is damaged, movements requiring on-line programs and continuous visual monitoring would be more impaired than movements dependent on prepared motor programs. We chose a modified version of a Fitts (1954) circle-tapping task to test this second hypothesis.

CASE REPORT

Patient

TB, a 26-year-old, right-handed African-American woman with 10 years of education, was admitted to Shands Hospital at the University of Florida following the sudden onset of inability to speak and right face, arm, and hand weakness. Her past medical history was significant for a spontaneous abortion, but there were no symptoms suggestive of collagen-vascular disease, and no previous neurological disorders. Her answers to a handedness questionnaire (Raczkowski, Kalat, & Nebes, 1974) revealed that she had used her right hand for all functions, and her husband confirmed this information.

TB was first seen by the Behavioural Neurology Service 10 days post-stroke. She was mute but able to obey complex auditory commands and to answer yes/no questions gesturally with complete accuracy. Cranial nerves were normal except for mild right lower face weakness. There was arm (4/5), leg (4/5), and hand weakness (4/5), and she was unable to perform any voluntary movements of the mouth to command, including protrusion of the tongue. However, spontaneous movements of the mouth, tongue, and face occurred normally in activities such as laughing and eating. When performing skilled learned purposive limb movements, the left nonparetic hand was clumsy. On initial assessment, her left-handed productions of gestures to command and imitations of gestures were unrecognisable. There were no sensory abnormalities, including stereognosis and graphesthesis, in either the left or right hand. Reflexes were mildly increased on the right and there was a right extensor plantar response, with the left flexor.

A computed tomographic scan was done on the day following onset of her symptoms, and magnetic resonance imaging done 1 month later revealed a similar finding: a left dorsolateral frontal infarct (see Fig. 1) affecting Brodmann areas 44, a small portion of area 45, the lateral portion of area 6, and the insular cortex. The medial and parasagittal portions of Brodman area 6, usually defined as supplementary motor cortex, were completely spared. There was also a small right posterior parietal infarct that appeared old (Brodman area 40). Evaluation for underlying causes of thrombotic, embolic, or vasculitic infarct including serologic studies, cerebrospinal fluid examination, carotid Dopplers, four-vessel angiography, and cardiac workup revealed only a small patient foramen fovale seen on trans-oesophageal echocardiography. There were low levels of antiphosphatidyl serine antibody (IgM, no IgG) and antiphosphatidyl inositol antibody (IgM, no IgG), but erythrocyte sedimentation rate, antinuclear antibody, serological test for syphilis (VDRL), rheumatoid factor, and anticardiolipin antibody tests were negative. The patient and primary care physicians decided against surgical correction of her patent foramen ovale, and she started oral anticoagulation with warfarin.

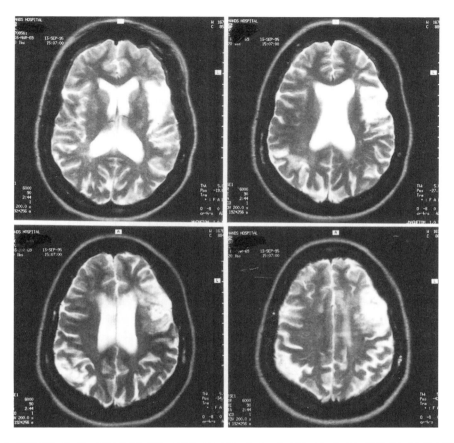

Fig. 1. MRI of brain for patient TB. T2-weighted images, 1 month after onset of symptoms. Acute left dorsolateral frontal stroke (right side of image) and small right parietal stroke, which was hypodense on T1-weighted images, consistent with an old lesion, are evident.

Speech and Language Assessment

On examination, TB was unable to produce any voluntary speech or sounds except the sound /ʃp/. She obeyed two- and three-step auditory commands easily, but clearly indicated uncertainty when questions were syntactically complex (e.g. in the passive form). She obeyed two- and three-step written commands, but her attempts to write (with either hand) resulted in unrecognizable scrawls or a few poorly formed letters, consistent with apractic agraphia. The Western Aphasia Battery (Kertesz, 1980) was administered 4 days post-stroke. Her score on the auditory comprehension subtest was 6.1 out of a possible 10, but her scores on other subtests (spontaneous speech, repetition, and naming) were 0 for a total Aphasia Quotient of 12.2 out of a possible 100 points. This is consistent with a profound nonfluent aphasia. She spontaneously gestured communicatively, with a reliable yes/no gesture and some simple symbolic as well as spontaneous emotion gestures.

METHODS AND RESULTS

Tests of Representational vs. Innervatory Praxis Systems

Ten days after onset of symptoms we evaluated praxis with the Florida Apraxia Battery (unpublished test, Rothi, Raymer, Ochipa, Maher, Greenwald, & Heilman, 1992). In praxis production tasks, TB produced pantomimes to verbal command and pantomimes of tool use while viewing a tool (held flat in the examiner's hand), performed transitive acts while holding tools, and imitated the examiner's pantomimes. Her performance on all of these tasks was videotaped and later scored by the consensus of three judges trained to use a system for qualitative/quantitative analysis of errors described by Rothi, Mack, Verfaellie, Brown, and Heilman (1988). We then examined her ability to select transitive pantomimes that matched a presented tool, to pick a tool from an array of 15 when shown a pictured object ordinarily acted upon by that tool, and to match tool-used pantomimes to pictured objects typically acted upon with that tool. The pictured objects were chosen from the Florida Action Recall Test (unpublished test, Schwartz, Crosson, Rothi, Nadeau, & Heilman, 1996) to match the pantomime-to-command items.

As shown in Fig. 2, TB displayed a dramatic impairment in all tasks involving gesture output, but tasks involving gesture comprehension were relatively spared. On the gesture to command task, her productions were rated by the system described by Rothi et al. (1988). Productions were scored correct or incorrect, and incorrect productions were analysed for error types. In her 30 productions, TB made 62 total errors, 20 of which were classified as unrecognisable (32%). There were 18 (29%) movement errors (moving the incorrect joint or not correctly coordinating joint movements), 10 errors involved incorrect hand internal configurations or postures (16%), 5 errors in which the movement was directed with an incorrect external spatial configuration (8%), 3 perseverative errors (5%), 3 no-response errors (5%), 2 content errors (e.g. hammer for screwdriver) (3%) and 1 error where she used her hand as a tool (2%). The difficulty with movement generation in the presence of spared tool concepts suggests that TB's apraxia was induced by damage to the production system rather than action semantics (Rothi & Heilman, 1996). Because gesture output but not comprehension was impaired it would appear that TB's apraxia may have been induced by an impaired ability to develop innervatory patterns.

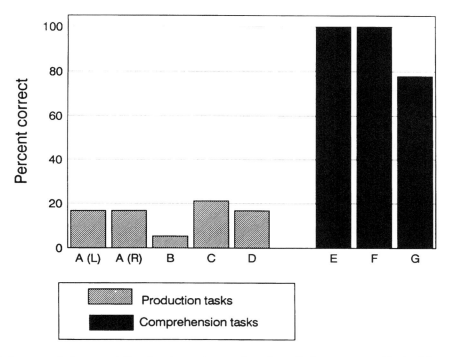

Fig. 2. Assessment of representational vs. innervatory praxis systems. Performance on subtests of the Florida Apraxia Battery (Rothi et al., 1992, unpublished) is graphed as percentage performed correctly. Gesture production tasks are as follows: A(L) = tool pantomimes to command, left hand; A(R) = tool pantomimes to command, right hand; B = attempts to mime use while viewing tool; C = attempts to mime use while holding tool; D = imitation of examiner's pantomime. Gesture comprehension tasks are as follows: E = select correct pantomime to match viewed tool; F = select correct tool from array to match tool pantomime; G = select correct tool from array to match pictured object tool would ordinarily act upon. It can be seen that TB's performance was better on gesture comprehension tasks than on gesture production tasks, supporting a deficit in gesture output.

Tests of Single-, Multi-joint and Sequential vs. Simultaneous Movement

Unlike many patients with ideomotor apraxia, who make spatial errors but whose gestures still have signal value and can be recognised for their intended content, TB made many gestures that were unrecognisable. Clinically, it appeared that her movements were degraded because she could not simultaneously coordinate multiple joint movements. To test this clinical observation formally, we asked TB and four right-handed, age-matched controls to perform nonmeaningful single- and multi-joint movements to command. Some multi-joint movements were combined sequentially (e.g. "open and close your hand and then bend your elbow") and some were combined simultaneously with the second movement added to the first (e.g. "open and close your hand; keep doing that while bending your elbow"). More examples of commands in all categories, and summary information for controls, are listed in Table 1. There were 38 trials of single-joint movements (e.g. "lift your shoulder") and 46 trials of multiple-joint movements, that included 25 sequential and 21 simultaneous combinations.

TB's performance of nonmeaningful movements was videotaped and scored later by the consensus of three raters trained in the scoring method of Rothi et al. (1988). Movements were scored "correct" (+) or "incorrect" (–). Movements were scored for accuracy (the correct joint moved in the specified direction) and for sequence and synchrony, as specified by the instructions. When judging synchrony, the raters viewed the movements in slow motion.

Table 1. Single- vs. Multiple-joint Movements

Sample Items Testing Single- vs. Multiple-joint Movements

Single joint movements
"Raise your elbow."
"Extend your arm."
"Turn your hand over."
"Raise your index finger."

Multi-joint sequential movements
"Straighten out your arm and make a fist."
"Lift your first finger and bend your elbow."
"Lift your thumb in the air and lift your shoulder."

Multi-joint simultaneous movement
"Turn your wrist; keep turning your wrist while bending your elbow."
"Open and close your hand; keep doing that while bending your elbow."
"Tap your fingers; keep tapping and move your wrist in a circle."

Summary Information for Four Normal Controls and Patient TB (in parentheses)

	Age			Sex			Education		
Mean	*Range*	*(TB)*	*Female*	*Male*	*(TB)*	*Mean*	*Range*	*(TB)*
30.25 years	23–40	(26 years)	3	1	(F)	13.5 years	11–16 years	(10 years)

		Multiple-joint Movements	
	Single-joint Movements	Sequential	Simultaneous
Mean	97.2%	98.2%	97.6%
Range	94.0–100%	92.9–100%	90.5–100%
(TB)	(71.4%)	(68.0%)	(28.6%)

Controls performed consistently at > 90% mean accuracy. As shown in Fig. 3, TB's performance was inferior to that of control subjects for all the movements, but her performance on the single-joint (71.5% correct) vs. multi-joint sequential movements (68% correct) did not significantly differ (t-test, $P = .77$). In contrast, her performance on the multi-joint simultaneous movements (28.6% correct) was significantly worse than that for multi-joint sequential movements or single-joint movements (t-test, $P < .05$).

Sequential and simultaneous movement commands may have differed in complexity and in working memory demands. Rothi and Heilman (1985) have shown that people with apraxia may have impairment in their working memory for gestures. Therefore, independent of the different demands that these tasks place on the control of motor programs, performance differences that we observed may have been related to language or working memory impairments. However, we doubt that the difference in commands could account for our results because the difference in syntactic complexity between sequential and simultaneous commands consistently favoured simultaneous movements. It can be seen when reviewing the commands for both categories (Table 1) that the sequential commands involved processing a syntactic message of "First A, then B." The simultaneous commands, in contrast, included no sequential specification ("A, B together"), and imposed less burden on gestural working memory. This conclusion is supported by the fact that we have since repeated this protocol on another patient with apraxia, who had a large stroke affecting the entire left middle cerebral artery territory (Barrett, Hughes, Beversdorf, Rothi, & Heilman, unpublished data). This patient could not remember the second part of more than half of the sequential movement commands but performed all of the simultaneous movements to command easily and correctly.

Tests of On-Line Movement Programs vs. Prepared Movement Programs

Fitts (1954) demonstrated that the speed of a movement is dependent on the size of the target of the movement and the distance traversed. Repeated movements to a large target can be performed without visual feedback. However, movements to small targets require on-line feedback. To test Goldberg's (1985) hypothesis about the convexity premotor cortex's role in feedback-dependent motor perform-

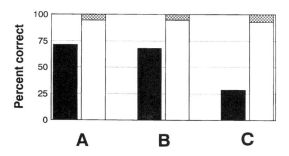

Fig. 3. Assessment of single- and multi-joint sequential and simultaneous movements. Performance on each condition, for TB (dark bar) and for four normal controls (white bar), graphed as percentage correct. Hatched area indicates standard error for controls. A = single-joint movements; B = multi-joint sequential movements; C = multi-joint simultaneous movements. See text for full interpretation of results.

ance, we administered a modified version of a Fitts (1954) tapping task to our patient. When the circles are large, it is assumed that a minimum of visual monitoring is required for the task, and that this movement is relatively controlled by prepared rather than on-line motor programs. When the circles are small, on-line programs are used to incorporate visual feedback monitoring so that subjects can place dots in the two circles accurately. Implementation of on-line motor programs integrating visual feedback typically reduces the rate at which movements proceed. TB and four right-handed, age- and gender-matched normals (recruited separately from the previous control subjects) alternately touched two target circles of 2cm, 5.5cm, 7.5cm and 8.5cm with a pen (see Fig. 4 for example), making dots in both circles. All circles were 19cm apart from centre to centre. Each trial continued for 20sec and the number of hits were recorded for 5 trials at each circle size.

As can be seen in Fig. 5 and Table 2, TB made fewer dots (was slower) than normal subjects

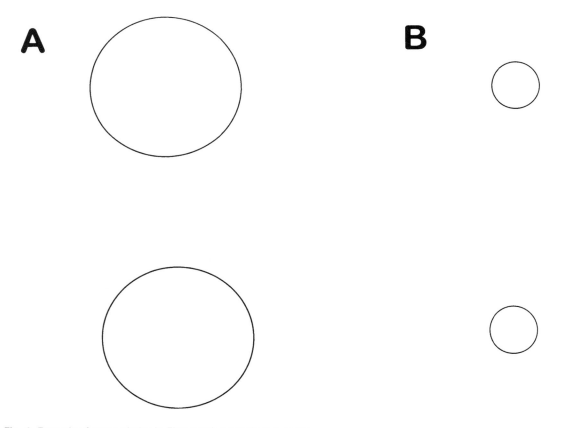

Fig. 4. Example of target circles in Fitts tapping task (not to scale).

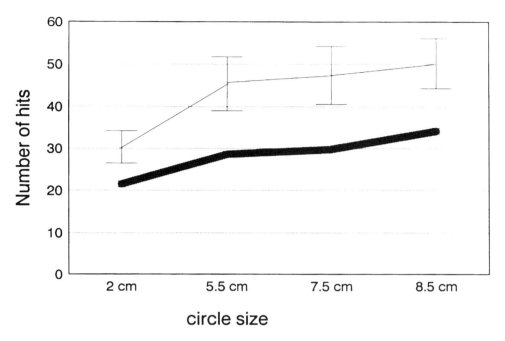

Fig. 5. Assessment of prepared vs. on-line movement programs using Fitts circle task. TB (thick line) and four normal controls (thin line), total number of hits per circle size on Y-axis and increasing circle diameter on X-axis, with standard error bars for control performance. It can be seen that TB made fewer hits at each circle size than normal controls.

Table 2. Summary Information for Four Normal Controls and TB (in parentheses): Fitts (1954) Tapping Task

	Mean	Range	(TB)
Age	27.3	22–37	(26)
Education	13.75	12–16	(10)
All female (F)			

	No. of Hits	% Reduction[a]
Circle Size	Mean (TB)	Mean (TB)
8.5cm	50.2 (34.2)	—
7.5cm	47.4 (29.8)	5.6% (12.9%)
5.5cm	45.4 (28.6)	9.6% (16.4%)
2.0cm	30.4 (21.5)	39.4% (37.1%)

[a] % Reduction from largest size.

for all circle sizes (below 95% confidence interval for the 4 normal controls). This replicates the findings of Carmon (1971), who found both left- and right-hemisphere-damaged patients to be slowed on a fast-paced tapping task, at similar percentages of normal performance. Carmon found that movement efficiency, however, measured by the ratio of correct responses over possible correct responses, discriminated left- from right-hemisphere-damaged patients, with left-hemisphere-damaged patients clearly impaired at the fast-paced (more preprogrammed) condition. Based on Carmon's work, we intended a priori to examine a similar efficiency parameter as well as overall speed.

TB's tapping speed, as measured by the number of hits per 20sec interval, increases slightly but probably not significantly over decreasing circle size when expressed as a percentage of normal performance (68.1%, 62.9%, 63%, and 70.7% for the 8.5cm, 7.5cm, 5.5cm, and 2cm circles, respectively). This does not explain, however, why her performance curve is flatter over different circle sizes (why her speed changes less) than that of controls. Like Carmon's left-hemisphere-damaged patients, it is possible that she may have been less efficient at larger circle sizes. When efficiency was coded as a percentage decrease in hits from the fastest (presumably maximum speed) condition (the largest circles), her percentage decrease in total number of hits from the largest circle size to the smallest circle size is comparable to normal subjects. However, her percentage decrease in hits from the 8.5cm, largest circle size to the 7.5cm large circle is greater than that observed in normal controls (Fig. 6). Overall, this dropoff in speed at larger circle sizes is consistent with a deficit in prepared movement programs, which Haaland and co-workers (Haaland & Harrington, 1989, 1994; Haaland et al., 1987) and Winstein and Pohl (1995) as well as Carmon (1971) have demonstrated to be impaired after left-hemisphere damage. This finding is, however, contrary to Goldberg's (1985) model of convexity premotor cortex function, which predicts that the smaller circles (on-line program condition) would be more impaired, as this condition is more visual-feedback dependent.

The means of calculating times for each circle size as a percentage increase from the fastest time was determined a priori and based on determining efficiency. However, mean performance of normal subject differs more from TB's performance on the middle-sized 5.5cm and 7.5cm circles than on the smallest and largest circles (see Fig. 5). In order to evaluate whether our finding could be an artefact of the percentage calculation, we recalculated means and standard errors using percentage increase in hits from the slowest condition (a method that does not reflect efficiency). Our finding of relative impairment in the large circle over the small circle condition was not reversed using this method of calculation.

DISCUSSION

Geschwind (1965) argued that the convexity premotor cortex was necessary as a "motor association area" by which language and visual areas were connected to the motor cortex of both the right and left hemispheres. However, case reports reviewed by Faglioni and Basso (1985) of patients with apraxia and lesions in this area, and the comparison of frontal- and parietal-damaged patients by De-Renzi, Faglioni, Lodesani, and Vecchi (1983) do not make a strong case for an apraxic syndrome accompanying left frontal damage. Tzavaras (1978) reported a patient with a left frontal haemorrhage, right hemiparesis, and apraxia involving movements to command as well as to imitation, but the exact location of the patient's lesion is unclear. Morlaass (1928)

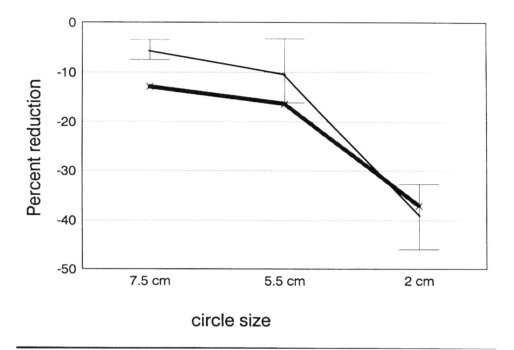

Fig. 6. Assessment of prepared vs. on-line movement programs using Fitts circle task. Results for TB (thick line) and four controls (thin line) are illustrated, with percent decrease in total hits from fastest condition over circle sizes on Y-axis and decreasing circle size on X-axis, with standard error bars for control performance. It can be seen that when performance is equated relative to fastest performance, TB returns to within control range for the smaller, closed-loop-movement circles, and is deficient compared with control range only at the larger, open-loop movement circles.

reported bilateral apraxia in a patient with damage to the pes of inferior frontal gyrus, but this patient's lesion also included inferior precentral gyrus, the pes of middle frontal gyrus, as well as occipital and parietal areas. Brun (1921) reported a patient with damage to the inferior frontal gyrus and the anterior pars of precentral gyrus that had the "incapacity to use limbs for more complex actions". This patient also had involvement of superior middle frontal gyrus, but the lesion spared pars triangularis. Both Kolb and Milner (1981) and Kimura (1982) reported that patients with frontal premotor ablations or damage had difficulty with motor sequences, but it is unclear if these patients had the type of spatial errors described by Liepmann as ideomotor apraxia.

TB's right parietal lesion probably contributes trivially to her apraxia and preprogramming deficit. This lesion was old and she had no history of impaired praxis prior to her current stroke. Heilman, Rothi, Mack, Feinberg, and Watson (1986) reported a right-handed patient with a right superior parietal lesion whose performance of gestures with the left hand was marred by spatial errors, especially with the eyes closed. This patient's apraxia was limited to the left had, and felt to be due to a failure of somesthetic integration in executing the movements. This is clearly different from TB's incorrect movements, which were mostly unrecognisable. Foix (1916) felt that the right parital area had a principal role in praxis. However, Maher (1995) and her co-workers studied the praxis of right-brain-damaged subjects and found that they did not differ in their total praxis scores from normal controls. Maher's study supports the finding in other large studies (de Ajuriaguera, Hecaen, & Angelergues, 1960; Kolb & Milner, 1981) that damage to the right post-Rolandic area does not produce severe apraxia.

TB's left hand and arm movements meet both the exclusion and inclusion criteria for ideomotor apraxia, with what appears to be a specific impairment in praxis production systems developing innervatory patterns. In contrast to her severe impairment in producing gestures to verbal command, her performance was very good on all tasks meant to assess her ability to understand tool-object relationships and decode visually presented gestures. Thus, her movement representations and action semantics appear to be relatively spared. This dissociation supports a specific deficit in praxis production, and is consistent with the radiologic appearance of her stroke, which spared the parietal cortex, the region proposed to store praxis knowledge or movement representations.

Praxis production must involve the transcoding of representational information into specific innervatory patterns. Given that TB's stroke may have injured areas important to innervatory pattern development, we wanted to know whether prepared movement programs or on-line movement programs might be more affected. We have evidence from two experimental tasks that prepared (off-line) movement programs were more deficient than on-line, continuously monitored movement programs and therefore this innervatory pattern subtype may depend upon left convexity premotor cortex activity.

In tests of single-, multi-joint, and sequential vs. simultaneous movements, TB was unable to coordinate multiple joint movements of the left hand and arm simultaneously, compared with her ability to execute the same movements sequentially and to execute isolated movements (Fig. 3). Thus, synchronous movements ipsilateral to her left dorsolateral premotor cortex lesion were disrupted. We suspect that her impairment in programming and coordinating simultaneous movements was responsible for her poor performance when attempting to gesture and pantomime to command with either hand (Fig. 2). This observation suggests that the left frontal cortex may be dominant for preprogramming both meaningful and nonmeaningful synchronous movements.

Our finding also suggests that linking preprograms and coordinating simultaneous movement among multiple joints may be especially reliant on the left frontal convexity cortex. Sequential preprograms, which coordinate movements to follow each other in a specified order, may not be as affected by lesions in this area, though these programs are undoubtedly also prepared and can be impaired after left-hemisphere damage (Harrington & Haaland, 1992).

When we tested prepared movement program vs. on-line movement programs specifically, our patients exhibited relatively more impairment on a larger circle target condition, relying on prepared movement programs, than on a small circle target condition. Sequential tapping in the small circle target condition relies upon longer periods of deceleration, a movement component felt to be visually determined (Jackson, Jones, Newport, & Pritchard, 1997; Saling, Stelmach, Mescheriakov, & Berger, 1996). Our patient therefore had a relatively selective impairment affecting prepared movement programs more than motor programs requiring continuous visual monitoring. It can be argued that some on-line visual monitoring of movement occurs when performing the large circle condition with eyes open. Similarly, the small circle condition is likely to incorporate prepared motor programs in planning the initial trajectory or reach component of the movement (Marteniuk, Leavitt, MacKenzie, & Athenes, 1990). Although performance in the small and large circle conditions may require elements of both types of processing, the difference in performance that TB demonstrated relative to controls suggests that both conditions are *not* dependent on the same neuroanatomic systems.

A review of neuroanatomic studies of the dorsolateral and medial premotor cortex in nonhuman primates suggests that medial and lateral zones may be functionally distinct. Previously, movement was felt to be hierarchically organised in the primate cortex, with most, if not all, corticospinal output originating from the primary motor cortex, and other areas having their influence either via modulation of the primary motor cortex or via the extrapyramidal system. There is now evidence, however, that the premotor areas project directly to the spinal cord and have the capacity to initiate and control movements directly (Dum & Strick, 1991b, 1996; Tanji & Shima, 1996; Wise, 1996). Although nomenclature and boundaries of the premotor areas in the monkey vary depending on investigators and techniques (see Wise, 1996, for a review), Dum and Strick (1991a, 1991b), define six premotor areas: two are on the lateral convexity, the lateral ventral and dorsal premotor areas, and four are medial. The medial areas include the SMA, the rostral cingulate motor area, and the caudal dorsal and ventral motor areas. This schema is considerably more complex than that based on Brodmann areas. The parietal primary somatosensory cortex and somaesthetic association areas appear to project much more strongly to the medial zone, although kinaesthetic-tactile "haptic" information may project to the lateral zone as well. Parietal visual association areas project primarily to the lateral and not the medial zone (Dum & Strick, 1996; Rizzolatti, Luppino, & Matelli, 1996).

On the basis of architectonic and phylogenetic analyses, anatomic connections, and lesion studies, Goldberg (1985) hypothesised that the supplementary motor area (roughly equivalent to the medial zones, see earlier) and the convexity premotor areas participate differently in internally vs. externally triggered movements. According to Goldberg, the SMA and its connections form a system "necessary for extended, internally-dependent, predictive or *projectional* action ... more concerned with ... navigating the limb through space rather than the more focal problem of accurately acquiring ... objects in 'peripersonal' or reachable space". The lateral premotor cortex, more closely connected with the piriform cortex and the object recognition system, "is utilised in the production of interactive, externally-contingent responsive action ... driven by the presence and identified nature of specific objects in the ... immediate environment". Goldberg specifically referred to the medial system as "feedforward" and the lateral system as "feedback". This would predict that damage to the lateral premotor cortex should produce a deficit in feedback-dependent movements requiring on-line visual monitoring.

Our findings do not support Goldberg's prediction. Other investigators have similarly found evidence against his hypothesis. Tanji and Shima (1996), citing four studies showing that externally triggered movements activated medial (SMA) neurons in monkeys in a similar fashion to neurons in lateral premotor cortex (Kurata & Wise, 1988; Okano & Tanji, 1987; Romo & Schulz, 1987; Thaler, Rolls, & Passingham, 1988), wrote "the view that the SMA is primarily related to self-initiated movements and that the PM [premotor] is related to visually or sensorially guided ... movements must be discarded". Passingham (1993, 1996), based on PET scanning data from humans doing motor tasks, also emphasised that the SMA's role is likely not to be limited to self-initiated movement.

When a patient is required to pantomime a transitive movement, there are no spatial cues to guide the movement. All of the movement parameters must be self-generated. Therefore, given our patient's severe apraxia, our patient's abnormal movements when she was operating relatively independent of visual cues should not be surprising. Patients who were unable to execute a gesture to command correctly often improve when they are shown the tool to be used, or given the tool to hold while gesturing (Heilman & Rothi, 1993). TB's gestures and pantomimes of tool use (Fig. 2) improved when she was given tools to hold. When she held the tool, visual and tactile stimulation could help support continuous, on-line movement programs.

Although these studies of TB suggest that the left frontal convexity cortex may be specialised for linking preprograms governing movement synchrony, we have not studied patients with discrete lesions to other portions of the brain. We cannot, therefore, conclude that left convexity frontal cortex is unique for mediating this process. Additional studies are needed to both replicate our observations and determine if these functions are unique to this part of frontal cortex.

REFERENCES

Ajuriaguera, J. de, Hecaen, H., & Angelergues, R. (1960). Les apraxies. Varietes cliniques et lateralization lesionelle. *Revue Neurologique, 102*, 566–594.

Annett, J., Annett, M., Hudson, P.T.W., & Turner, A. (1979). The control of movement in the preferred and non-preferred hands. *Quarterly Journal of Experimental Psychology, 31*, 641–652.

Brun, R. (1921). Klinische und anatomische Studieren über Apraxie. *Schweitzer Archiv für Neurologie und Psychiatrie, 9*, 29–64, 194–226.

Brust, J.C.M. (1996). Lesions of the supplementary motor area. In H.O. Lüders (Ed.), *Advances in neurology, Vol. 70: Supplementary sensorimotor area* (pp. 237–248). Philadelphia, PA: Lippincott-Raven Publishers.

Carmon, A. (1971). Sequenced motor performance in patients with unilateral cerebral lesions. *Neuropsychologia, 9*, 445–449.

DeRenzi, E., Faglioni, P., Lodesani, M., & Vecchi, A. (1983). Performance of left brain-damaged patients on imitation of single movements and motor sequences, frontal and parietal-injured patients compared. *Cortex, 19*, 333–343.

Dick, J.P.R., Beneke, R., Rothwell, J.C., Day, B.L., & Marsden, C.D. (1986). Simple and complex movements in a patient with infarction of the right supplementary area. *Movement Disorders, 1*, 255–266.

Dum, R.P., & Strick, P.L. (1991a). Premotor areas: Nodal points for parallel efferent systems involved in the central control of movement. In D.R. Humphrey & H.J. Freund (Eds.), *Motor control: Concepts and issues* (pp. 231–295). London: Wiley.

Dum, R.P., & Strick, P.L. (1991b). The origin of corticospinal projections from the premotor areas in the frontal lobe. *Journal of Neuroscience, 11*, 667–689.

Dum, R.P., & Strick, P.L. (1996). The corticospinal system: A structural framework for the central control of movement. In L.B. Rowell & J.T. Shepard (Eds.), *Handbook of exercise physiology: Integration of motor circulatory, respiratory and metabolic control during exercise, Section A: Neural control of movement* (pp. 68–113). New York: Oxford University Press.

Faglioni, P., & Basso, A. (1985). Historical perspectives on neuroanatomical correlates of limb apraxia. In E.A. Roy (Ed.), *Neuropsychological studies of apraxia and related disorders* (pp. 3–44). New York: Elsevier.

Fitts, P.M. (1954). The information capacity of the human motor system controlling the amplitude of movements. *Journal of Experimental Psychology, 47*, 381–391.

Foix, C. (1916). Contribution a l'etude de l'apraxie ideo-motrice de son anatomie pathologique et de ses rapports avec les syndromes qui, ordinairement, l'accompagnent. *Revue Neurologique, 23*, 283–298.

Geschwind, N. (1965). Disconnexion syndromes in animals and man. *Brain, 88*, 237–294, 585–644.

Goldberg, G. (1985). Supplementary motor area structure and function: Review and Hypotheses. *Behavioral Brain Sciences, 8*, 567–616.

Haaland, K.Y., & Harrington, D.L. (1989). Hemispheric control of the initial and corrective components of aiming movements. *Neuropsychologia, 27*, 961–969.

Haaland, K.Y. & Harrington, D.L. (1994). Limb-sequencing deficits after left but not right hemisphere damage. *Brain and Cognition, 24*, 104–122.

Haaland, K.Y., Harrington, D.L. & Yeo, R. (1987). The effects of task complexity on motor performance in left and right CVA patients. *Neuropsychologia, 25(5)*, 783–794.

Harrington, D.L., & Haaland, K.Y. (1992). Motor sequencing with left hemispheric damage. Are some cognitive deficits specific to limb apraxia? *Brain, 115*, 857–874.

Heilman, K.M., & Rothi, L.J.G. (1993). Apraxia. In K.M. Heilman & E. Valenstein (Eds.), *Clinical neuropsychology* (3rd Edn.) (pp. 141–164). New York: Oxford University Press.

Heilman, K.M., Rothi, L.G., Mack, L., Feinberg, T., & Watson, R.T. (1986). Apraxia after a superior parietal lesion, *Cortex, 22*, 141–150.

Heilman, K.M., Rothi, L.J.G., & Valenstein, E. (1982). Two forms of ideomotor apraxia. *Neurology, 32*, 342–346

Jackson, S.R., Jones, C.A., Newport, R., & Pritchard, C. (1997). A kinematic analysis of goal-directed prehension movements executed under binocular, monocular and memory-guided viewing conditions. *Visual Cognition, 4(2),* 113–142.

Kertesz, A. (1980). *Western Aphasia Battery.* London, Ontario: University of Western Ontario Press.

Kimura, D. (1982). Left-hemisphere control of oral and brachial movements and their relation to communication. *Philosophical Transactions of the Royal Society of London B298,* 135–149.

Kolb, B., & Milner, B. (1981). Performance of complex arm and facial movement after focal brain lesions. *Neuropsychologia, 19(4),* 491–503.

Kurata, K., & Wise, S.P. (1988). Premotor and supplementary motor cortex in rhesus monkeys: Neuronal activity during externally and internally instructed motor tasks. *Experimental Brain Research, 72,* 237–248.

Laplane, D., Talairach, J., Meininger, J., Bancaud, J., & Orgogozo, J.M. (1977). Clinical consequences of corticectomies involving the supplementary motor area in man. *Journal of Neurological Science, 34,* 301–314.

Lauritzen, M., Henriksen, L., & Lassen, N.A. (1981). Regional cerebral blood flow during rest and skilled movements by Xenon-133 inhalation and emission computerized tomography. *Journal of Cerebral Blood Flow and Metabolism, 1,* 385–389.

Liepmann, H. (1905). The left hemisphere and action. *Münchener medizinische Wochenschrift,* 48–49.

Liepmann, H. (1920). Apraxia. *Ergebenisse der Gesampten Medizin, 1,* 516–543.

Maher, L. (1995). *Praxis and the right hemisphere.* A doctoral dissertation presented to the Graduate School of the University of Florida.

Marteniuk, R.G., Leavitt, J.L., MacKenzie, C.L., & Athenes, S. (1990). Functional relationships between grasp and transport components in a prehension task. *Human Movement Science, 9(2),* 149–176.

Morlaas, J. (1928). *Contribution a l'etude de l'apraxie.* Paris: Amedee Legrand.

Okano, K., & Tanji, J. (1987). Neuronal activities in the primate motor fields of the agranular frontal cortex preceding visually triggered and self-paced movements. *Experimental Brain Research, 66,* 155–166.

Orgogozo, J.M., & Larsen, B. (1979). Activation of the supplementary motor area during voluntary movement in man suggests it works as a supramodal area. *Science, 206,* 847–850.

Passingham, R.E. (1993). *The frontal lobes and voluntary action.* New York: Oxford University Press.

Passingham, R.E. (1996). Functional specialization of the supplementary motor area in monkeys and humans. In H.O. Lüders (Ed.), *Advances in neurology, Vol. 70: Supplementary sensorimotor area* (pp. 105–116). Philadelphia, PA: Lippincott-Raven Publishers.

Penfield, W., & Welch, K. (1951). The supplementary motor area of the cerebral cortex. *Archives of Neurology and Psychiatry, 66,* 289–317.

Raczkowski, D., Kalat, J.W. & Nebes, R. (1974). Reliability and validity of some handedness questionnaire items. *Neuropsychologia, 12(1),* 43–47.

Rao, S.M., Binder, J.R., Bandettini, P.A., Hammeke, T.A., Yetkin, F.Z., Jesmanowicz, A., Lisk, L.M., Morris, G.L., Mueller, W.M., Estkowski, L.D., Wong, E.D., Haughton, V.M. & Hyde, J.S. (1993). Functional magnetic resonance imaging of complex human movements. *Neurology, 43,* 2311–2318.

Rizzolatti, G., Luppino, G., & Matelli, M. (1996). The classic supplementary motor area is formed by two independent areas. In H.O. Lüders, (Ed.), *Advances in neurology, Vol. 70: Supplementary sensorimotor area* (pp. 45–56). Philadelphia, PA: Lippincott-Raven Publishers.

Roland, P.E., Larsen, B., Lassen, N.A., & Skinhøj, E. (1980). Supplementary motor area and other cortical areas in organization of voluntary movements in man. *Journal of Neurophysiology, 43,* 118–136.

Roland, P.E., Skinhøj, E., Lassen, N.A., & Larsen, B. (1980). Different cortical areas in man in organization of voluntary movements in extrapersonal space. *Journal of Neurophysiology, 43,* 137–150.

Romo, R., & Schulz, W. (1987). Neuronal activity preceding self-initiated or externally timed arm movements in area 6 of monkey cortex. *Experimental Brain Research, 67*, 656–662.

Rothi, L.J.G., & Heilman, K.M. (1985). Ideomotor apraxia: Gestural learning and memory. In E.A. Roy (Ed.), *Neuropsychological studies in apraxia and related disorders* (pp. 65–74). New York: Oxford University Press.

Rothi, L.J.G., & Heilman, K.M. (1996). Liepmann (1900 and 1905): A definition of apraxia and a model of praxis. In C. Code, C.-W. Wallesch, Y. Joanette, & A.R. Lecours (Eds.), *Classic cases in neuropsychology*. Hove, UK: Psychology Press.

Rothi, L.J.G., Heilman, K.M., & Watson, R.T. (1985). Pantomime comprehension and ideomotor apraxia. *Journal of Neurology, Neurosurgery and Psychiatry, 48*, 207–210.

Rothi, L.J.G., Mack, L., Verfaellie, M., Brown, P., & Heilman, K.M. (1988). Ideomotor apraxia: Error pattern analysis. *Aphasiology, 2(3/4)*, 381–388.

Rothi, L.J.G., Ochipa, C., & Heilman, K.M. (1991). A cognitive neuropsychological model of praxis. *Cognitive Neuropsychology, 8*, 443–458.

Rothi, L.J.G., Ochipa, C., & Heilman, K.M. (1997). A cognitive neuropsychological model of limb praxis and apraxia. In L.J.G. Rothi & K.M. Heilman (Eds.), *Apraxia: The neuropsychology of action* (pp. 29–50). Hove, UK: Psychology Press.

Saling, M., Stelmach, G.E., Mescheriakov, S., & Berger, M. (1996). Prehension with trunk assisted reaching. *Behavioral Brain Research, 80(1–2)*, 153–160.

Tanji, J., & Shima, K. (1996). Contrast of neuronal activity between the supplementary motor area and other cortical motor areas. In H.O. Lüders (Ed.), *Advances in neurology, Vol. 70: Supplementary sensorimotor area* (pp. 95–104). Philadelphia, PA: Lippincott-Raven Publishers.

Thaler, D.E., Rolls, E.T., & Passingham, R.E. (1988). Neuronal activity of the supplementary motor area (SMA) during internally and externally triggered wrist movements. *Neuroscience Letters, 93*, 264–269.

Todor, J.I., & Cisneros, J. (1985). Accommodation to increased accuracy demands by the right and left hands. *Journal of Motor Behavior, 17*, 355–372.

Tzavaras, A. (1978). Les apraxies unilaterales. In H. Hecaen & M. Jeannerod (Eds.), *Du control moteur a l'organisation de geste*. Paris: Masson.

Watson, R.T., Fleet, W.S., Rothi, L.J.G., & Heilman, K.M. (1986). Apraxia and the supplementary motor area. *Archives of Neurology, 43*, 787–792.

Winstein, C.J., & Pohl, P.S. (1995). Effects of unilateral brain damage on the control of goal-directed hand movements. *Experimental Brain Research, 105*, 163–174.

Wise, S.P. (1996). Corticospinal efferents of the supplementary sensorimotor area in relation to the primary motor area. In H.O. Lüders (Ed.), *Advances in neurology, Vol. 70: Supplementary sensorimotor area* (pp. 57–70). Philadelphia, PA: Lippincott-Raven Publishers.

THE PERCEPTION OF SPATIAL RELATIONS IN A PATIENT WITH VISUAL FORM AGNOSIA

Kelly J. Murphy
Rotman Research Institute of Baycrest Centre, and University of Toronto, Ontario, Canada

David P. Carey
University of Aberdeen, Old Aberdeen, UK

Melvyn A. Goodale
University of Western Ontario, London, Canada

It is generally believed that neuropsychological patients presenting with visual agnosia, a deficit on object perception/recognition, have suffered damage to the ventral visual cortical pathway (Milner & Goodale, 1995). Rarely has the ability of such patients to perceive the spatial location of objects been investigated—perhaps because "spatial vision" is thought by some researchers to be mediated exclusively by the dorsal visual cortical pathway. Here we present data on spatial perception in a patient DF, who has a profound visual form agnosia. DF and two control subjects were required to make a copy of the spatial arrangement of a target display of five differently coloured circular tokens using a duplicate set of the same tokens. Spatial performance was analysed in two ways: (1) relative location measured the ability to reconstruct the relative spatial relations between the tokens such as left versus right, above versus below, and nearer versus farther; (2) absolute location measured the exact displacement in millimetres of each token's copied position relative to its true location. DF was able to copy some of relative location relations between the tokens although her ability to do so was not nearly as accurate as that of the control subjects. Nevertheless, DF's limited appreciation of relative location was enough to enable her to discriminate rather well between spatial patterns of tokens. She could not, however, reconstruct the absolute distance relations between the tokens and showed large displacements of token position compared to the control subjects. Interestingly, although Df was not "normal" in her ability to appreciate the allocentric spatial relations between the locations of the tokens relative to one another, she could accurately process token location egocentrically (i.e. relative to her own body and hand

---

Requests for reprints should be addressed to K.J. Murphy, PhD, Rotman Research Institute of Baycrest Centre, 3560 Bathurst Street, Toronto, Ontario, Canada M6A 2E1 (Tel: 416-7852500 x 2593; Fax: 416-7852862; E-mail: kmurphy@rotman-baycrest.on.ca)

This research was supported by Medical Research Council of Canada Grant MT-7269 and a postdoctoral fellowship from the Rotman Research Institute of Baycrest Centre. The authors thank DF and the control subjects for their willingness to participate in the investigation presented here. Thanks also to Angela Haffenden and AnnaLee Kruyer for their assistance with administration of some of the spatial analysis tasks and to Malcolm Binns for advice on statistical analyses.

position). Thus, like controls, she was perfectly able to point to and touch all the tokens in an array. These results demonstrate deficits in the ability to perceive spatial relations between objects in a patient with visual form agnosia and suggest that the ventral steam also plays a functional role in spatial vision, particularly allocentric spatial vision.

INTRODUCTION

On the basis of evidence marshalled from both human and monkey experiments, Goodale and Milner (1992; Milner & Goodale, 1993, 1995) put forward a new interpretation of the division of labour between the ventral and dorsal cortical visual pathways. The traditional account, advanced by Ungerleider and Mishkin (1982) over 15 years ago, describes the functions of the two cortical visual systems as a distinction between "object" versus "spatial" vision. In their reformulation, Goodale and Milner (1992; Milner & Goodale, 1993) advanced the argument that the distinction is perhaps more parsimoniously described as one between visual "perception" and the visual control of "action." In this new account the ventral steam of visual projections, extending from striate into inferotemporal cortex, mediates the perception of objects and their relations. The dorsal steam, extending from striate to posterior parietal cortex, mediates the visual control of actions directed at those objects.

The role of the dorsal steam in visuomotor control is now well established and even those scientists who subscribe to the spatial vision theory of dorsal stream function acknowledge the contributions of the dorsal stream to sensorimotor control (e.g. Boussaoud, Ungerleider, & Desimone, 1990; Haxby et al, 1993). Proponents of the spatial vision story view visuomotor behaviour as a component of dorsal stream function that is subsumed under the broader function of spatial vision. For those who subscribe to the spatial vision account, the primary function of the dorsal pathway is the *perception* of the spatial relations between, and locations of, objects in the environment (for review see Milner & Goodale, 1995). The view that optic ataxia (disordered reaching to visual targets that the subject can "see") is a consequence of problems in the perception of space is an example of this kind of reasoning—this is the so-called "visual disorientation" account of the disorder.

In Goodale and Milner's (1992: Milner & Goodale, 1993) original interpretation of ventral stream function, the term *perception* is used in a restricted sense to refer to our phenomenological visual experience. This use of the term perception is different from the more general meaning of perception as sensory processing. In Milner and Goodale's (1995) view the construction of a perceptual representation of the world, which can act as the foundation for cognitive operations, is the domain of mechanisms in the ventral stream of visual processing. In this context, the processing of the spatial relations of objects (or the parts of objects) is as central to perception as is the processing of the intrinsic features of the objects.

In contrast to the perception of spatial layout, provided by the ventral stream, the computation of spatial location, carried out by the dorsal stream, is entirely related to the guidance of specific visuomotor actions, such as grasping an object, locomoting around obstacles, or gazing at different objects in a scene. As a consequence, the dorsal stream mechanisms do not compute the "allocentric" location of a target object, i.e. its location relative to other objects in the scene, but rather the "egocentric" coordinates of the location of the object with respect to the observer. Indeed, the egocentric coordinates are in the particular frame of reference for the action to be performed. For example, to control a grasping movement the dorsal stream must eventually transform visual information about the object into arm- and hand-centred coordinates. To reiterate, allocentric spatial information about the layout of objects in the visual scene is computed by the ventral stream mechanisms, which mediate perception, while precise egocentric spatial information about the location of an object in a body-centred frame of reference is computed by the dorsal stream mechanisms, which mediate the visual control of action. This viewpoint represents a marked departure from traditional spatial vision accounts, which appear to assume that both kinds of spatial analysis (allocentric and egocentric) are the domain of the dorsal stream.

Consider the following argument. Perceptual systems are biased towards stability: Objects in the world retain their identity and their spatial relations in spite of dramatic changes in the retinal stimuli that are produced when, for example, the observer moves from one viewpoint to another. Thus, our coffee cup looks like a coffee cup independent of viewpoint, and appears to remain in the same place on the desk in spite of dramatic rotations about the axes of our favourite swivelling chair. Visuomotor systems, on the other hand, must compute the precise location of objects in the environment in a metrically accurate fashion with respect to the effector being used. Information about a coffee cup's relative position on the corner of the desk is not adequate for guiding an accurate and efficient grasping movement to a cup's handle. The visuomotor system must have the position of the cup in precise body-centred coordinates. In short, both dorsal and ventral visual systems compute information about spatial location, but in very different ways.

Goodale, Milner, and colleagues have conducted experiments with neurological patients with either dorsal stream (e.g. optic ataxia) or ventral stream (e.g. visual agnosia) damage (e.g. Carey, Harvey, & Milner, 1996; Dijkerman, Milner, & Carey, 1996; Goodale et al., 1994; Jakobson, Archibald, Carey, & Goodale, 1991; Milner et al., 1991; Murphy, Racicot, & Goodale, 1996). For example, the visual form agnosic patient DF, who has been extensively studied by this group, can produce well-calibrated grasping movements towards objects that she cannot visually discriminate. This result suggests that object attributes (i.e. dimensions or shape) are processed by two relatively independent visual mechanisms, one specialised for the perception of objects and the other specialised for the control of actions directed at objects. This work with DF led Goodale and Milner (1992; Milner & Goodale, 1993) to hypothesise that both the dorsal and ventral stream use information about object size, shape, and orientation, but that the two streams use this visual information in different ways and for different purposes.

Although the work with DF, briefly described here, provides support for differential processing of object features for perception and action, it says nothing about how spatial information might be handled by perceptual and action systems. In fact there are only a handful of studies that have reported on both allocentric and egocentric spatial processing abilities in brain-damaged patients presenting with visual disturbances. One such study was conducted by Perenin and Vighetto (1988) on patients with optic ataxia following damage to the human homologue of the dorsal stream. Perenin and Vighetto noted that even though their optic ataxic patients were unable to reach accurately to objects presented in different positions in their contralesional visual field they could nonetheless make normal judgements about the relative locations of these same objects in their affected visual field. In other words, Perenin and Vighetto's patients could use spatial information for the purposes of making perceptual judgements about the locations of objects in their environment but they could not accurately use spatial information about object location for the purposes of making accurate visually guided movements. These optic ataxic patients are thought to have damage to the dorsal stream system but not the ventral stream. DF, on the other hand, is presumed to have an intact dorsal but damaged ventral cortical visual system, based upon Goodale and colleague's examination of her visuomotor control abilities and her visual object perception impairments. How, then, might one expect someone like DF to perform on tasks of spatial perception? If one subscribes to the "spatial vision" theory of the visual functions of the dorsal stream, then DF should do well at tasks requiring her to analyse the spatial relations among objects in the environment. In contrast, a strict interpretation of Milner and Goodale's (1995) theory about ventral steam function leads one to anticipate that DF would show deficits in her ability to perceive the relative location of objects in a visual scene even though she can use information about their location to direct her actions towards them.

In the current investigation we examined DF's ability to use spatial information about a small group of objects in order to make decisions about where those objects were located relative to her (egocentric) and relative to one another (allocentric) in space. For these tasks we presented small groups of differently coloured circular tokens that could easily be discriminated on the basis of colour but which could not be distinguished on the basis of shape or edge-based cues. Thus, relatively pure spatial responses to the group of tokens could be examined. In our egocentric task DF was required to point to all available tokens in the target array in a specified sequence. In our allocentric task DF was required to copy the spatial arrangement of tokens in the target array as precisely as possible. In addition, DF was asked to perform an allocentric judgement task in which she was asked to make same–different judgements about the spatial configuration of tokens in different pairs of token arrays. Given the fact that DF appears to show normal visually guided movements, which require, egocentric spatial processing, we expected that she would show normal sensitivity to the spatial locations of all of the tokens in a given array when she had to point to them. We expected, however, that she would do poorly when the tasks required allocentric frames of reference and she was asked to process information about the positions of targets relative to one another.

METHODS

Subjects

The patient DF. DF is a right-handed woman of 41 years of age who suffered brain damage as a result of accidental carbon monoxide poisoning 8 years ago. The accident left DF with a profound visual form agnosia. Magnetic Resonance Imaging (MRI) of her brain 1 year after the accident indicated that the densest area of cortical damage was in the ventral portion of the lateral occipital region (areas 18 and 19 bilaterally), with apparent sparing of primary visual cortex (area 17) (for further neuroanatomical details see Milner et al.,

1991). Although DF has relatively intact basic visual function (e.g. visual acuity, flicker fusion, colour perception, visual fields out to 30° are normal; Milner et al., 1991), she has great difficulty visually recognising or identifying objects on the basis of their size, shape, or orientation. Interestingly, she is adept at using these same object properties to make accurate visually guided reaching and grasping movements (e.g. Goodale, Milner, Jakobson, & Carey, 1991). Her preserved visuomotor abilities, and MRI findings, indicate that the dorsal stream of projections travelling from V1 into posterior parietal cortex must be intact. In contrast, the pattern of impaired visual abilities in DF, coupled with the locus of greatest damage as seen on MRI, indicates that she has damage close to the origin of the ventral stream of visual cortical projections extending from V1 into inferotemporal cortex[1].

DF's impaired visual recognition abilities cannot be explained on the basis of anomia, because she has no trouble identifying objects haptically. Moreover, her difficulties in visual perception cannot be accounted for by an inability to comprehend or remember task instructions because her language and memory function is well within normal range (Milner et al., 1991). Praxis to verbal command is also intact and, as mentioned previously, Goodale and colleagues have repeatedly demonstrated her intact visuomotor behaviour (see Milner & Goodale, 1995, for review).

---

[1] Recent functional MRI confirms ventral stream damage in DF. Unlike control subjects, DF does not show inferotemporal activation during object perception despite having normal patterns of V1 activation to visual stimulation (T. James & M.A Goodale, personal communication of an unpublished observation, November 1997).

Control subjects. Two control subjects matched for sex, handedness, and approximate age were also tested. Control subjects (C1 and C2) were aged 40 and 42 years and had no history of visual, neurological, or psychiatric disturbance. The participation of all subjects was voluntary.

Apparatus

Two sets of five differently coloured (yellow, green, blue, black, and red) wooden circular tokens, each measuring 3cm in diameter and 0.9cm in height, were used. Two identical sets of 20 white paper target templates measuring 27.5×27.5cm were also used. These target templates contained the spatial coordinates for the tokens. Four of the templates contained the position coordinates for one, two, three, and four tokens respectively, The remaining 16 templates were test templates and each contained the position coordinates for five tokens. Eight of the 16 test templates contained "linear" position coordinates for the tokens (i.e. the tokens were placed in a straight line). Relative to the subject's viewpoint, two of the eight linear displays contained position coordinates forming a horizontal line, two formed a vertical line, two formed a diagonal line extending left to right, and two formed a diagonal line extending right to left. In the first four of the eight linear arrays the tokens were arranged equal distances apart. In the second four linear displays the tokens were arranged at unequal distances apart. The remaining eight templates contained "non linear" position coordinates for the tokens (i.e. the pattern of tokens was not orderly and the tokens appeared to be positioned randomly relative to one another). All token distances were unequal in the nonlinear displays. Forty blank sheets of white paper with the same dimensions as the templates (27.5×27.5cm) were used as response backgrounds upon which the positions of the tokens arranged by the subjects could be recorded and later measured. A Sony ccd-F401 video camera, (Sony, Tokyo) was used to record subject performance.

Procedure

Because DF has difficulties in making perceptual judgements about the identity and orientation of objects we chose stimuli that were devoid of these properties. The objects we chose were circular tokens, all of which had the same dimensions but which could easily be distinguished on the basis of colour. Prior to conducting the spatial analysis experiments with DF it was established that she could distinguish each of the 5 tokens (in a given set) by colour, at 100% accuracy. In addition, it was established that DF showed no evidence of simultagnosia. She could report the number of tokens arranged in front of her whether or not the display contained 1, 2, 3, 4 or all 5 tokens with 100% accuracy. It was important to use stimuli that she could reliably discriminate from one another; otherwise, any potential errors in her performance could be accounted for by nonspatial processes. It was also important that the objects did not possess identity and orientation information in order to ensure that, aside from colour, the primary visual attribute of the target was its position in space.

Both the target array (positioned on a square white background) and a blank white background of the same dimensions as the background of the target array were presented together on a black tabletop 120cm wide and 75cm deep. On a given trial one of the square backgrounds was positioned 1cm to the left of the subject's sagittal axis and the other 1cm to the right. The leading horizontal edge of both backgrounds was 15cm from the edge of the table nearest the subject. The left–right position of the target array background and the blank response background was completely counterbalanced across trials. On a given trial the subject was instructed to close her eyes while the experimenter set up the token array. The subject opened her eyes upon verbal command and proceeded with the spatial analysis task.

The experiment involved three tests of spatial analysis. The first was an egocentric spatial pointing test, which involved having the subject point to and touch each token in the target array according to a sequence order specified by the experimenter. The number of items in the sequence was always the number of items (five) in the array. This was followed by an allocentric spatial copy test that involved having the subject copy the spatial configuration of tokens in the target array, using a duplicate set of tokens, onto the blank response background positioned to the left/right of the square background containing the target token array. Both the egocentric pointing test and the allocentric spatial copy test were administered together on two separate test occasions. Thus, there were 2 trials for each of the 16 test templates conducted on 2 separate days. During the second test session 2 trials of 4 additional token arrays, containing less than 5 tokens, were also administered. These 4 token arrays contained 3, 2, 1 and 4 tokens respectively and were administered once at the start of the second session and again at the end of the second session after the 16 test arrays had been completed. These four arrays of fewer than five tokens were administered in order to determine if performance at reproducing the spatial configuration of token(s) in a target array would be improved if fewer tokens were employed. Finally, an allocentric spatial judgement test was administered. This last task required subjects to visually examine two token arrays, positioned side by side, and to state whether or not the arrangement of the coloured tokens was the same in the two displays or different.

Egocentric spatial pointing test. In order to determine that subjects could spatially locate each item in the array egocentrically, that is relative to their own body position, they were required to reach out with their right hand and point to and touch, with their index finger, each item in the test array according to a sequence order specified by the experimenter. After the experimenter named out loud all five tokens by colour, the subject's task was to execute one pointing movement to each token in the array according to the order in which each colour had been named. Pointing accuracy was grossly assessed as one correct contact between the subject's index finger and each token in the array in the order specified by the experimenter. The ability to perform this exercise established the following: (1) that subjects could egocentrically localise all tokens in the array; and (2) that they could see and attend to all of the token items in the array. Pointing performance was videorecorded and the time taken to execute the pointing sequence was measured in seconds.

Allocentric spatial copy test. Following the egocentric pointing test subjects were instructed to use the duplicate set of tokens provided to produce an exact copy of the spatial position of the tokens in the target array onto the white response background. Subjects were told to try and position their tokens on the blank response background such that they matched the token positions of the target array. No time limit was imposed on how long subjects could take to produce their copy. Performance was videorecorded. The time it took to complete each copy was measured in seconds by the experimenter. Timing began with the placement of the first token and ended when the subject verbally indicated that they had finished constructing their copy. Once the subject had completed their copy the experimenter recorded the position of each token on the response background by tracing the outline of each token with a pencil and identifying the token according to its colour by placing the first one or two letters of the colour name of the token within its traced outline.

Two types of measures, *relative location* and *absolute location*, were utilised to assess the accuracy of subject's token display copies/reproductions. The relative location measures were similar in principle to Kosslyn's (1987) concept of "categorical space relations" in which the relative positions of objects or parts of an object are analysed in terms of imprecise relations such as above or below, to the left or right, closer or further. The absolute location measure was similar to Kosslyn's concept of "coordinate space relations" in which the precise distance relations between objects or parts of an object are analysed. Thus, we were able to use our allocentric spatial copy test to examine both kinds of spatial perception abilities in our subjects. (Note, however, that both the relative and absolute location measures used are measures of allocentric spatial perception.)

Relative location accuracy was examined using two measures, ordinal position and proximal (or nearest neighbour) position. Ordinal position was examined by determining how accurately subjects reproduced the order of the array from left to right and from top to bottom. Each reproduction was scored for both horizontal and vertical orders with the exception of the four instances where the templates for the copies were only horizontal or vertical. Each horizontal and/or vertical sequence was scored out of four. Take the case of a sequence of items in a horizontal target display ordered as follows: Yellow to Blue, Blue to Green, Green to Red, and Red to Black. If the subject reproduced exactly the same horizontal sequence, she would receive a score of 4 out of 4 possible points. If she reproduced the horizontal sequence, Blue to Yellow, Yellow to Red, Red to Green, and Green to Black, where two of the steps in the sequence were correct and two were incorrect, the score would be 2/4. For example, in Fig. 1a, DF received a score of 4/4; in Fig. 1b, a score of 3/4; and in Fig. 1c, a score of 2/4 for the horizontal component and a score of 2/4 for the vertical component. The examples of copies produced by the normal controls, depicted in Fig. 1d–i, received scores of 4/4 on all of the above measures. As stated earlier, 2 trials were run for each of the 16 test arrays. A perfect cumulative score would be 224 (12 displays with horizontal and vertical components multiplied by 2 trials multiplied by 8 possible points equalled 192 points and 4 displays with only a horizontal or vertical component multiplied by 2 trials multiplied by 4 possible points equalled 32 points).

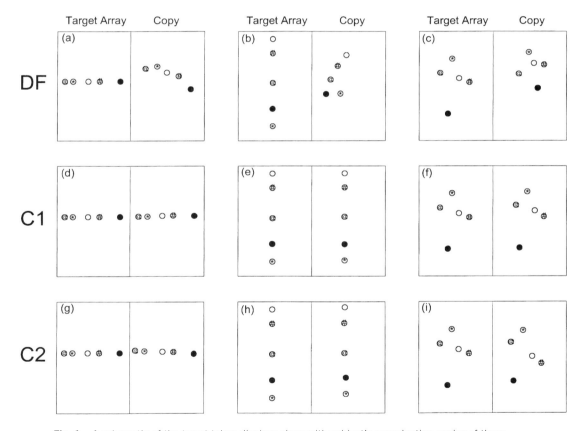

Fig. 1. A schematic of the target token displays along with subject's reproduction copies of those same displays. The three trials depicted in this figure were chosen because they represent the range of performance in DF. Columns one and two show examples of linear target arrays and column three shows an example of a nonlinear target array.

Relative proximal position (e.g. the relative distance relations between tokens in the array) was examined for each of the 2 trials of the 12 test displays in which the distances between tokens in the arrays were unequal. This included the four linear displays and eight non-linear displays. The identity and number of nearest neighbour pairs was determined for each test array. This involved measuring the distances between the tokens in a given test array to determine which pairs of tokens were closest together and then repeating this procedure for subjects' copy arrays. The subjects' copy arrays were then scored according to the number of nearest neighbour pairs produced that were identical in kind to those present in the test array. Five of the test displays contained only three nearest neighbour pairs (e.g. the target arrays depicted in columns one and two of Fig. 1) and seven contained four possible nearest neighbour pairs (e.g. the target arrays depicted in column three of Fig. 1). Thus, the maximum number of points that could be achieved on 2 trials of these 12 test arrays was 86. In the examples provided in Fig. 1, DF scored 1/3 for her copy in Fig. 1a; 2/3 for her copy in Fig. 1b: and 2/4 for her copy in Fig. 1c. The normal control subjects both scored 3/3, 3/3, and 3/4 respectively on their reproductions of the same target arrays (see Fig. 1d–i).

Absolute location was assessed by measuring the vector displacement of each token in the subject's array copy with respect to its true position in the target array. This was accomplished by simply marking the centre of each token's true position on the response background containing the traced locations of where subjects had positioned the tokens in their copy. The distance of the centre of each traced token to the location of its true centre was measured in millimetres.

Allocentric spatial judgement test. In this final task the examiner arranged 2 token displays side by side using 2 of the target templates from the 2 sets of 16 test arrays. Once the displays were in position the subject was instructed to open her eyes and to look carefully at both token arrays and state whether or not the arrangement of the coloured tokens was the same on the 2 displays or different. There were 32 trials in total. On 16 of the trials the 2 token displays were the same and on 16 they were different. Same and different trials were randomised. Responses were scored as number of correct judgements out of 32.

RESULTS

Egocentric Spatial Pointing Test

DF and the two control subjects, C1 and C2, had no difficulty pointing directly towards and touching each token in the arrays. Further, none of the subjects had any difficulty remembering the sequence in which they were instructed to touch each of the five tokens in a given display (with the minor exception that both DF and C1 made 1 sequence error out of 32 trials and C2 made 3). Thus, DF, like the control subjects, had no difficulty attending to and egocentrically locating all five token items in a target array. DF, though no different from controls in her ability to localise each token in a given array, tended to be slightly longer in completing her pointing movement sequence ($M = 4.3$sec; SD = 1.2sec) than the two control subjects (C1, $M = 2.2$sec; SD = 1.0sec; C2, $M = 3.0$sec; SD = 0.8sec). (Due to equipment malfunction the last 8 of the 16 trials of the second session were not videorecorded and pointing time could not be measured from the videotape for these trials for control subject C1).

Allocentric Spatial Copying Test

Relative location for both ordinal and proximal position spatial relations was examined using a series of proportions analyses in which a z score was computed to examine the difference in the *proportions* of overall correct token placements, between DF and the control subjects, relative to the inherent variability associated with the proportion of hits (number of correct token placements). More specifically, the z score was calculated based on the difference between the independent proportions of obtained points (hits), out of the total number of possible points, achieved on each of the subjects' reproduction copies according to the scoring criteria outlined above. These z scores represent a quantification of the degree of difference between the performance of DF and the two control subjects. As Fig. 2a shows, DF was considerably poorer than controls at

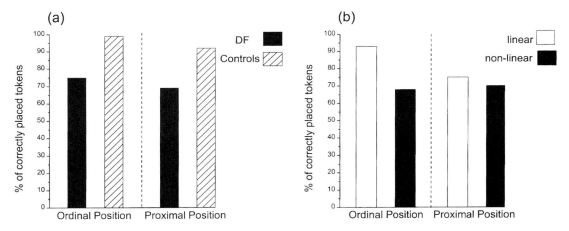

Fig. 2. The performance of DF and the control subjects on two measures of relative location accuracy: ordinal position, which represents the accuracy with which the horizontal (i.e. left to right) and vertical (i.e. top to bottom) organisational sequence of the tokens was reproduced; and proximal position, which represents the accuracy with which distance relations (i.e. closer versus further) were preserved in the subject's reproductions. (a) DF was considerably poorer ($z = -7.2$ on ordinal; $z = -4.3$ on proximal) than control subjects at reproducing the relative spatial relations between the tokens. (b) DF was more accurate at reproducing ordinal position when the target displays were linear (i.e. tokens formed an orderly straight line) as opposed to nonlinear (i.e. tokens formed a random pattern) ($z = 4.9$). There was little difference between DF's copies of linear versus nonlinear target arrays with respect to her ability to reproduce proximal position.

reproducing both relative ordinal position ($z = -7.2$) and relative proximal position ($z = -4.3$).

Reproducing the ordinal and proximal token positions was a trivial task for controls who achieved near-ceiling performance on both these measures (see Fig. 2a). To understand exactly what the spatial attributes of the arrays were that posed difficulty for DF, we examined her accuracy at reproducing relative location in the linear versus the nonlinear arrays. DF did achieve better accuracy ($z = 4.9$) at reproducing the ordinal position in the linear as compared to the nonlinear arrays. A slight trend in the same direction was noted in her ability to reproduce the proximal spatial relations in the linear versus the nonlinear displays (see Fig. 2b). No obvious differences were noted in DF's ability to reproduce vertical versus horizontal ordinal sequences in her reproductions of the target displays.

Consistent with the relative location results, absolute location measures revealed that DF was considerably less accurate than the control subjects at reproducing the true location of the tokens in her array copies (see Fig. 3). Indeed, when the standardised difference between means was computed across test sessions, DF's mean performance was found to be *six* standard deviations away from the overall mean of the two control subjects.

As Fig. 3 shows, there was no difference in absolute location performance across test sessions (the same was true for relative location performance). In order to examine for practice effects, paired t-tests were conducted to compare sessions one versus session two performance for each individual subject and to compare linear versus nonlinear reproduction performance. None of these t-tests proved significant in any of the subjects. No difference was observed in the magnitude of absolute displacement error for the extra trials conducted in the second session using displays with fewer tokens (i.e. the four displays containing either one, two three or four tokens) for either DF or the normal control subjects.

Control subject one (C1) was slightly more accurate at reproducing absolute location than C2. One reason C1 was a little better than C2 on this measure may have been because C1 took longer ($M = 45.6$sec; SD = 21.4sec) to complete her copy arrays than C2 ($M = 15.6$sec; SD = 4.7sec). DF took the greatest amount of time to complete her array copies ($M = 70.4$sec; SD = 23.2sec); however, this extra time did not appear to improve her accuracy.

Allocentric Spatial Judgement Test

The relative location results indicated that DF can to some extent appreciate the spatial relations between objects and in particular their ordinal positions (see Fig. 2a,b). We believe that her ability to perceive ordinal position may account for why she did so well at judging whether or not the spatial configuration of two token arrays was the same or different. DF made 29 correct judgements out of 32 trials. This spatial judgement task was trivial for someone with normal object perception abilities and both control subjects easily obtained perfect scores. Unfortunately, performance on this task was not videorecorded and the time taken to make the judgements for each pair of

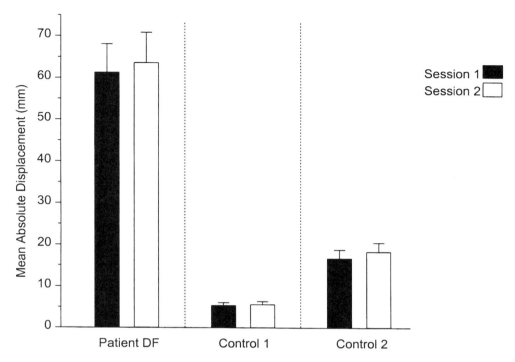

Fig. 3. The performance of DF and the two control subjects on the absolute location accuracy of their token placements in their array copies. The absolute distance between each token in the subject's reproduction copies and the token's true spatial location was measured in millimetres. DF was far less accurate (6 SD different) than control subjects at reproducing the true/absolute spatial locations of the tokens. No practice effects in location accuracy were noted between the first and second test administrations. Error bars represent standard error of the mean.

spatial arrays was not measured. Qualitative impression was that DF took considerably longer to render her same/different judgement decision than did the normal controls, who made their judgements immediately after the arrays were presented.

We re-examined the stimulus pairs used in the allocentric spatial judgement test and determined that the ability to appreciate even one part of the ordinal sequence position information in the arrays would be likely to provide enough cues to confidently render a correct decision. The colour cues provided by the different tokens could have allowed DF to perceive differences between the two arrays without any appreciation of their spatial patterns. The judgement task did not require subjects to analyse the relative proximal (distance) relations between the tokens in the array and it could be solved based upon ordinal cues alone. In other words, we had made the task too easy for DF. One of us (DPC) has recently begun to explore the results of this same–different judgement task with DF a little further.

These efforts have indicated that when the ordinal relations in the two arrays are the same but the proximal relations are different, DF is at chance when making her same–different decision. We plan to explore this preliminary finding in future experiments examining spatial perception abilities in DF.

DISCUSSION

DF showed no difficulty whatsoever in pointing to each of the targets in turn. Except for the fact that she was slightly slower, her performance was indistinguishable from the control subjects. This means that DF can use egocentric spatial coding of target location to control visuomotor acts such as manual aiming movements. Her normal visuomotor performance on this task is consistent with her past performance on a broad range of visually guided behaviours (for review, see Milner & Goodale, 1995).

Her normal performance on a task where egocentric coding is demanded can be contrasted with her poor performance on tasks where allocentric coding must be used. This is clearly evident in the fact that she shows large displacements of the position of objects with respect to one another when she attempts to reproduce the target arrays. Her copies look quite different from the actual target patterns she was presented with. Nevertheless, DF is not completely devastated in her ability to make spatial judgements. In general, she seems capable of discriminating some of the relative (ordinal and proximal) relations between objects in space, albeit not nearly as well as control subjects. Further, her limited appreciation of relative spatial relations, especially ordinal position, appears to provide her with enough information to discriminate between different spatial patterns. Her deficit is most evident when she tries to reproduce the patterns. But again it is important to emphasise that DF's poor performance in spatial perception cannot be ascribed to an over-arching deficit in "spatial vision" since she has no problem directing her hand to any of the targets. Her problem is with allocentric, not egocentric, coding of spatial position.

It is important to note that DF's poor performance on the copying task cannot be explained by suggesting that this task was more "cognitively challenging" than the other tasks that we used. The copying task, for example, did not require much working memory, since the model was always available. In contrast, the pointing task, one which DF did as well as normal subjects, did put a load on working memory, since subjects were required to point to the tokens in a predetermined order.

Although DF's copies tended to be displaced slightly towards the top half of the square white background, this shift in placement is unaccompanied by any evidence for a neglect of or inattention to the bottom part of the target display. She always copied the entire array of five tokens. More importantly, she reproduced the vertical arrangement of the tokens as well (or as poorly) as she did the horizontal.

It is generally believed that patients with a visual agnosia (like DF) have suffered damage to the ventral visual cortical pathway (Milner & Goodale, 1995). Rarely has the ability of such patients to perceive the spatial location of objects been investigated. This may be the case for two reasons. First, "spatial vision" is thought by some researchers to be mediated exclusively by the dorsal visual cortical pathway. Second, the severity of the object perception/recognition deficits overshadows any possible observations of deficits in space perception, particularly when there is an absence of visuomotor disturbance. Neurological patients who present with spatial perceptual impairments as a primary feature (i.e. patients with hemispatial neglect, topographical agnosia, dressing dyspraxia, and constructional apraxia) typically have damage in the parietal lobe, in particular the inferior parietal region. We have not observed any of the classically described visuospatial impairments, often reported after parietal lobe damage, in DF's behaviour and this is consistent with the fact that her MRI scan does not show evidence of parietal lobe injury. Nonetheless, we feel it is important to address the issue of constructional apraxia (i.e. the inability to perform familiar movement sequences when engaged in the act of making or preparing something) and explain why we feel that DF's difficulty at reproducing the allocentric location of an object cannot be accounted for by appealing to some sort of constructional apraxic deficit. First, DF's drawings of objects from memory are remarkably intact and do not show any evidence of constructional apraxia (Servos, Goodale & Humphrey, 1993). Second DF, can perform image construction tasks and is able to use visual imagery to manipulate objects in order to fashion new objects in her "mind's eye" just as well as normal subjects (Servos & Goodale, 1995). In short, DF does not have constructional apraxia; but she does have a fundamental deficit in allocentric spatial coding.

As outlined in the Procedure, our relative and absolute location measurements for evaluating allocentric spatial perceptual abilities in our subjects share some features in common with Kosslyn and colleagues' categorical and coordinate concepts of spatial perception (Kosslyn, 1987; Kosslyn, Chabris, Marsolek, & Koening, 1992; Kosslyn et al., 1989). Kosslyn et al. identify two broad classes of spatial processing: categorical, which can be associated with our relative location measurements because it is concerned with the relative positions of targets in space (i.e. closer versus further, above versus below, left versus right); and coordinate, which can be associated with our absolute location measurement because it is concerned with the precise metric relations between objects. However, an important distinction between our relative and absolute location measurements and the concepts of categorical and coordinate spatial processing is that our location measurements are restricted to allocentric coding only and do not include egocentric coding. Bearing this distinction in mind, Kosslyn and colleagues' categorical and coordinate distinction can provide a useful framework within which to view our allocentric spatial copy task. In these terms, our allocentric task requires relatively intact categorical spatial processing to solve the ordinal and proximal aspects of the relative location problem. DF is unable to copy the relative locations of the targets with the same degree of accuracy as the normal control subjects—although, by adopting a slavish one-by-one element approach when making her copies of the target arrays, she seems to have encoded some of the categorical relations amongst elements in the target array. Nevertheless, to produce accurate copies, subjects also had to solve the absolute location problem and make coordinate or metrical transformations about target positions on the model and the copy. (It is important to note again here that, even though the judgements involve metrical scaling, they are still fundamentally allocentric). One strategy for making these judgements would be to compare the distance vectors between each pair of elements; another would be to compare vectors of each element with the edges of the array. DF seemed incapable of using either of these two strategies.

Regardless of the exact interpretation of the nature of the spatial deficits, DF's spatial abilities should be intact if all forms of visuospatial processing, and not just visuomotor control, depended on mechanisms in the dorsal stream. The results of these investigations demonstrate that this is clearly not the case. Instead, as argued elsewhere (Goodale & Milner, 1992; Milner & Goodale, 1995), the characterisation of dorsal stream function as visuomotor rather than visuospatial predicts the obtained data.

As was mentioned in the Introduction, other accounts of single cases support the dissociation of visuospatial perceptual encoding from visuomotor encoding. Recently, Stark, Coslett, and Saffran (1996) described a patient, with normal visual object recognition abilities, who was impaired at making spatial judgements about the locations of visual targets but who could accurately localise these targets with a pointing movement. Balint's patient had poor visuomotor control when he used his right hand but not his left; a finding that is hard to reconcile with the idea of an over-arching visuospatial deficit (Harvey & Milner, 1995). Perenin and Vighetto (1988) have argued that poor manual localisation of targets in patients with dorsal stream damage is largely unrelated to the degree of visuospatial dysfunction measured in perceptual tests.

DF's problems with allocentric spatial perception, while clearly evident, are not as profound as her deficits in object or face recognition. Carey, Dijkerman, and Milner (in press) have found that DF's estimates of target distance were quite poor compared to controls, particularly with monocular vision, but were nevertheless correlated with the target's true distance. At the same time, of course, her visuomotor performance with respect to the distance of these objects (as estimated by peak velocity distance scaling) was indistinguishable from control subjects. As seen in our allocentric tasks here, the perceptual spatial judgements reveal capacities that appear to be somewhere between the profound deficits in object perception and the remarkably intact visuomotor performance. Further investigations will be necessary to uncover whether or not these intermediate levels of performance are due to strategies that exploit her spared visuomotor skills, or depend on less damaged cortex in zones between occipitotemporal and superior parietal cortex. It has been suggested by Milner and Goodale (1995), for example, that the more inferior regions of the posterior parietal lobe may contain areas where the dorsal and ventral streams interact. This region of confluence between the dorsal and ventral systems may contain mechanisms that mediate operations in which object-based and allocentric representations interact with egocentric frames of reference. These mechanisms could support visuocognitive operations such as mental rotation and action planning (Milner & Goodale, 1995; Turnbull, Carey, & McCarthy, 1997). Whatever the neural substrates of allocentric coding might be, the results of this investigation make it clear that such coding is separate from the egocentric spatial coding mediating the control of skilled movements—and point to the involvement of ventral stream mechanisms in the perception of spatial relations of objects as well as in the perception of their intrinsic features.

REFERENCES

Boussaoud, D., Ungerleider, L.G., & Desimone, R. (1990). Pathways for motion analysis: Cortical connections of the medial superior temporal and fundus of the superior temporal visual areas in the macaque. *Journal of Comparative Neurology, 296,* 462–495.

Carey, D.P., Dijkerman, H.C., & Milner, A.D. (in press). Perception and action in depth. *Consciousness and Cognition.*

Carey, D.P., Harvey, M., & Milner, A.D. (1996). Visuomotor sensitivity for shape and orientation in a patient with visual form agnosia. *Neuropsychologia, 34,* 329–337.

Dijkerman, H.C., Milner, A.D., & Carey, D.P. (1996). The perception and prehension of objects oriented in the depth plane. I. Effects of visual form agnosia. *Experimental Brain Research, 112,* 442–451.

Goodale, M.A., Meenan, J.P., Bulthoff, H.H., Nicolle, D.A., Murphy, K.J., & Racicot, C.I. (1994). Separate neural pathways for the visual analysis of object shape in perception and prehension. *Current Biology, 4,* 604–606.

Goodale, M.A., & Milner, A.D. (1992). Separate visual pathways for perception and action. *Trends in Neurosciences, 15,* 20–25.

Goodale, M.A., Milner, A.D., Jakobson, L.S., & Carey, D.P. (1991). A neurological dissociation between perceiving objects and grasping them. *Nature, 349,* 154–156.

Harvey, M., & Milner, A.D. (1995). Bálint's patient. *Cognitive Neuropsychology, 12,* 261–281.

Haxby, J.V., Grady, C.L., Horwitz, B., Salerno, J., Ungerleider, L.G., & Mishkin, M. (1993). Dissociation of object and spatial visual processing pathways in human extrastriate cortex. In B. Gulyás, D. Ottoson, & P.E. Roland (Eds.), *Functional organisation of the human visual cortex* (pp. 329–340). Oxford: Pergamon Press.

Jakobson, L.S., Archibald, Y.M., Carey, D.P., & Goodale, M.A. (1991). A kinematic analysis of reaching and grasping movements in a patient recovering from optic ataxia. *Neuropsychologia, 29,* 803–809.

Kosslyn, S.M. (1987). Seeing and imagining in the cerebral hemispheres: A computational approach. *Psychological Review, 94,* 148–175.

Kosslyn, S.M., Chabris, D.F., Marsolek, C.J., & Koening, O. (1992). Categorical versus coordinate spatial relations: Computational analyses and computer simulations. *Journal of Experimental Psychology: Human Perception and Performance, 18,* 562–577.

Kosslyn, S.M., Koening, O., Barrett, A., Cave, C.B., Tang, J., & Gabrieli, J.D.E. (1989). Evidence for two types of spatial representations: Hemispheric specialisation for categorical and coordinate relations. *Journal of Experimental Psychology: Human Perception and Performance, 15*, 723–735.

Milner, A.D., & Goodale, M.A. (1993). Visual pathways to perception and action. In T.P. Hicks, S. Molotchnikoff, & T. Ono (Eds.), *Progress in brain research, Vol. 95*. Amsterdam, Elsevier.

Milner, A.D., & Goodale, M.A. (1995). *The visual brain in action* Oxford: Oxford University Press.

Milner, A.D., Perrett, D.I., Johnson, R.S., Benson, O.J., Jordan, T.R., Heeley, D.W., Bettucci, D., Mortara, F., Mutani, R., Terazzi, E., & Davidson, D.L.W. (1991). Perception and action in "visual form agnosia". *Brain, 114*, 405–28.

Murphy, K.J., Racicot, C.I., & Goodale, M.A. (1996). The use of visuomotor cues as a strategy for making perceptual judgements in a patient with visual form agnosia. *Neuropsychology, 10*, 396–401.

Perenin, M.T., & Vighetto, A. (1988). Optic ataxia: A specific disruption in visuomotor mechanisms. I. Different aspects of the deficit in reaching for objects. *Brain, 111*, 643–674.

Servos, P., & Goodale, M.A. (1995). Preserved visual imagery in visual form agnosia. *Neuropsychologia, 33*, 1383–1394.

Servos, P., Goodale, M.A., & Humphrey, G.K. (1993). The drawing of objects by a visual form agnosic: Contribution of surface properties and memorial representations. *Neuropsychologia, 31*, 251–259.

Stark, M., Coslett, H.B., & Saffran, E.M. (1996). Impairment of an egocentric map of locations: Implications for perception and action. *Cognitive Neuropsychology, 13*, 481–523.

Turnbull, O.H., Carey, D.P., & McCarthy, R.A. (1997). The neuropsychology of object constancy. *Journal of the International Neuropsychological Society, 3*, 288–298.

Ungerleider, L.G., & Mishkin, M. (1982). Two cortical visual systems. In D.J. Ingle, M.A. Goodale, & R.J.W. Mansfield (Eds.), *Analysis of visual behaviour*. Cambridge, MA: MIT Press.

EVIDENCE FOR MOVEMENT PREPROGRAMMING AND ON-LINE CONTROL IN DIFFERENTIALLY IMPAIRED PATIENTS WITH PARKINSON'S DISEASE

Catherine L. Reed
University of Denver, CO, USA

Ian M. Franks
University of British Columbia, Vancouver, Canada

Investigated were those aspects of motor planning and execution underlying movement dysfunction in patients with Parkinson's disease (PD). Specifically examined was the effect of disease severity on these processes. An experiment is reported that dissociates preprogramming processes from on-line programming processes in a simple motor task that varies in movement complexity. Dependent measures included reaction time, movement time, as well as kinematic measures of peak velocity, peak acceleration, peak deceleration, and their respective time values, plus inter-trial variability and EMG activation. While PD patients as a whole were able to pre-program movements, inter-trial variability for these measures was increased for more severely affected PD patients. Nonetheless, evidence for on-line programming occurred for all PD patients in later intervals of more complex movements. Further, EMG impulses correspond with acceleration trace deviations. The data as a whole support the hypothesis that disrupted basal ganglia function influences the consistency of cortical activation and the selection of motor program components.

INTRODUCTION

In this paper we investigate motor planning and programming, aspects of motor control thought to be cognitive in nature, in differentially impaired patients with Parkinson's disease. Research has indicated that motor planning is strongly associated with premotor

---

Requests for reprints should be addressed to Catherine L. Reed, Department of Psychology, University of Denver, 2155 S. Race St., Denver, CO 80208, USA (E-mail: creed@du.edu).

The authors gratefully acknowledge the help of Dr. Donald Calne, Susan Calne, and the Parkinson's Association of Vancouver for sharing their knowledge and for helping to recruit participants. We also wish to thank Paul Nagelkerke for his technical wizardry, and members of the University of British Columbia Motor Learning Laboratory for their technical and intellectual contributions.

The authors were supported, in part, by the Social Science Foundation of the University of Denver and the Natural Sciences and Engineering Research Council of Canada.

cortical areas, including the supplementary motor area (SMA), and that the basal ganglia influence motor planning processes through their thalamocortical connections (Alexander & Crutcher, 1990; Chesselet & Delfs, 1996; Cunnington, Bradshaw, & Iansek, 1996). Parkinson's disease (PD), produced by impaired functioning of the basal ganglia and their related neural circuits, provides a window into how cortical and subcortical processes interact to produce coordinated, accurate movement. We use a paradigm designed to separate motor planning, or preprogramming, and on-line programming components of the motor control process in PD patients. To relate these findings to possible neural mechanisms and determine whether these processes are affected by different degrees of basal ganglia dysfunction, we investigated the effects of disease severity on motor programming.

Parkinson's disease is a progressive disease that affects approximately 1% of the population (O'Brien & Shoulson, 1992). The average age of onset is 60 years and incidence increases with age. It is characterised by tremor at rest, rigidity (increased muscle tone simultaneously in extensor and flexor muscles), difficulty in initiating movement, paucity of spontaneous movements (akinesia), and slowness in the execution of movement (bradykinesia).

Although motor execution deficits are characteristic of PD, researchers disagree whether they result from impaired motor programming or planning (e.g. Flowers, 1978; Marsden, 1982; Robertson & Flowers, 1990). In situations where information is provided about the required movement, PD patients can plan and prepare simple movements relatively normally (Stelmach, Worringham, & Strand, 1987). Longer movement times are more commonly associated with PD than are prolonged reaction times (Phillips, Bradshaw, Iansek, & Chiu, 1993). However, researchers report mixed findings about PD patients' ability to plan complex movements or to change movements midstream (Harrington & Haaland, 1991, 1996; Jennings, 1995; Rafal, Inhoff, Friedman, & Bernstein, 1987; Stelmach et al., 1987). For example in a delayed reaction time task, normal participants produce increases in reaction time (RT) with the number of movement components in a sequence but PD patients may not (Stelmach et al., 1987). Other studies have shown that PD patients were able to prepare a sequence of finger taps in advance only when there was a compatible relationship between stimulus and response (Harrington & Haaland, 1991). This difficulty in preparing complex movement sequences in advance suggests that the basal ganglia have a role in engaging subsequent movements in a movement sequence (Marsden, 1982). The different results may stem from the fact that different researchers use different measures to indicate preprogramming and by ignoring important information available in alterations of ongoing movement.

Differing results regarding motor preprogramming in PD may also arise from the patient groups' disease severity. PD-related deficits in motor function are associated with a degeneration of neurons in the substantia nigra with subsequent reductions of dopamine

(DA) in the substantia nigra and its projection sites (e.g. striatal and limbic regions) as well as other neurochemical abnormalities. Recent research has indicated that it is not only the presence and utilisation of DA that is critical for planning accurate movement, but also the timing of its release (Graybiel, 1990). Although disease progression corresponds to decreasing dopamine levels, PD symptoms vary widely among patients. This variability in the patient population implies a complexity in the underlying neural circuitry that organises the components of movement.

One way that disease severity and decreased DA production can influence movement is through the underactivation of cortical regions involved in motor programming and planning. Decrements in DA production in the basal ganglia affect motor planning through the activation of premotor brain regions, including the SMA and ventral premotor cortex (PMv). The SMA appears to be involved in motor programming, especially for internally generated movements (Decety, Kawashima, Gulyas, & Roland, 1992; Decety et al., 1994; Goldenberg et al., 1992; Jeannerod, 1995; Roland, Larsen, Lassen, & Skinhoj, 1980). The SMA has been implicated in the programming of motor subroutines prior to movement initiation (Roland et al, 1980), in the timing and initiation of submovements in terms of selecting the correct movements, and in the integration of these submovements into ongoing movement sequences (Kornhuber et al., 1989; Cunnington et al., 1996). A major cortical input to the basal ganglia is from SMA and a major cortical output of the basal ganglia is to SMA (Wickens, 1993). The basal ganglia modulate motor program outflow via the SMA and motor cortex via a complex interplay of excitatory and inhibitory pathways (e.g. Albin, Young, & Penney, 1989; Alexander & Crutcher, 1990; DeLong, 1990). Basal ganglia dysfunction alters activation in the SMA. In PD patients, decreased dopamine production decreases activation of SMA. Data from PET neuroimaging studies of PD patients confirm hypoactivation of the SMA (Shinotoh & Calne, 1995; Snow, 1996). As a result, underactivation of the SMA may lead to disrupted motor planning.

The basal ganglia also have strong thalamocortical connections to PMv (Hoover & Strick, 1993; Matelli, Luppino, Fogassi, & Rizzolatti, 1989; Passingham, 1993). The PMv has been implicated in the programming of externally generated movements. PET studies of PD patients performing motor tasks have found underactivation in this area as well as (Playford et al., 1992).

Alternatively, disease severity and DA production may affect motor programming and the functioning of these cortical areas by introducing noise or variability into the motor system through inconsistent premotor cortex activation. PD selectively impairs the activation or switching of motor sequences rather than the individual elements of the movement sequence. Thus, PD patients have difficulty in consistently activating the required sequence. Noise in the motor system would express itself, over the course of many movements, in terms of variability in both motor planning and movement accuracy (Sheriden & Flowers, 1990). For example, PD patients may be able to generate sufficient force to execute the movement but be unable to activate the correct parameters for the movement. As a result, movements are inaccurate as the force varies erratically from trial to trial (Sheridan & Flowers, 1990). This unpredictability of limb placement forces patients to monitor each action while it is being made and to perform on-line corrections. Further, greater basal ganglia dysfunction would increase variability in PD patients' ability to plan and control movements.

Both inconsistent basal ganglia output and cortical underactivation correspond with the movement deficits observed in PD patients. These patients have relatively intact perceptual processes and do not exhibit obvious problems with the selection and sequencing of muscle activity (Hallett & Khoshbin, 1980). However, they tend to have variable delays in motor initiation and increased movement times. Further, kinematic analysis of movement execution in PD patients suggests that, relative to controls, these patients tend to have lower initial acceleration, lower overall velocity, and less accuracy in most movements, particularly lower-amplitude movements (e.g. Stelmach, Teasdale, & Phillips, 1992). They are also impaired when producing ballistic movements, varying movement acceleration and velocity, and executing movements without visual information.

Although it is clear from the literature that PD patients have motor execution problems that increase with disease severity, it is not clear whether motor programming processes are similarly affected. In the present study, we investigated whether disease severity influenced motor programming and, if so, what aspects of programming were disrupted. We modified a paradigm developed by Franks and colleagues to isolate motor processes involved in the preprogramming, on-line programming, and production of fast, accurate movements (van Donkelaar & Franks, 1991). In normal populations, these types of fast movements require little on-line monitoring. The separation of motor programming processes is achieved through multiple, kinematic-dependent variables that are sensitive to differences between preprogrammed and on-line prepared movements: movement initiation time (RT), acceleration traces, and EMG data from the triceps and biceps muscles of the arm. Although previous studies have used these measures to study preprogramming in PD, few have incorporated all of these measures in the same study to provide converging evidence for separate preprogramming and on-line programming processes.

The detailed level of analysis available from this paradigm and these measures permits insight into those motor programming mechanisms affected by disease severity and the role of the basal ganglia in motor control. For example, if disrupted basal ganglia function influences the consistency of selecting accurate motor components, then several explicit predictions in terms of motor control can be proposed: (1) variable motor parameter selection suggests evidence of on-line programming, but not necessarily a lack of evidence for motor preprogramming (i.e. one can plan an incorrect movement); (2) on-line programming should increase with sequence length because inaccurate movements early in a movement sequence affect subsequent movement segments; and (3) increased disease severity and corresponding basal ganglia dysfunction will produce increased inter-trial variability of movement parameters.

In addition, specific predictions can be made regarding the effect of disease severity on specific dependent variables.

Preprogramming

Do patients with Parkinson's disease preprogram stimulus-triggered movements? Is this ability affected by disease severity? Problems in programming a movement prior to execution can be revealed through the examination of the time necessary to initiate a movement. In this experiment, maximal speeds of movement were required. If participants can prepare upcoming actions in advance, then more complex movements should take longer to prepare and initiate than simple movements (Fischman, 1984; Henry & Rogers, 1960; Rosenbaum, Inhoff, & Gordon, 1984; Rosenbaum, Kenny, & Derr, 1983; Sternberg, Monsell, Knoll, & Wright, 1978). Thus, an increase in RT that is associated with an increase in movement complexity would be evidence for preprogrammed movement whereas a lack of increase in reaction time may imply that the movement is being controlled on-line (during its execution). However, inferences about movement programming from simple reaction time (SRT) have their problems due to the imprecise nature of the dependent variable. It is possible to partial out the programming aspects of the movement variable. It is also possible to partial out the programming aspects of movement preparation from the mechanical aspects of movement initiation using electromyography (EMG). SRT can be fractionated into premotor and motor time components (Botwinick & Thompson, 1966). Premotor RT is the time from the onset of the imperative stimulus to the beginning of EMG and has typically been associated with delays in central programming activity. On the other hand, motor RT is the time from the beginning of EMG activity to the start of external limb movement. This period is believed to reflect the duration of nonprogramming events, such as the electromechanical delay in the muscle and development of sufficient torque to initiate movement (Anson, 1982). Thus the use of premotor time as an indicator of movement programming enables valid interpretations of SRT differences across levels of movement complexity (Anson, 1989; Christina & Rose, 1985). If PD patients can preprogram stimulus-triggered actions, then simple movements should be faster to prepare and initiate than complex movements (Henry & Rogers, 1960; Sternberg et al., 1978). Normal preprogramming in PD patients has been found under some conditions (Harrington & Haaland, 1991; Rafal et al., 1987; Stelmach et al., 1987). Alternatively, if RT is invariant across levels of complexity, then on-line preparation is inferred; that is, the sequence is prepared in parts throughout the movement as opposed to entirely beforehand. In other words, if PD patients can only plan and initiate movements one at a time and not in combination (Marsden,

1982; Robertson & Flowers, 1990), then we should find no increase in RT or PMT with increases in movement complexity. PD has been found to disrupt the ability to preprogram a single movement or a series of movements (e.g. Flowers, 1978; Harrington & Haaland, 1991, 1996; Jennings, 1995; Stelmach et al., 1987).

Whether evidence for preprogramming is found or not, evidence for inconsistent parameter selection may be found in the variability of preprogramming as well as movement measures. If the extent of subcortical dysfunction is related to parameter inconsistency, then within-subject variability should increase with disease severity.

Movement Execution

Does the severity of the disease affect movement execution? Execution variables reflect movement and movement monitoring processes. Execution deficits are characteristic of all PD patients. We expect that disease severity will influence these variables given that severity is largely defined in terms of execution difficulties. In addition, we predict that movement execution becomes more variable with increased disease severity, because it is the output of variable motor parameter selection. Previous studies indicate that the movement of PD patients differs from that of normal patients in terms of lower initial acceleration, lower overall velocity, and lower accuracy, particularly in larger amplitude movements (e.g. Flowers, 1975, 1976). PD patients are also impaired when executing movements without visual information, suggesting that they are more dependent on "on-line" visual guidance to achieve a spatial target.

On-line Programming

Does the severity of PD affect the degree of on-line movement control? Kinematic profiles of the acceleration trace are used to parse aiming movements into two phases: the initial impulse and the error correction phase. The initial impulse is assumed to be preprogrammed and is characterised by a rapid, continuous change in the position of the limb. In normal populations, fast movements performed under 200msec were predominantly programmed in advance with no use of feedback to detect and correct errors (Abrams, Meyer, & Kornblum, 1990; Meyer, Abrams, Kornblum, Wright, & Smith, 1988; Woodworth, 1899). The initial impulse does not contain movement modifications and is therefore comprised of one submovement. The correction phase begins if the endpoint of the initial impulse misses the target. Error corrections are indexed by discontinuities in the position, velocity, and acceleration of the moving limb, which are said to reflect the presence of on-line adjustments to a movement, or error reduction. Error correction phases may consist of only a single submovement or they may contain multiple submovements (see Crossman & Goodeve, 1983; Meyer et al., 1988; Meyer, Smith, Kornblum, Abrams, & Wright, 1990, for more in-depth discussions on the number of submovements). Parsing of movements into their initial impulse and error correction phases has been accomplished by locating the first moment at which one of the following movement modifications occurs: a positive to negative zero-line crossing velocity, a negative to positive zero-line crossing in acceleration, or a significant deviation in acceleration. Positive-to-negative transitions in velocity correspond to reversals in the direction of the movement, going from a forward to backward direction. Zero-line crossings in the acceleration trace represent an increase in the velocity of the movement after it was slowing down. Significant deviations are relative minimums in the absolute value of acceleration while the acceleration is negative. In contrast to zero-line crossings, significant deviations represent abrupt changes in the acceleration trace that reflect a decrease in the net braking force of the limb without an increase in velocity.

If disease severity affects movement planning and accuracy, we would expect to find evidence for increased on-line control, as indicated by an increased incidence of significant deviations of the acceleration trace for later segments of the more complicated movement overall, and in particular, for more severely affected patients. Few zero-line crossing differences are expected given that the data analysis is constrained to accurate movements. On-line processing may be reflective of a movement strategy employed by patients with PD, among others. If one cannot be certain of the endpoint of a limb movement, a good strategy would be to visually monitor movements. In addition, inaccuracies in the endpoints of early movement segments may warrant additional monitoring of later movement segments. As a result, early movement segments may preprogrammed, but later segments may be subject to on-line control.

METHOD

Participants

Twelve participants diagnosed with idiopathic PD (9 male, 3 female; age range 48–76 years; mean age = 57.5 years; mean years from diagnosis = 6.75) volunteered to participate in this study. All were tested during their normal medication cycle, that is, they were taking medication for their PD symptoms and participated in the experiment during periods of good dose effectiveness. All participants were right-handed and had no history of stroke, heart attack, or serious medical problems. All were able to follow verbal instructions. They were classified into two groups, based on the severity of PD symptoms indicated by their Hoehn-Yahr motor rating: PD1 = less severe (Stages I–II), PD2 = more severe (Stages III–IV) (Table 1).

In addition, three participants from the Vancouver community (1 male, 2 female; aged range 47–71 years; mean age = 59.67 years), matched for age and education, volunteered to participate as control participants. All control participants were right-handed and had no history of stroke, heart attack, or serious medical problems. Given that the primary predictions concern differences among PD patients, the data from control participants is reported primarily for broad comparison purposes. In

Table 1. Patient Information

Gender	Age	Years Since Diagnosis	Medication
PD1[a]			
F	48	5	Sinemet, Eldypyrl
M	49	3	Sinemet, Tolcapone
F	49	4	Artane, Amantadine
M	52	6	Sinemet
M	52	3	Sinemet
M	66	9	Sinemet, Amantadine, Selegiline
PD2[b]			
M	48	7	Sinemet, Lisuride
M	57	9	Sinemet, Bromocryptine
F	58	8	Sinemet, Amitriptyline, Deprenyl, Permax
M	63	8	Sinemet
M	72	9	Sinemet, Deprenyl, Bromocryptine
M	76	10	Sinemet, Tolcapone

[a]PD1: Less affected by PD.
[b]PD2: More affected by PD.

addition, the performance of normal participants on these tasks is reported in van Donkelaar and Franks (1991).

Task, Apparatus, and Procedure

Participants performed two simple motor tasks: horizontal arm extension and horizontal arm extension and flexion. They gripped a vertical handle at the free end of a horizontal level attached to a vertical shaft such that the elbow was coaxial with the axis of rotation. The movement of the frictionless level controlled a cursor on an oscilloscope screen. On the screen were two target boxes located ± 22.5° (± 5cm) from the centre of the screen for a total movement of 45° or 10cm. The left start target was ± 1.225° for a total of 2.25° or 0.5cm on the screen. The right end target was ± 2.25° for a total of 4.5° or 1cm on the screen. Before each trial, participants heard a 200msec warning tone at 100Hz and saw a brightening of the targets to indicate that the trial was about to begin. The foreperiod between the warning tone and "go" tone varied from 500–2500msec. Upon hearing a 300msec "go" tone at 700Hz, participants moved the cursor (and their arm) as quickly and accurately as possible to the right target. Catch trials, during which participants heard a warning tone without a subsequent "go" tone, occurred on 10% of the trials to ensure that movement onset was not anticipated. Data was collected for 5000msec.

The within-subject independent variable was complexity of movement (task complexity), as indicated by the number of changes in direction of movement: extension and extension-flexion. In the extension task, participants extended their arm at the elbow joint to move the cursor to the right target, without exceeding its bounds. For the extension-flexion task, participants extended then flexed their arm to move the cursor within the bounds of the right target and then move the cursor within the bounds of the left target. Trials were blocked by task. This manipulation provides evidence of preprogramming in that RTs should increase with additional movement segments being programmed.

To distinguish between central and on-line processing components of movement, only fast and accurate movements were acceptable for analysis. Participants received feedback regarding RT, movement time (MT), and accuracy; they were encouraged to "beat their best times." After completing 10 accurate practice trials, participants completed 10 accurate experimental trials.

During each trial four components of data were collected at a sample rate of 1000Hz: angular displacement, acceleration, triceps EMG, and biceps EMG. After data collection, the angular displacement was calculated. All data were stored on-line during a trial, analysed, and reviewed; they were then saved in a data file on the computer hard disk. Angular displacement was sampled from a custom optical encoder interface card (Nagelkerke & Franks, 1996). Angular acceleration traces were obtained from an accelerometer filtered at 50Hz, then sampled via a 12-bit Analogue-to-Digital Converter. EMG activity was sampled using surface electrodes placed on the biceps brachii and triceps brachii muscles. The raw EMG signal was amplified, and then sampled via the 12-bit Analogue-to-Digital Converter. Angular velocity was calculated by copying the displacement data into a temporary array, low-pass filtering the data at 10Hz and then differentiating into the velocity data array.

Dependent Variables

Reaction time (RT). RT, or movement initiation time, is the time between the onset of the imperative stimulus tone and the subject's first movement. If the ability to preprogram upcoming actions are relatively intact in PD, then simple movements should be faster to prepare and initiate than complex movements. A less accurate motor program/plan would mean that RTs would be more variable for more complex movements and for more severe PD patients.

EMG recordings. From EMG recordings, RT may be divided into premotor time, associated with central planning activity, and motor time, associated with the electromechanical delay of the muscle overcoming the inertia of the limb. Thus, like RT, an increase in premotor time with movement complexity would be indicative of movement preprogramming (e.g. Kasai & Saki, 1992). A possible outcome of a noisy motor program/plan is the finding of more variable premotor times for more complex movements and for more severe PD patients.

Acceleration traces. Two aspects of the acceleration trace indicate on-line adjustments to movement: the number of zero-line crossings within a movement segment, and significant deviations in the shape of the trace. Whereas preprogrammed movements produce acceleration traces that cross the zero-line only once during each movement segment, movements in which adjustments are prepared on-line produce acceleration traces with multiple zero-line crossings. Zero-line crossings in the acceleration trace represent an increase in the velocity of the movement after it was slowing down and vice versa.

In addition, on-line prepared adjustments can be inferred from an acceleration trace through significant deviations in its shape. In contrast to zero-line crossings, significant deviations represent abrupt changes in the acceleration trace which reflect a decrease in the net force of the limb without an increase in velocity (Kahn, Franks, & Goodman, 1996). A significant deviation is a peak or valley in the data between successive maximum points of positive or negative acceleration, that are preceded or followed by a minimum of 10 data points (e.g. 20msec) (van Donkelaar & Franks, 1991). Typically, preprogrammed movements result in smooth acceleration traces; however, adjustments made during movement appear in the acceleration trace as deviations from the normally smooth curve. In normal participants these fast movements show little on-line adjustments; they are only seen when moving at slower rate (van Donkelaar & Franks, 1991). If the motor program is often inaccurate, one would expect evidence of increased on-line processing, especially for more complex movements.

Movement Analysis

Only data from accurate trials were analysed. Despite the many errors made by PD patients in these tasks, movements of accurate length were required to compare the dependent measures for simple and complex movements in which the initial movement was equated. Trials exceeding 1sec or 3 standard divisions (SDs) from the individual's mean were also eliminated from the analyses. One patient's performance exceeded 3 SDs from mean patient performance and, thus, was eliminated from numerical analyses.

Mean data for each subject was collected using an interactive graphics display. The graphic display included displacement as a function of time, velocity traces, acceleration traces, and biceps and triceps raw EMG traces. To determine the beginning of movement, the displacement profile, which was recorded by an optical encoder, was searched for the point in which the displacement value exceeded zero. This point was defined as the beginning of movement. The end of movement was defined as the point in time following peak velocity in which the absolute angular velocity of the change fell below 10 deg/sec for 150msec. A search was then performed from the acceleration profile for the possible initiation of an error correction phase, that is the occurrence of zero-line crossings and significant deviations. To qualify as a significant deviation the difference in the absolute values of acceleration between minimas and maximas had to be at least 100 deg/sec. EMG traces were analysed for the time corresponding to the first significant burst of activity.

Data across participants was subsequently analysed using MANOVAs for Severity (PD1, PD2) and Task Complexity (extension, extension-flexion). Alpha was set at the .05 level unless otherwise reported.

RESULTS AND DISCUSSION

In the literature, PD patients are reported to reliably undershoot targets (e.g. the movement stops short of hitting the target). An analysis of error trials in this task indicated that both control and Parkinson's patients did not reliably undershoot or overshoot the target, $F(1,12) = 1.64$, $P > .10$. No significant group differences were found between control patients and Parkinson's patients when testing for a Group (control, PD) × Task (extension, extension-flexion) × Error Direction (undershoot, overshoot) interaction, $F(1,13) = 2.9$, $P > .10$. These results are different from the prevalent undershoot errors observed in many clinical settings and experimental studies. It is possible that the incidence of undershoot errors in our study was minimised by the external movement cue and by trial-by-trial error feedback. Typically, undershoot errors were immediately followed by overshoot errors.

Do Patients with PD Preprogram Stimulus-triggered Movements? Is This Ability Affected by Disease Severity?

To address the questions of whether patients with Parkinson's disease are able to preprogram stimulus-triggered movements and whether this ability is affected by the severity of the disease progression, we examined reaction time (RT) data and premotor time (PMT).

Reaction time (RT: time from the auditory signal to movement initiation as indicated by displacement, msec). An analysis of RT data revealed no significant effects of severity, $F(1,9)$, < 1 (PD1: 256.77; PD2: 276.50; Ctrl: 222.4) and no interaction with task complexity. However, it did produce a significant effect of task complexity, $F(1,9) = 7.62$, $P > .03$ (Extension Task: 251.03; Extension-Flexion Task: 276.50). For tasks of different movement complexity, holding perceptual features constant and changing only the task requirement, PD participants are able to show evidence of preprogramming. This ability is not affected by the progression of the disease. PD participants' RTs are in the same range as those of the control participants.

Premotor time (PMT: time from stimulus to onset of EMG, msec). No effect of disease severity, $F(1,9 < 1$ (PD1: 200.66; PD2: 212.20; Ctrl: 177.32) or interaction was found. A main effect of task complexity was significant, $F(1,9) = 9.86$, $P < .02$ (Extension Task: 189.64; Extension-Flexion Task: 222.17). Consistent with the RT data, PMT data provides converging evidence of preprogramming. Research with normal participants suggests that PMT represents delays associated with central programming (Anson, 1982; 1989).

Within-subject variability for preprogramming measures (inter-trial variability). Less severe patients tended to have less inter-trial variability than more severe patients. There was a significant severity effect for RT, $F(1,9) = 7.00, P < .03$, and a marginal effect for premotor time, $F(1,9) = 4.31, P < .07$. These findings imply inconsistent functioning of the basal ganglia, even in accurate movements.

In sum, PD patients can execute the tasks correctly and are able to preprogram at least some part of the movement. Disease severity did not appear to affect preprogramming process in terms of mean measures. However, disease severity did influence the within-subject variability of the preprogramming measures, suggesting increasing inconsistencies in pre-programming processes.

Does the Severity of the Disease Affect Movement Execution?

To address the question of how the severity of the disease affects movement execution, we examined the following kinematic variables related to execution: movement time (MT), peak velocity (first peak: deg/sec), time to peak velocity, peak acceleration (deg/sec/sec), time to peak acceleration (msec), acceleration and deceleration intervals for the extension task (msec), and within-subject variability (inter-trial variability).

Movement time (MT: time from movement initiation to end of first movement, to permit comparison between extension and extension-flexion tasks, msec). MT is reliably slowed in Parkinson's patients as a result of rigidity and bradykinesia inherent in the disease symptoms. As expected, we found a severity effect, $F(1,9) = 14.11$, $P < .005$ (PD1: 345.00; PD2: 548.75; Ctrl: 363.72), but no task complexity effect, $F(1,9) = 3.48$, $P < .10$, (Extension Task: 411.79; Extension-Flexion: 463.86) or interaction.

Peak velocity (first peak; deg/sec). Peak velocity has also been found to be lower in patients with PD. The significant severity effect, $F(1,9) = 21.10$, $P < .002$ (PD1: 250.23; PD2: 159.29; Ctrl: 239.05) suggests that the lowering of peak velocity increases with greater basal ganglia dysfunction. There was also a task complexity effect, $F(1,9) = 7.22$, $P < .03$ (Extension Task: 224.20; Extension-Flexion: 193.59), with the single movement task producing higher peak velocities than the complex movement. There was no interaction.

Time to peak velocity (msec). The time to peak velocity reflected the peak velocity data. There was a severity effect, $F(1,9) = 19.57$, $P < .002$ (PD1: 156.13; PD2: 242.39; Ctrl: 173.78) as well as a task complexity effect, $F(1,9) = 6.57, P < .04$ (Extension Task: 180.65; Extension-Flexion Task: 210.04). There was no interaction.

Peak acceleration (deg/sec/sec). Peak acceleration data mirrored peak velocity data despite different measuring instruments. There was a group effect $F(1,9) = 21.96$, $P < .002$ (PD1: 270.45; PD2: 113.56; Ctrl: 227.58) and task complexity effect, $F(1,9) = 5.98$, $P < .04$ (Extension Task: 228.52; Extension-Flexion Task: 169.77). There was no interaction.

Time to peak acceleration (msec). Time to peak acceleration showed a group effect, $F(1,9) = 10.95$, $P < .01$ (PD1: 97.90; PD2: 146.21; Ctrl: 100.40), and a task complexity effect, $F(1,9) = 5.37$, $P < .05$ (Extension: 112.84, Extension-Flexion: 126.89), but no interaction.

Acceleration and deceleration intervals for extension task (msec). The acceleration and deceleration intervals for the extension task produced effects of severity, $F(1,9) = 18.54, P < .002$ (PD1: 212.3; PD2: 312.43; Ctrl: 225.77) and interval, $F(1,9) = 55.03$, $P < .0001$ (Acceleration Interval 1: 205.89; Acceleration Interval 2: 309.74), but no interaction. Teasdale, Stelmach, and Mueller (1991) emphasise that the goal of most tasks is to initiate a movement and to decelerate a limb accurately to produce a skilled movement. In normal populations, equal acceleration and deceleration intervals are commonly found. Here we find greater deceleration intervals for the PD group.

Motor time (time from EMG onset to movement initiation, msec). No effect of severity, $F(1,9) = 1.18$, $P > .10$ (PD1: 56.09; PD2: 64.30; Ctrl: 45.08), or task complexity, $F(1,9) = 2.41, P > .10$ (Extension Task: 61.39; Extension-Flexion Task: 58.26), was found for motor time. The Severity × Task Complexity effect was significant, $F(1,9) = 12.64$, $P > .01$ (Extension: PD1 = 53.54; PD2 = 70.80; Extension-Flexion Task: PD1 = 58.64; PD2 = 57.80). This interaction does not have a clear interpretation. The longer motor time for the PD2 group in the extension task appears to be related to coactivation of both agonist and antagonist muscles at the beginning of the extension movement, thus requiring more time for the agonist to overcome inertia. Proportionately more coactivation was found for PD2 than for PD1. Figures 1a, 1b, 2a and 2b provide visual examples of this finding for typical patient performances on the extension and extension-flexion task. Both biceps and triceps are concurrently active at the initiation of movement.

Qualitative Analysis of EMG Activity

For normal participants performing a forearm extension movement, the characteristic pattern of EMG activity from triceps and biceps muscles is triphasic (Enoka, 1994): The triceps agonist activates to accelerate the arm to the target then the biceps antagonist provides a breaking force to the arm as it nears the target, followed by triceps reactivation to secure the arm in place. A slightly different pattern of activity is typically displayed during a continuous extension-flexion arm movement. The triceps initiate arm movement and the biceps slow the arm as it nears the target; however, in this case, the antagonists become agonists to move the arm in the opposite direction. The biceps remain active until the triceps stop the arm and then the biceps secure the arm at the start position (van Donkelaar & Franks, 1991).

This normal pattern of activity was not typically present in PD patients, as depicted in Figs. 1a, 1b, 2a, and 2b. Both within and between individuals, there was great variability of muscular activation patterns. Figures 1a and 1b illustrate an example of a participant from the less affected PD1 group, which displays EMG patterns similar to those of normal young adults. This participant's performance was included for comparison purposes. This example displays triphasic muscular activity, although the amount of activation appears to be smaller than that of normal participants. Muscular activation is maintained to allow the switch from agonist to antagonist function. In contrast, Figs. 2a and 2b illustrate an example of a participant from the more affected PD2 group that displays very different EMG patterns. Participants in the PD2 group tended to show multiple weak phasic bursts of both agonist and antagonist muscles, in contrast to stronger, more sustained firing found in the perform-

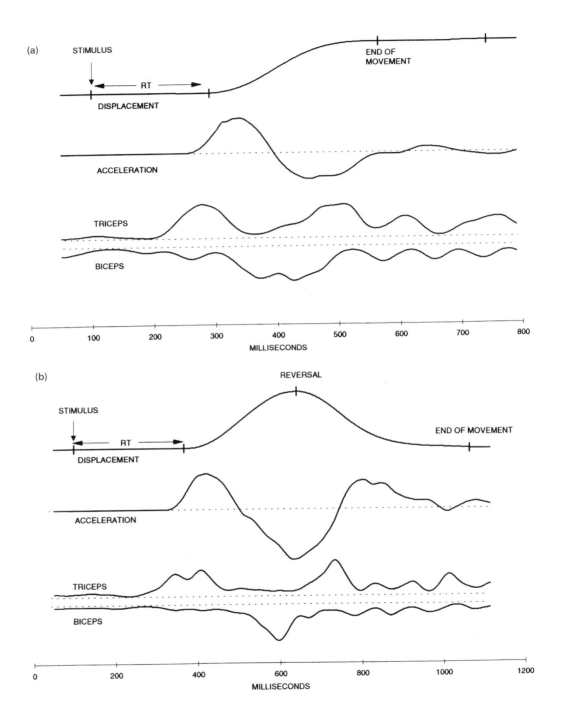

ance of normal participants. Additional coactivation of biceps and triceps muscles is found as well as significant deviations in the positive acceleration portions of the acceleration curve, a finding never found in normal performance.

PD patients seem unable to maintain the long burst of the biceps as its role changes from antagonist to agonist at the reversal point of the movement. Previous research from Berardelli, Accornero, Argenta, Meco, and Manfredi (1986) and Hallett and Khoshbin (1980) suggests that PD patients require additional bursts of muscle activity (especially agonist muscle activity) to produce both faster and longer movements. This is consistent with our data. However, the multiple bursts of muscle activity also corresponds with a strategy for accurate movement strategies with PD. Compare the bursts of EMG activity with changes in the acceleration curves in the earlier figures. The largest bursts correspond with major maxima and minima of the trace. Note, however, how the additional bursts correspond in time with deviations of the acceleration trace that are indicative of on-line control. It appears that patients with PD are trying to impose control throughout the movement to produce an accurate movement. If PD patients have difficulty predicting the endpoint of a limb movement, a good strategy for producing an accurate movement would be to use smaller, more frequent bursts of muscle activity to control the movement.

Within-subject variability (inter-trial variability). Less severe patients were generally less variable than more severe PD patients in terms of movement time, $F(1,9) = 16.16$, $P < .003$, RT to peak velocity, $F(1,9) = 4.95$, $P < .06$, marginal, RT to peak acceleration, $F(1,9) = 4.90$, $P < .06$, marginal, and motor time, $F(1,9) = 16.76$, $P < .003$. However, there were two variables in which less severe patients were more variable than more severe patients: peak velocity, $F(1,9) = 8.12$, $P < .02$, and peak acceleration, $F(1,9) = 6.47$, $P < .04$.

To summarise, both severity and movement complexity affect motor control in PD. First, the severity of PD affects all of the above variables in predictable ways. People with more severe PD are slower to move, producing movements of lesser velocity and acceleration. They also tend to be more variable in their movements. The task effects show increasing effects of dopamine deficits with complexity. Less severe patients with PD may have a

Fig. 1 (opposite). Performance data from patient no.3 of the less affected PD1 group. Figure 1(a) illustrates performance on the extension task. Figure 1(b) illustrates performance on the extension-flexion task. The top graph represents lever displacement in degrees as a function of time. The acceleration trace (degrees/sec/sec), and the rectified and filtered EMG activation trace (volts) for biceps and triceps follow. All traces correspond to the same function of time. Note that RTs are longer for the extension-flexion task than for the extension task and that MT for the first movement is similar for both tasks. In addition, note that although biceps activation appears to be smaller than that for normal participants, there is a triphasic pattern of EMG activity between biceps and triceps. Last, muscular activation is maintained as triceps switch from agonist function to antagonist function in the extension-flexion task. Overall, these examples are similar to EMG patterns of a normal, young person.

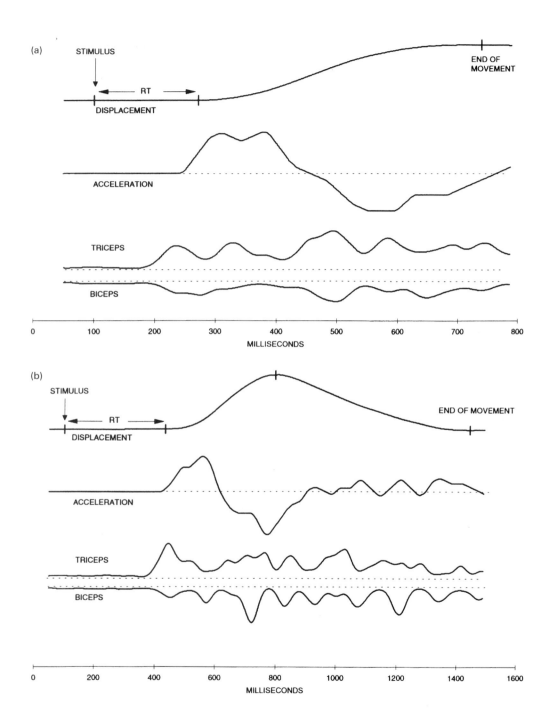

greater variability for faster movements (e.g. Sheridan & Flowers, 1990) because they have a larger repertoire of movement speeds available to them. The variability of movement would decrease if there are fewer movement speeds available, as seen in more severe patients with PD. In other words, the less severe group's greater within-subject variability can be explained by the range of velocities available to the control of patients more severely affected by PD. People in the less severe group may have greater within-subject variability because they have a larger range available to them given the status of the dopamine in the system. However, patients with less dopamine in the system have fewer options for movement velocity and acceleration.

Second, movement complexity (the difference between extension and extension-flexion tasks) affects execution variables differently. Movement times for the same segment are increased for the more complex task. In addition, peak velocity and peak acceleration are lower and time to peak velocity and acceleration are increased for the more complex task. Unlike the other measures, motor time was not significantly different for the two tasks. These effects do not interact with severity. This is also of interest because PD patients as a whole have performance that is in a similar range of our control participants.

Does the Severity of PD Affect the Degree of On-line Movement Control?

Zero-line crossings. Because we restricted our analyses to fast and accurate movements, we did not expect to find extraneous zero-line crossings. No severity effects were found for either the extension task, $F < 1$ (PD1 = 1; PD2 = 1) or the extension-flexion task, $F < 1.0$ (PD1 = 2.07; PD2 = 2.17. This result is to be expected given that we only analysed accurate movements that appeared to be, on the basis of smooth and bell-shaped displacement and velocity curves, single movements for each direction of movement. In addition, participants were instructed to move as fast as possible, which should have precluded zero-line crossings. These analyses confirmed that participants primarily programmed one or two movements, depending upon the task.

Significant deviations in the acceleration curve. For the extension task, there was no severity effect [$F(1,9) < 1$ (PD1 = .403, PD2 = .350; Ctrl = .067; PD$_{overall}$ vs. Ctrl, $P < .08$], interval effect [$F = 2.57, P > .10$], or interaction. However, for the

Fig. 2 (opposite). Performance data from patient no.6 of the more affected PD2 group. Figure 2(a) displays performance on the extension task. Figure 2(b) displays performance on the extension-flexion task. The top graph represents level displacement in degrees as a function of time. The acceleration trace (degrees/sec/sec) and the rectified and filtered EMG activation trace (volts) for biceps and triceps follow. All traces correspond to the same function of time. In the PD2 group, note the coactivation of biceps and triceps. Multiple bursts of weak activity are evident in the EMG traces. As illustrated here, PD patients have difficulty maintaining muscular activation at the point at which the agonist muscle becomes the antagonist. Note that there are significant deviations in the positive acceleration portion of the acceleration curve that are never seen in normal performance.

extension-flexion task, there was an interval effect, $[F = 14.19, P < .0002$ (Interval 1 = .259; Interval 2 = .988; Interval 3 = .512]. The interval effect suggests that more on-line adjustments occur when a second movement is required. No effects were found for severity $[F(1,9) < 1$ (PD1 = .569; PD2 = .644; Ctrl; = .211; $PD_{overall}$ vs. Ctrl; $F(1,9) = 6.94$, $PD < .03)]$ or interaction. This general interval effect is not due to the severity of PD; however, there were significant differences between the PD patients and the controls; the PD group produced a greater number of significant deviations in the acceleration curve, especially on the extension-flexion task. As movements became more complex, PD patients overall made additional on-line adjustments.

Mean significant deviations for extension and extension-flexion tasks. No effect of severity $[F(1,9) < 1$ (PD1: .503; PD2: .505; Ctrl: .150)] was found. A task effect was significant $[F(1,9) = 5.42, P < .05$ (Extension: .399; Extension-Flexion: 6.09]. In contrast, there was a significant difference between the PD patients and the control group $[F(1,9) = 5.53, P < .04]$. The performance of the control group was significantly different from that of the PD groups for mean number of significant deviations, suggesting increased on-line control for people with PD. The greatest difference occurred at the end of the more complex movement. We would expect to find evidence for increasing on-line programming as movement because more complex if one postulates a noisy motor system.

In summary, patients with PD can preprogram movements, but on-line monitoring of movement is evident as well, especially in the more complex task. Disease severity appears to influence within-subject variability of the preprogramming measures, movement, and on-line processing despite evidence of preprogramming.

GENERAL DISCUSSION

Parkinson's disease can provide insight into the role of the basal ganglia in the planning and production of coordinated movement. The main goal of the present study was to examine motor planning processes in PD patients with varying degrees of severity and to infer neural function from behavioural measures. Specifically we investigated the differences between preprogrammed movements and those prepared on-line. In normal young populations movement initiation time, acceleration traces, and EMG data provide evidence as to the conditions under which preprogramming and on-line control processing occur within a movement sequence. Using a paradigm designed to separate motor control processes, we found that patients with PD are indeed able to preprogram movement, as evidenced through increased RT and PMT with movement complexity. However, there was also evidence for variable motor planning processes. All patients with PD showed increased on-line adjustments to their movements as the movement complexity increased. Furthermore, there were effects of disease progression that were consistent with a hypothesis of a variable motor system producing an inappropriate scaling of movement. Both preprogramming measures and execution measures showed an increase in within-subject (or intertrial) variability.

How do the current results address hypotheses regarding the role of the basal ganglia in motor control? The basal ganglia modulate motor program outflow via the supplementary motor cortex (SMA), premotor cortex, and motor cortex via a complex interchange of excitatory and inhibitory pathways. The basal ganglia receive excitatory input from all areas of cortex. In particular, the striatum (the caudate and putamen) receive cortical input from the SMA. In the direct, facilitory pathway, cortical activity through the basal ganglia disinhibits thalamic activity, potentially facilitating one motor program while suppressing activation of other motor programs. An indirect, inhibitory nigrostriatal side loop modulates this activity. Dopamine further modulates thalamocortical activity through both facilitatory and inhibitory mechanisms. Research has indicated that the phasic release of dopamine in addition to its presence is critical for smooth voluntary motor control (Graybiel, Aosaki, Flaherty, & Kimura, 1994). Thus, the loss of dopamine and the dopamine cell activity produces the loss of smooth motor control Thus, PD resulting from decreased dopamine from SNpc produces an overactive indirect pathway and an underactive direct pathway. Both decrease thalamic and cortical activation. The net result is decreased activation to SMA and other premotor areas (Chesselet & Delfs, 1996).

This study and clinical observations support the contention that the basal ganglia do more than simply modulate the activation of premotor areas. If PD simply produced an underactivation of these areas, then the preprogramming of movement should be increasingly disrupted with disease severity. However, that prediction was not born out by the data. Alternatively, if the basal ganglia are specialised to perform parallel processing in which certain movements are enhanced and others are suppressed, then their function may be to amplify the signal-to-noise ratio in the selection of appropriate parameters of movement. Many of the motor symptoms found in PD may arise from the dysfunctional operation of the basal ganglia that introduces additional noise into the motor system (Sheridan & Flowers, 1990). Noise in the motor system would manifest itself in increased performance variability and this is supported by our data.

A neurally based model of basal ganglia function in Parkinson's disease has been proposed by Wickens and colleagues (Wickens, 1993; Wickens, Hyland, & Anson. 1994). They describe a model in which neurons in cortical motor areas have functional groupings or cell assemblies. Complex movements are represented in cortical motor areas as a set of closely linked cell subassemblies, each representing a different element in the sequence of movements. Inhibitory connections between cells create a winner-take-all mechanism in which only the most strongly activated cell in the network fires. The process of motor planning would involve bringing each different subassembly to partial activation so that only a small amount of additional activation would bring the assembly to threshold and "ignite" it. The planning of more complex movements requires the activation of additional subassemblies and thus the activation of additional inhibitory neurons. Thus, longer sequences would produce more inhibition and slow the "ignition" time of the first subassembly of the sequence. Thus, this model would predict longer RTs for complex movements.

An additional consequence of this competitive network of cell assemblies is that the assembly that most closely matches the constraints of the input receives the most activation and ignites. Overall insufficient activation produced by excessive inhibition of the thalamus and premotor areas from the basal ganglia would produce consistently longer RTs. However, inconsistent basal ganglia operation could have two implications. First, it could affect the selection of motor cell assemblies in that no one motor cell assembly has dominance over several others, thereby allowing less optimal assemblies to reach threshold first and fire. Second, it could affect the time to ignite the correct subassemblies in that additional input is needed to reach threshold. Thus, basal ganglia dysfunction in PD patients could affect cell assembly firing rates. Inconsistent igniting could manifest itself in behaviour as variability in motor preparation and on-line programming where activation is insufficient to activate the rest of the subassemblies for complex movements. This type of mechanism is consistent with our findings that PD patients demonstrated increased RTs for more complex movements, that disease severity affected the variability of their increased RTs, and that more on-line programming was required for complex movements.

In addition, evidence for the on-line programming, as well as preprogramming of movement, provides support for the explanation that PD patients may initiate movement before programming is finished (Jennings, 1995). The initiation of movements before programming is complete is often found in the performance of normal populations when various loads are placed on performance (Ketelaars, 1995). If disease severity increases the likelihood of incomplete programming, then we would expect to find and do find corresponding differences in the variability of preprogramming measures with disease severity. Also, both a noisy motor system and incomplete programming would produce our finding of additional on-line adjustments at the end of complex movements. However, incomplete programming does not necessarily predict significant increases in preprogramming time for more complex movements. If movements are reliably initiated before programming is complete, then RT and PMT could be similar for simple and complex movements. Further, incomplete programming should produce corresponding increases in motor time. Our data do not support either of these outcomes.

In conclusion, the examination of Parkinson's disease subpopulations provides insight into the influence of the basal ganglia on various aspects of motor control. The investigation of disease severity permits the formulation of hypotheses about subcortical influences on various motor control functions. Further investigation using such patient groups and sensitive paradigms, however, is necessary to confirm and support our conclusions as to the specific neural mechanisms involved in motor programming.

REFERENCES

Abrams, R.A., Meyer, D.E., & Kornblum, S. (1990). Eye-hand coordination: Oculomotor control in rapid aimed limb movements. *Journal of Experimental Psychology: Human Perception and Performance, 16*, 248–267.

Albin, R.L., Young, A.B., & Penney, J.B. (1989). The functional anatomy of basal ganglia disorders. *Trends in Neuroscience, 12*, 366–375.

Alexander, G.E., & Crutcher, M.D. (1990). Functional architecture of basal ganglia circuits: Neural substrates of parallel processing. *Trends in Neuroscience, 13*, 266–271.

Anson, J.G. (1982). Memory drum theory: Alternative tests and explanations for the complexity effects on simple reaction time. *Journal of Motor Behaviour, 14*, 228–246.

Anson, J.G. (1989). Effects of moment of inertia on simple reaction time. *Journal of Motor Behaviour, 21*, 60–71.

Berardelli, A., Accornero, N., Argenta, M., Meco, G., & Manfredi, M. (1986). Fast complex arm movements in Parkinson's disease. *Journal of Neurology, Neurosurgery, and Psychiatry, 49*, 1146–1149.

Botwinick, J., & Thompson, L.W. (1966). Premotor and motor components of reaction time. *Journal of Experimental Psychology, 71*, 9–15.

Chesselet, M., & Delfs, J.M. (1996). Basal ganglia and movement disorders: An update. *Trends in Neuroscience, 19*, 417–422.

Christina, R.W., & Rose, D.J. (1985). Premotor and motor reaction time as a function of response complexity. *Research Quarterly for Exercise and Sport, 56*, 306–315.

Crossman, E.R.F.W., & Goodeve, P.J. (1983). Feedback control of hand-movement and Fitts' law. *Quarterly Journal of Experimental Psychology, 35A*, 251–278.

Cunnington, R., Bradshaw, J.L., & Iansek, R. (1996). The role of supplementary motor area in the control of voluntary movement. *Human Movement Science, 15*, 627–647.

Decety, J., Kawashima, R., Gulyas, B., & Roland, P.E. (1992). Preparation for reaching: A PET study of the participating structures in the human brain. *Neuroreport, 3*, 761–764.

Decety, J., Perani, D., Jeannerod, M., Bettinardi, V., Tadary, B., Woods, R., Mazziotta, J.C., & Fazzio, F. (1994). Mapping motor representations with positron emission tomography. *Nature, 371*, 600–602.

DeLong, M.R. (1990). Primate models of movement disorders of basal ganglia origin. *Trends in Neuroscience, 13*, 281–285.

Enoka, R.M. (1994). *Neuromechanical basis of kinesiology* (2nd edn.). Champaign, IL: Human Kinetics.

Fischman, M.G. (1984). Programming time as a function of number of movement parts and changes in movement direction. *Journal of Motor Behaviour, 16*, 405–423.

Flowers, K.A. (1975). Ballistic and corrective movements on an aiming task. Intention tremor and Parkinsonian movement disorders compared. *Neurology, 25*, 413–421.

Flowers, K.A. (1976). Visual "closed-loop" and "open-loop" characteristics of voluntary movement in patients with Parkinsonism and intention tremor. *Brain, 99*, 269–310.

Flowers, K.A. (1978). Lack of prediction in the motor behaviour of Parkinsonism. *Brain, 101*, 35–52.

Goldenberg, G., Prodreka, I., Steiner, M., Willmes, K., Suess, E., & Deecke, L. (1989). Regional cerebral blood flow patterns in visual imagery. *Neuropsychologia, 27*, 641–664.

Graybiel, A.M. (1990). Neurotransmitters and neuromodulators in the basal ganglia. *Trends in Neuroscience, 13*, 244–253.

Graybiel, A.M., Aosaki, T., Flaherty, A.W., & Kimura, M. (1994). The basal ganglia and adaptive motor control. *Science, 265*, 1826–1831.

Hallett, M., & Khoshbin, S. (1980). A physiological mechanism of bradykinesia. *Brain, 103,* 301–314.

Harrington, D.L., & Haaland, K.Y. (1991). Sequencing in Parkinson's Disease: Abnormalities in programming and controlling movement. *Brain, 114,* 99–115.

Harrington, D.L., & Haaland, K.Y. (1996). *The basal ganglia and motor control.* Paper presented at the 3rd Annual Meeting of the Cognitive Neuroscience Society, San Francisco.

Henry, F.M., Rogers, D.E. (1960). Increased response latency for complicated movements and a "memory drum" theory of neuromotor reaction. *Research Quarterly, 31,* 448–458.

Hoover, J.E., & Strick, P.L. (1993). Multiple output channels in the basal ganglia. *Science, 259,* 819–821.

Jeannerod, M. (1995). Mental imagery in the motor context. *Neuropsychologia, 33,* 1419–1432.

Jennings, P.J. (1995). Evidence of incomplete motor programming in Parkinson's disease. *Journal of Motor Behaviour, 27,* 310–324.

Kahn, M., Franks, I.M., & Goodman, D. (1996). *The effects of practice on the visual control of rapid aiming movements: Evidence for an interdependency between programming and feedback processing.* Manuscript under review.

Kasai, T., & Saki, H. (1992). Premotor reaction time (PMT) of the reversal elbow extension-flexion as a function of response complexity. *Human Movement Science, 11,* 319–334.

Ketelaars, M.A.C. (1995). *On-line programming of simple movement sequences: An application of the probe reaction time paradigm.* Unpublished Masters thesis, School of Human Kinetics, University of British Columbia.

Marsden, C.D. (1982). The mysterious functions of the basal ganglia: The Robert Wartenberg Lecture. *Neurology, 32,* 514–539.

Matelli, M., Luppino, G., Fogassi, L., & Rizzolatti, G. (1989). Thalamic input to inferior area 6 and area 4 in the macaque monkey. *Journal of Comparative Neurology, 280,* 468–488.

McClelland, J.L., & Rumelhart, D.E. (1987). *Explorations in parallel distributed processing.* Cambridge, MA: MIT Press.

Meyer, D.E., Abrams, R.A., Kornblum, S., Wright, C.E., & Smith, J.E.K. (1988). Optimality in human motor performance: Ideal control of rapid aimed movements. *Psychological Review, 95,* 340–370.

Meyer, D.E., Smith, J.E.K., Kornblum, S., Abrams, R.A., & Wright, C.E. (1990). Speed-accuracy tradeoffs in aimed movements: Toward a theory of rapid voluntary action. In M. Jeannerod (Ed.), *Attention and performance XIII* (pp. 173–226). Hillsdale, NJ: Lawrence Erlbaum Associates Inc.

Nagelkerke, P., & Franks, I.M. (1996). An optical encoder and XY oscilloscope interface for the IBM PC. *Behaviour Research Methods, Instruments and Computers, 28,* 404–410.

O'Brien, C.F., & Shoulson, I. (1992). Cognitive dysfunction in Parkinson's disease and related disorders. In L.J. Thal, W.H. Moos, & E.R. Gamzu (Eds.), *Cognitive disorders: Pathophysiology and treatment.* New York: Marcel Dekker.

Passingham, R.E. (1993). *The frontal lobes and voluntary action.* New York: Oxford University Press.

Phillips, J.G., Bradshaw, J.L., Iansek, R., & Chiu, E. (1993). Motor functions of the basal ganglia. *Psychological Research, 55,* 175–181.

Playford, E.D., Jenkins, I.H., Passingham, R.E., Nutt, J., Frackowiak, R.S., & Brooks, D.J. (1992). Impaired mesial frontal and putamen activation in Parkinson's disease: A positron emission tomography study. *Annals of Neurology, 32,* 151–161.

Rafal, R.D., Inhoff, A.W., Friedman, J.H., & Bernstein, E. (1987). Programming and execution of sequential movements in Parkinson's disease. *Journal of Neurology, Neurosurgery, and Psychiatry, 50,* 1267–1273.

Robertson, C., & Flowers, K.A. (1990). Motor set in Parkinson's disease. *Journal of Neurology, Neurosurgery, and Psychiatry, 53,* 583–592.

Roland, P.E., Larsen, B., Lassen, N.A., & Skinhoj, E. (1980). Supplementary motor area and other cortical areas in the organization of voluntary movements in man. *Journal of Neurophysiology, 43,* 118–136.

Rosenbaum, D.A., Inhoff, A.W., & Gordon, A.M. (1984). Choosing between movement sequences: A hierarchical editor model. *Journal of Experimental Psychology: General*, 113, 372–393.

Rosenbaum, D.A., Kenny, S., & Derr, M.A. (1983). Hierarchical control of rapid movement sequences. *Journal of Experimental Psychology: Human Perception and Performance*, 9, 86–102.

Sheridan, M.R., & Flowers, K.A. (1990). Movement variability and bradykinesia in Parkinson's disease. *Brain*, 113, 1149–1161.

Shinotoh, H., & Calne, D.B. (1995). The use of PET in Parkinson's disease. *Brain and Cognition*, 28, 297–310.

Snow, B.J. (1996). Fluorodopa PET scanning in Parkinson's disease. *Advances in Neurology*, 69, 449–457.

Stelmach, G.E., Teasdale, N., & Phillips, J. (1992). Response initiation delays in Parkinson's disease patients. *Human Movement Science*, 11, 37–45.

Stelmach, G.E., Worringham, C.J., & Strand, E.A. (1987). The programming and execution of movement sequences in Parkinson's disease. *International Journal of Neuroscience*, 36, 55–65.

Sternberg, S., Monsell, S., Knoll, R.L., & Wright, C.E. (1978). The latency and duration of rapid movement sequences: Comparisons of speech and typewriting. In G.E. Stelmach (Ed.), *Information processing in motor control and learning* (pp. 117–152). New York: Academic Press.

van Donkelaar & Franks, I.M. (1991). The effects of changing movement velocity and complexity on response preparation: Evidence from latency, kinematic and EMG measures. *Experimental Brain Research*, 83, 618–633.

Wickens, J.R. (1993). Corticonstriatal interactions in neuromotor programming. *Human Movement Science*, 12, 17–35.

Wickens, J., Hyland, B., & Anson, G. (1994). Cortical cell assemblies: A possible mechanism for motor programs. *Journal of Motor Behaviour*, 26, 66–82.

Woodworth, R.S. (1899). The accuracy of voluntary movements. *Psychological Review Monograph Supplements*, 3, No. 3.

Gesture Imitation in Autism I: Nonsymbolic Postures and Sequences

Isabel M. Smith
Dalhousie University, Halifax, Canada

Susan E. Bryson
York University, Toronto, Canada

This study investigated the ability of children and adolescents with autism to imitate nonsymbolic manual postures and sequences. The controls were children with receptive language delays (matched to the autistic group for age and language level), and typically developing children (matched for language level). Control tasks assessed gesture memory and manual dexterity. Imitation tasks were videotaped for blind scoring of overall accuracy and specific errors. Children with autism performed relatively poorly on posture imitation, but not imitation of simple posture sequences. Reduced manual dexterity contributed to, but did not entirely account for the autistic imitative deficit. An error that was significantly more common in the autistic group suggests that their difficulty in assuming another's perspective may be apparent at the level of simple actions.

INTRODUCTION

Individuals affected by autism share distinctive difficulties in social understanding and forming relationships with others, in reciprocal communication, and in thinking and behaving flexibly in response to environmental demands. The specific manifestations of these difficulties vary according to both age and developmental level. The conceptualisation of autism has changed considerably since this developmental syndrome was first described in 1943 by Kanner (1973). Discussion of the boundaries of the diagnosis and of the validity of subgroups within the spectrum of autistic disorders is ongoing (Volkmar, Klin, & Cohen, 1997).

Requests for reprints should be addressed to Dr. Isabel M. Smith, IWK Grace Health Centre, PO Box 3070, Halifax, Nova Scotia B3J 3G9, Canada (E-mail: ISmith@iwkgrace.ns.ca).

This research was conducted in partial fulfilment of the requirements for the degree of Doctor of Philosophy at Dalhousie University, and was supported by a Medical Research Council of Canada studentship to IMS. A portion of the data was presented at a meeting of the Society for Research in Child Development, Indianapolis, IN, March 1995. The authors acknowledge with thanks the contributions of committee members Drs Raymond Klein, John Barresi, and John Fisk.

Efforts to define further the neuropsychological parameters of autistic spectrum disorders have been intense in recent years. Implicit in this approach is the belief that clearer definition of the characteristics that are specifically associated with autism will aid in efforts to discover causes, identify subgroups, predict outcome, and plan effective interventions. Given that autism is a congenital disorder, considerable focus has been on pivotal social-communicative abilities that develop early in life. Several theoretical accounts have identified impairment in the ability to imitate as critical in the development of other autistic symptoms (Dawson & Lewy, 1989a, b; Meltzoff & Gopnik, 1993; Rogers & Pennington, 1991).

The experimental literature on imitation in autism dates at least to the work of DeMyer et al. (1972), who proposed that children with autism showed a form of apraxia, characterised by difficulty in imitating both symbolic and nonsymbolic gestures. Although the early literature strongly suggested that deficient imitation was associated with autism, a review by Smith and Bryson (1994) identified a number of limitations in the data, and recommended several avenues of investigation. Specifically, Smith and Bryson concluded that much the literature addressed only the question of whether individuals with autism have relatively more difficulty imitating the actions of others, rather than exploring the nature of the difference. They also pointed out that there was an overemphasis on symbolic processes in imitation, in the absence of sufficient information about the ability to imitate arbitrary (nonsymbolic) gestures. Smith and Bryson argued for further investigation of the way in which individuals represent and reproduce movements as a possible key to understanding the nature of their neuropsychological deficits, especially as manifested in imitation.

The present paper is a first step towards addressing the call for a more analytic approach to the study of imitation in autism, and presents data collected as part of a larger project (Smith, 1996). Here we report the results of nonsymbolic gesture tasks, whereas a companion report (Smith & Bryson, 1997) describes data on symbolic gestures and integrates the two sets of results. A discussion of the general issues to be addressed can be found in Smith and Bryson (1994). Briefly, the study reported here provides more descriptive data about nonsymbolic gesture imitation in autism than do previous investigations. In addition to varying the types of tasks (e.g. symbolic vs. nonsymbolic gestures) and request conditions (e.g. verbal instructions vs. imitation), there were other manipulations designed to inform us as to the nature of the imitative deficit. Variables derived from the literature on imitation in autism, as well as from research on adult apraxia and developmental dyspraxia, were incorporated into the study. Most importantly, the study involved analyses of imitation errors by individuals with autism.

The nonsymbolic gestures in this study are restricted to simple manual (hand and finger) postures and sequences. In addition, the availability of visual feedback while imitating the gestures was manipulated by screening the child's view of his/her hands. This manipulation was described by Bergès and Lézine (1965) in their classic monograph on the development of gesture imitation. They reported that obstructing the child's view during imitation resulted in approximately a 2-year old performance decrement. However, Bergès and Lézine also found an age-related increase in the accuracy of imitation in the absence of visual feedback over the 6- to 10-year-old range. Thus, in normal children, manual imitation is facilitated when both visual and kinaesthetic information about hand position is available, but with increasing age, children can compensate for the lack of visual input. Research by Hermelin and her colleagues using nonimitative perceptual-motor tasks has suggested that visual and kinaesthetic input may not be integrated in a normal manner by children with autism (Frith & Hermelin, 1969; Hermelin & O'Connor, 1970; see also Masterton & Biederman, 1983). In these studies, the performance of children with autism appeared not to be adversely affected by the lack of visual feedback, as was the case for both normally developing and mentally handicapped controls. In previous studies of imitation skills in autism, the availability of visual feedback either has not been analysed as a variable (as in all studies using the Uzgiris-Hunt Scales, Uzgiris & Hunt, 1975; e.g. studies by Abrahamson & Mitchell, 1990, and Curcio, 1978), or has been confounded with manual versus facial imitation (e.g. Rogers, Bennetto, McEvoy, & Pennington, 1996), in which different processes may be at work (Dawson, Warrenburg, & Fuller, 1983). Evidence from adults similarly suggests that acquired facial and limb apraxias are determined by somewhat independent mechanisms (Raade, Rothi, & Heilman, 1991).

Bergès and Lézine (1965) reported that complex hand and arm gestures are more likely to be performed by the dominant limb. The lateralisation of praxic processes in autism is an important issue that has received little attention (with the exception of Dawson et al., 1983), despite evidence of atypical patterns of hand preference (Bryson, 1990; Soper et al., 1986). In the present study, spontaneous hand use was observed in each task. Also, manual dexterity was assessed using a standardised measure to evaluate the possible contribution of problems of fine motor coordination to performance on imitation tasks.

Sequencing of manual gestures was assessed using very simple gestures; both sequence length and context were varied. The intent was to minimise the demands of reproducing the gestures themselves in order to assess possible group differences in gesture sequencing ability, since claims for specific deficits in autism have been made (e.g. Hermelin, 1976). When children with autism have been described as having deficits in the sequencing of movements (e.g. Hughes, 1996; Rogers et al., 1996), length of the movement sequences has not been manipulated. Therefore, there was no opportunity for any interaction to arise between sequence length and diagnostic group, as would be observed if groups were differentially affected by sequencing per se. Moreover, problems in the sequencing of information in both perception and production have also been associated with nonautistic language disorders (Tallal, Miller

& Fitch, 1993). Perseverative movement errors might be expected in the group with autism, given the association between autism and repetitive behaviours, and the evidence of such errors on the Wisconsin Card Sorting Test by people with autism (e.g. Ozonoff, Pennington, & Rogers, 1991). However, motor perseveration has also been reported to be frequent in the language-impaired population (Roy, Elliott, Dewey, & Square-Storer, 1990). In the present study, special attention was paid to control for language level, as this is clearly critical in studies of motor and praxic functions in individuals with autism (Smith & Bryson, 1994).

METHOD

Participants

Three groups participated in the research: 20 children and adolescents with autism, 20 with language impairment, and 20 typically developing children. Group characteristics are outlined in Table 1. The participants with autism met Bryson, Clark, and Smith's (1988) operational diagnostic criteria used in their epidemiological study; these are similar to the DSM-IV criteria (American Psychiatric Association, 1994). Each of the children obtained Autism Behaviour Checklist (ABC; Krug, Arick, & Almond, 1980) scores above the cutoff that distinguished autistic from nonautistic children in the epidemiological study (Wadden, Bryson, & Rodger, 1991). As seen in Table 1, these children and adolescents (aged between 7 and 19 years) represent a relatively high-functioning autistic group, with nonverbal ability generally in the mildly impaired to normal range, despite lower levels of receptive language ability.

The Language Impaired control group consisted of 20 children who were matched to the group with autism for sex, and as closely as possible on a case-control basis for chronological age and standard scores on the Peabody Picture Vocabulary Test-Revised (PPVT-R; Dunn & Dunn, 1981). These children were selected based on history (i.e. of significant receptive language delay or severe reading disability), and/or on current relative discrepancy between verbal and nonverbal abilities (i.e. low PPVT-R or verbal IQ relative to nonverbal IQ). Matching on receptive language ability was the primary goal for the present purpose; the clinical groups were also well matched on the basis of Wechsler Block Design scores (see Table 1), although we recognise that this is typically a peak skill for children with autism (Bartak, Rutter, & Cox, 1975). The second control group consisted of 20 normally developing children, matched by sex, chronological age, and PPVT-R mental age equivalents to the children with autism.

Materials

Picture stimuli used in the nonsymbolic posture tasks were black-and-white photographs of the hands of author IMS, who tested all of the children. Each photograph measured 15.3 × 11.2cm, and was mounted on 21.2 × 15.2cm card and laminated with plastic. A 36 × 67cm tabletop screen was used in the No Visual

Table 1. Characteristics of Participants

	Autistic (17 M, 3 F)	Language Impaired (17 M, 3F)	Normal (17 M, 3 F)
Age			
M (SD)	11:4 (41.8)[a]	11:9 (33.6)[a]	6:6 (39.7)
Range	7:0–18:5	6:10–17:8	3:4–13:7
Receptive Language PPVT-R[c]			
Mental age			
M (SD)	7:8 (55.7)[b]	8:7 (46.5)[b]	7:5 (49.3)[b]
Range	2:10–19:9	3:0–17:10	2:10–16:5
Standard score			
M (SD)	68.7 (30.1)	75.7 (22.8)	108 (10.8)
Range	39–120	39–110	90–124
Nonverbal Wechsler Block Design			
M (SD)	8.74 (2.96)	9.73 (2.60)	
Range	5–13	5–14	

[a] $F(1,38) = 0.21$, $P = .65$.
[b] $F(2,57) = 0.45$, $P = .64$.
[c] Peabody Picture Vocabulary Scale-Revised.

Feedback imitation condition. The frame had a 21.5 × 52cm aperture, covered with black fabric in which there were openings for the child's arms. Three 3- to 4-year-old normal control children were too small to use the screen comfortably when seated. Manual preference was evaluated using 10 household objects (hammer, scissors, pencil, comb, toothbrush, spoon, screwdriver, key, straw, razor).

Procedure

Most ($N = 4$) of the children and adolescents were tested in their own residences; the remainder were tested at Dalhousie University, or at their own schools. The tasks were administered in two 30- to 45-minute sessions for all but two children (who required three and four sessions, respectively). The order of tasks within each session is outlined in Table 2. For half of the children in each group, the nonsymbolic action conditions (Postures and Sequences) were presented first; for the remainder, a series of symbolic actions (Actions with Objects and Communicative Gestures; Smith & Bryson, 1997) was presented first. Manual dexterity was evaluated with the Grooved Pegboard (Trites, 1997).

Table 2. Outline of Tasks

Tasks	Session 1: Recognition/Comprehension	Session 2: Imitation/Production
Nonsymbolic tasks[a]		
Postures	Photo recognition	Imitation
Sequences	Photo arrangement	Imitation
Symbolic Tasks[b]		
Actions with objects	Pantomime identification	Pantomime production
		Object use control
		Unconventional object use
Communicative gestures	Gesture identification	Gesture production
	Gesture recognition (context task)	

[a] This report. Nonsymbolic tasks were presented first in Session 1 to half of the participants in each group, to whom symbolic tasks would then be presented first in Session 2.
[b] Smith & Bryson, 1997

Each child was seated opposite the experimenter. Gestures were modelled with the right hand a maximum of twice per item, and each gesture was held for approximately 5 seconds. Children received generous verbal and nonverbal encouragement for their efforts, but no systematic feedback on performance was given. No instructions regarding hand use were given to the children except in the case of the Grooved Pegboard, which was administered first to the dominant hand. Hand dominance for this purpose was based on observation of the child's spontaneous use of a pencil during the Actions with Objects condition (Smith & Bryson, 1997). Each child's performance on the imitation tasks was videotaped. Due to the constraints of the individual settings, the video records were not standardised in terms of angle or distance of recording, or of lighting conditions.

Gesture Memory

Postures. Eight unimanual postures from the deaf alphabet (a, b, e, f, g, h, v, w) and eight bimanual postures adapted from Bergès and Lézine (1965) were selected. Two children in the Autistic group had some prior exposure to signed letters, but neither was able to identify reliably the names of the letters used in this study. Four of the bimanual postures were symmetrical, and the other four were asymmetrical.

For both unimanual and bimanual postures, recognition was assessed in a multiple-choice task, using photographs of the postures (target and two foils, one similar and one dissimilar). The score on this test was the number of items correct out of the 8 unimanual and 8 bimanual postures (i.e. maximum total score = 16). Other data collected included whether a similar or dissimilar foil was selected, and the number of correctly recognised symmetrical versus asymmetrical bimanual postures.

Sequences. Each sequence consisted of a combination of either two or three gestures, selected from three simple components ("fist", "palm", "chop"). Two different sequences were performed for each sequence length in each of three locations (on the table-top, in the air directly in front of the experimenter, on the experimenter's head). The series of six two-action sequences (i.e. two on table, two in air, two on head) was followed by the three-action series. Before the task began, recognition of the individual components (fist, palm, chop) was assessed by pointing to each correct photograph following a demonstration. Ability to reconstruct each sequence was assessed by placing photographs of the correct individual components in mixed order on the table in front of the child, and asking the child either to point to or to place the photographs in order. Photographs were shuffled between trials. Each trial was scored as correct if the child identified the correct sequences (maximum total score = 12).

Gesture Imitation

Postures. Each child was asked to imitate the eight unimanual postures and eight bimanual postures described earlier, immediately after each had been modelled by the experimenter (Model Absent). After all of the unimanual postures were presented, each was modelled a second time and held in view of the child while the child produced a second attempt (Model Present). The same procedure was then followed with the bimanual postures. Finally, the same sets of unimanual and bimanual postures were modelled, and the child asked to imitate with the view of his/her own hands obscured by a screen (No Visual Feedback), while the model remained within view.

There were two types of scoring for Nonsymbolic Postures. First, the videotape record was viewed by the investigator (i.e. nonblind) to locate the still frame that most closely approximated the modelled gesture for each trial; this frame was photographed and printed. A rater assigned similarity judgements comparing each posture to the model (using the photographs from the recognition task as the standard), without knowledge either of group membership or of the purposes of this study. Ratings were made on a scale from 1 to 10, with 1 representing minimal similarity to the model and 10 and exact copy of the model.

Reliability. For all tasks, a second blind rater scored 19 out of 60 records (31%) for reliability. These records were selected randomly, but with constraints in order to represent both diagnostic groups and levels of functioning approximately. For the Postures data, ratings were derived from the still photographs; for the Sequences task, judgements were made from the videotaped records. The same two raters provided judgements on all of the data reported here. They were trained by providing written descriptions of criteria, by discussion with the investigator, and by provision of feedback after practice scoring of videotapes. All final analyses were conducted using the data from the primary blind rater, who scored data for all participants.

The reliability measure for the Postures data was the extent to which the ratings of the two blind judges were correlated, regardless of the absolute value of the ratings (since the scale was arbitrary). Pearson product-moment correlations between the two ratings indicated that the total scores (i.e. summing ratings of all postures and conditions for each individual) were strongly correlated, $r = .90$; correlations between the subscores across conditions ranged from $r = .72$ to .92 (mean $r = .83$). Scores were strongly related for both bimanual ($r = .85$) and unimanual ($r = .92$) postures.

The second type of scoring for Nonsymbolic Postures involved enumeration of errors (see categories in Table 3). Reliability was calculated for 31% of the records scored by the second blind rater (19/60 participants, 48 trials each for a total of 912 judgements). For these data, the question of interest was the extent to which the two raters assigned errors to the same or to different categories. Therefore, kappa coefficients of agreement were used (Bakeman & Gottman, 1986). A preliminary review of the reliability data indicated that the raters' judgements of the larger class of Rotation errors were unreliable. Therefore, further analyses were restricted to the subclass consisting of 180° Rotations (54% of the primary rater's total Rotation errors). Kappas were calculated separately for each of five nonexclusive error types. The reliability of Location errors (kappa = .38) was considered inadequate for these analyses (Fleiss, 1981; Landis & Koch, 1977); these errors were therefore discarded. Kappa for Form errors was .52, considered "fair" (Fleiss, 1981) or "moderate" (Landis & Koch, 1977) agreement; these errors were retained for subsequent analyses. The kappa values for the remaining error types were "excellent" (Fleiss, 1981): Symmetry (kappa = .79), Left-Right (kappa = .95), and 180° Rotation (kappa = .82).

Table 3. Types of Errors for Scoring of Posture Data

Left-right	Left-right reversal of demonstrated hand use.
Form	Incorrect form of gesture (e.g. wrong number of fingers, incorrect position of fingers).
Location	Incorrect position of hands relative to each other (e.g. hands at wrong relative heights, too close together or too far apart).
Rotation	At least one hand rotated 45° through the wrist or elbow in any plane. Subclass: Rotation of hand 180° such that the child's view of his/her own hand is an approximation of the child's view of the model's hand (e.g. palmar rather than dorsal surface is presented to the observer).
Symmetry	Any of: (1) "mirroring" of unimanual posture by other hand, (2) asymmetrical bimanual posture formed symmetrically, (3) symmetrical bimanual posture formed asymmetrically, or (4) bimanual posture performed unimanually.
No response	No recognisable attempt to reproduce the posture.

Sequences. Each child was asked to produce each of the six two-action and then the six three-action Nonsymbolic Sequences (two trials per sequence) immediately following demonstration. Each correct reproduction (correct postures and order) of a sequence was given 1 point (maximum = 24). Liberal criteria were used in judging the correctness of the postures.

Reliability of correct/incorrect judgements was calculated between the two blind judges; kappa (calculated for 19/60 individuals over 335 trials) was .86 ("excellent", Fleiss, 1981).

Errors for Sequences were enumerated separately, and are described in Table 4. Due to the infrequent occurrence of several error types, two categories (Form and Location) were discarded, and several others were collapsed for analysis. A single kappa was calculated reflecting the agreement between the blind raters on the distribution of errors across the following mutually exclusive categories: Perseveration (types I and II combined); Element Change (Substitution, Addition, and Deletion combined); Sequencing; and No Error. Overall, kappa was .70 for these judgements ("good", Fleiss, 1981).

Table 4. Types of Errors for Scoring of Sequence Data

Form	A recognisable attempt was made to form one of the three postures, but was distorted in form (e.g. fingers spread vs. together).
Location	The gesture was reproduced in a location other than that modelled (e.g. on the table vs. in the air).
Addition	An extra gesture was inserted into the sequence (scored independently of the accuracy of other elements).
Deletion	An element of the sequence was missing.
Substitution	An incorrect element replaced an element of the sequence
Perseveration	(1) An immediately preceding sequence was reproduced; or (2) An element within a sequence was repeated.
Sequencing	The correct elements were produced in an incorrect order.

RESULTS

As described in the Method, the three diagnostic groups (Autistic, Language Impaired, and Normal) did not differ significantly in their receptive language mental age scores, as measured by the PPVT-R. However, because of the wide variability in mental ages within groups, and the difficulty of matching on a case-by-case basis, analyses of covariance were conducted using PPVT-R mental age as a covariate. The results of evaluations of assumptions for the various analyses were satisfactory, with the exception of error data for the Nonsymbolic Posture imitation tasks. Logarithmic transformations were applied to these error data to correct positive skewness, and to minimise the effects of one outlying score for 180° Rotation errors in the Posture task.

For most of the repeated measures analyses reported here, compound symmetry assumptions were not met (Mauchly's sphericity test). Therefore, multivariate F, rather than averaged F, results are reported for the repeated-measures effects (SPSS Inc., 1988; Tabachnick & Fidell, 1989). Pillai's criterion was used to evaluate the significance of multivariate Fs.

Gesture Memory

Postures. The univariate analysis of covariance conducted on total Posture recognition scores (maximum score = 16) yielded no significant Group differences, $F(2,55) = 1.35$, $P = .27$; Autistic $M = 12.90$ (SD = 2.0), Language Impaired $M = 13.95$ (SD = 2.3), and Normal $M = 12.63$ (SD = 2.6).

Recognition errors across groups were examined by looking at whether children erred by choosing foils similar to, or dissimilar from, the target posture. The proportions of similar foils chosen relative to the total number of errors made by each child were subjected to a univariate analysis of covariance. There were no significant Group differences on this measure, $F(2,55) = 1.05$, $P = .36$; Autistic $M = .87$ (SD = .27), Language Impaired $M = .89$ (SD = .28), Normal $M = .75$ (SD = .38). When recognition errors occurred, all groups tended to err by choosing similar, rather than dissimilar, foils.

Next, recognition errors across groups were examined with respect to symmetrical versus asymmetrical bimanual postures. An analysis of covariance was performed on ratio scores consisting of the number of errors made on symmetrical bimanual postures as the numerator, with the total number of errors (on symmetrical and asymmetrical postures combined) as the denominator. This analysis revealed significant Group differences, $F(2,55) = 3.57$, $P = .04$; Autistic $M = 0.48$, SD = 0.39, Language Impaired $M = 0.25$, SD = 0.41, Normal $M = 0.16$, SD = 0.33. Planned contrasts indicated that means for the Language Impaired and Normal groups did not differ, $F(1,55) = 0.66$, $P = .42$, but that children in the Autistic group made more errors in recognising symmetrical postures, $F(1,55) = 6.47$, $P = .01$. That is, children in the Autistic group erred equally often on symmetrical and on asymmetrical postures, while the Language Impaired and Normal groups recognised asymmetrical postures less often. Overall, however, recognition performance did not differ across groups.

Sequences. Posture Sequences data consist of the numbers of sequences correctly recalled, as indicated by ordering of the posture photographs, and are displayed in Table 5. The data were analysed in a repeated measure analysis of covariance, with Sequence Length (two or three postures) and Location (table, air, head) as the repeated measures, and Group as the between-subjects measure. There were no significant main effects of Group, $F(2,56) = 0.28$, $P = .76$, or Location, multivariate $F(2,56) = 0.77$, $P = .47$. There was a significant main effect of Sequence Length, $F(1,57) = 163.73$, $P < .001$, such that, for children in all three groups, sequences of two postures were correctly recognised more frequently than were sequences of three postures. There were no significant two- or three-way interactions (Group × Sequence Length × Location, multivariate $F(4,114) = 1.27$, $P = .29$; Group × Sequence Length, $F(2,57) = 0.03$, $P = .98$; Group × Location,

Table 5. Mean Numbers (SD) of Sequences Correctly Identified (Maximum = 2)

Sequences	Autistic Mean (SD)	Language Impaired Mean (SD)	Normal Mean (SD)
Two-posture			
Table	1.6 (0.5)	1.5 (0.7)	1.6 (0.7)
Air	1.5 (0.7)	1.6 (0.7)	1.4 (0.7)
Head	1.4 (0.7)	1.4 (0.8)	1.5 (0.6)
Three-posture			
Table	0.5 (0.7)	0.9 (0.8)	0.7 (0.8)
Air	0.7 (0.9)	0.5 (0.7)	0.6 (0.7)
Head	0.9 (0.9)	0.8 (0.9)	0.9 (0.8)
Total (max = 12)	6.4 (2.8)	6.7 (2.7)	6.5 (3.1)

multivariate $F(4,114) = 0.59$, $P = .67$). Therefore, the effects of the manipulations of Sequence Length and Location were the same on the Autistic group as on controls.

Gesture Imitation

Postures—Global ratings. The data for Nonsymbolic Posture imitation represent global ratings of the similarity of the gestures to the standards. These ratings are presented in Fig. 1 by Group and Condition (Model Absent, Model Present, No Visual Feedback). The data were analysed in a repeated measures Group × Condition analysis of covariance, with Condition as the repeated measure. There were significant main effects of both Group, $F(2,52)$

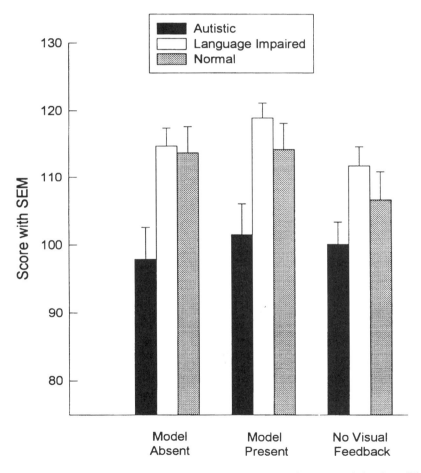

Fig. 1. Mean global ratings of posture imitation, as a function of diagnostic group and visual condition.

$= 5.88$, $P = .005$, and Condition, multivariate $F(2,52) = 6.44$, $P = .003$. Planned contrasts indicated that the effect of Group was due to significantly lower overall scores for the Autistic than for the two control groups, $F(1,52) = 11.1$, $P = .002$; whose scores did not differ, $F(1,52) = 0.66$, $P = .42$. The main effect of Condition was attributable to (1) lower scores obtained in the Model Absent condition, compared to the Model Present condition, $F(1,53) = 6.06$, $P = .02$, and (2) to lower scores in the No Visual Feedback condition, compared to the Model Present condition, $F(1,53) = 6.35$, $P = .02$. The Group × Condition interaction was not statistically significant, multivariate $F(4,106) = 1.44$, $P = .22$.

As the Autistic group erred equally frequently in *recognising* symmetrical and asymmetrical bimanual postures, while the controls erred more frequently on asymmetrical postures, the data were examined for corresponding *imitation* differences. Global ratings of Posture imitation by children in the three Groups for the bimanual symmetrical and asymmetrical postures (collapsed across Model Absent, Model Present, and No Visual Feedback conditions) are presented in Table 6. A repeated measures Group × Symmetry Condition (Symmetrical vs Asymmetrical) analysis of covariance revealed that, consistent with the analysis reported earlier for the imitation rating data, there was a significant main effect of Group, $F(2,43) = 4.95$, $P = .01$ [the Autistic group had lower scores than did either of the control groups, $F(1,43) = 9.78$, $P = .003$, whose scores did not differ, $F(1,43) = 0.13$, $P = .72$]. There was also a main effect of Symmetry Condition, $F(1,44) = 144.42$, $P = .001$, indicating that symmetrical postures were rated as being more accurately imitated than asymmetrical postures. However, there was no significant Group × Symmetry Condition interaction, $F(2,44) = 1.33$, $P = .28$, indicating that neither type of bimanual posture was relatively more difficult for the Autistic (or any other) group.

Postures—Errors. Posture imitation data were also analysed with respect to the presence of the four nonexclusive error types determined to be sufficiently reliable for analysis: Form, Left-Right Reversal, Symmetry, and 180° Rotation. Each error type was analysed in a repeated measures analysis of covariance, with Condition (Model Absent, Model Present, No Visual Feedback) as the repeated measure. Although all of these error types are more common in the Autistic group, there were no significant main effects or interactions for Form or Symmetry errors (all Fs approximately 1.00). The untransformed mean numbers of Left-Right Reversal errors are displayed in Table 7. Analysis of log Left-Right Reversal errors revealed that there was a nonsignificant main effect of Group, $F(2,55) = 2.08$, $P = .14$. There was a significant main effect of Condition, multivariate $F(2,55) = 3.87$, $P = .03$,

Table 6. Mean Global Ratings (SD) of Posture Imitation for Bimanual Postures

	Autistic Mean (SD)	Language Impaired Mean (SD)	Normal Mean (SD)
Symmetrical	81.1 (12.9)	92.8 (7.7)	91.3 (9.7)
Asymmetrical	73.5 (13.4)	82.5 (7.4)	80.7 (10.1)

Table 7. Mean Numbers (SD) of Left-right Reversal Errors

	Autistic Mean (SD)	Language Impaired Mean (SD)	Normal Mean (SD)
Model Absent	6.7 (3.4)	5.8 (3.2)	5.7 (3.6)
Model Present	7.2 (4.0)	5.5 (3.1)	5.2 (4.0)
No Visual Feedback	8.4 (3.1)	8.9 (3.8)	7.0 (5.0)
Total	22.3 (8.3)	20.2 (8.3)	17.9 (10.9)

but no significant Group × Condition interaction, multivariate $F(4,112) = 0.85$, $P = .50$. Planned contrasts indicated that the Model Absent and Model Present conditions did not differ, $F(1,56) = 0.93$, $P = .34$, while the No Visual Feedback condition differed significantly from both of these, $F(1,56) = 7.69$, $P = .008$. That is, significantly more Left-Right Reversal errors were made by children in all groups when they could not see their hands.

Table 8 presents mean numbers of 180° Rotation errors. Analysis of log 180° Rotation errors revealed that there was a significant main effect of Group, $F(2,56) = 4.11$, $P = .02$. Planned contrasts indicated that 180° Rotation errors were more common for the Autistic group than for the two control groups, $F(1,56) = 8.11$, $P = .006$, for whom there was no difference, $F(1,56) = 0.13$, $P = .72$. There was no significant main effect of Condition, multivariate $F(2,56) = 2.32$, $P = .11$, and no significant Group × Condition interaction, multivariate $F(4,114) = 0.50$, $P = .74$.

Sequences—Correct/Incorrect. Numbers of correctly imitated Nonsymbolic Sequences consisting of either two or three postures (Sequence Length) in each of three Locations are presented by Group in Table 9. These data were analysed in a repeated measures analysis of covariance with Sequence Length and Location as the repeated measures. For this analysis, in order for a sequence to be scored as correct, both the sequence and the location had to be reproduced correctly. There was a significant main effect of Sequence Length, $F(1,56) = 64.09$, $P < .001$, indicating that for all groups, fewer three-element sequences were

Table 8. Mean Numbers (SD) of 180° Rotation Errors

	Autistic Mean (SD)	Language Impaired Mean (SD)	Normal Mean (SD)
Model Absent	1.20 (2.38)	0.40 (0.82)	0.50 (1.05)
Model Present	1.50 (2.74)	0.20 (0.52)	0.50 (1.28)
No Visual Feedback	1.05 (2.33)	0.00 (0.00)	0.25 (0.55)
Total	3.9 (7.1)	0.6 (1.1)	1.3 (2.3)

Table 9. Mean Numbers (SD) of Sequences Correctly Imitated (Maximum = 4)

Sequences	Autistic Mean (SD)	Language Impaired Mean (SD)	Normal Mean (SD)
Two-posture			
Table	3.5 (0.9)	3.7 (0.9)	3.2 (1.2)
Air	3.7 (0.7)	3.1 (1.3)	3.0 (1.4)
Head	3.0 (1.2)	3.0 (1.3)	2.3 (1.6)
Three-posture			
Table	2.7 (1.2)	2.8 (1.0)	1.8 (1.7)
Air	2.3 (1.6)	2.1 (1.3)	1.5 (1.6)
Head	2.0 (1.3)	1.9 (1.2)	1.7 (1.8)
Total (max = 24)	16.8 (5.2)	16.5 (5.2)	13.4 (7.3)

correctly produced than two-element sequences. There was no significant main effect of Group, $F(2,55) = 2.29$, $P = .11$, or interaction of Group × Sequence Length, $F(2,56) = 0.29$, $P = .75$. There was a significant main effect of Location, multivariate $F(2,55) = 13.66$, $P < .001$. Planned contrasts indicated that sequences modelled on the table were more often correctly reproduced than those modelled in the air or on the head, $F(1,56) = 25.85$, $P < .001$, and that there was a trend toward fewer correctly imitated sequences when modelled on the head than in the air, $F(1,56) = 3.92$, $P = .053$. Both the Group × Location interaction, multivariate $F(4,112) = 0.77$, $P = .55$, and the Group × Sequence Length × Location interaction, multivariate $F(4,112) = 1.36$, $P = .25$, were nonsignificant.

Sequences—Errors. Three groups of errors were reliably identified in the sequencing task: Perseveration (of individual postures or sequences), Element Change (i.e. Substitution, Addition, or Deletion of postures), and Sequencing. The number of errors was prorated by dividing the actual number of errors of each type by the number of trials completed for each child, and multiplying by the maximum number of trials (24). Separate univariate analyses of covariance were conducted on the numbers of Perseverative, Element Change, and Sequencing errors for each Group, collapsed over Sequence Length and Location. There were no significant Group differences on any error type [all but one F less than 1.0; Group effect for Sequencing errors, $F(2,54) = 1.82$, $P = .17$; Autistic $M = 1.6$, SD = 1.2; Language Impaired $M = 1.9$, SD = 1.9; Normal $M = 1.1$, SD = 1.3].

Handedness and Dexterity

Manual preference was compared across groups in two ways. First, observations of each child's spontaneous use of nine household objects (one object, the screwdriver, was used bimanually by so many children that it was dropped from the preference score) resulted in a Handedness score ranging from 0 to 9, representing the number of objects for which the right hand was used. Handedness scores were analysed in a univariate analysis of covariance, with chronological age as the covariate. There was no significant effect of Group, $F(2,56) = 0.33$, $P = .72$; Autistic $M = 6.9$, SD = 3.2; Language Impaired $M = 7.2$, SD = 3.1; Normal $M = 7.3$, SD = 2.8.

Second, categorical scores were constructed such that individuals with Handedness scores of 9 were considered exclusive right-handers, those with Handedness scores of 0 were considered exclusive left-handers, and those with scores of 1 to 8 were considered to show mixed handedness. It is readily apparent that the proportions of children displaying right-, left-, or mixed-handedness did not differ between groups: 10 Autistic, 10 Language Impaired, and 9 Normal children were right-handers; 8 Autistic, 8 Language Impaired, and 9 Normal children were classed as mixed-handers, and 2 of each group were exclusive left-handers.

Analyses of manual dexterity (i.e. Grooved Pegboard) were restricted to the Autistic and Language Impaired groups. Grooved Pegboard times were available for only 14 children in the Normal group because the remainder were too young to complete the task. Analyses were conducted on a time-per-peg measure. The Grooved Pegboard times were analysed in a repeated measures analysis of covariance, with Hand (Dominant, Nondominant) as the repeated measure and PPVT-R mental age as the covariate. The means are displayed in Table 10. There was a significant main effect of Group, $F(1,36) = 7.45$, $P = .01$. Children in the Autistic group were significantly slower to place pegs than were those in the Language Impaired group. As expected, there were also a significant main effect of Hand, with the dominant hand being faster than the nondominant hand, $F(1,37) = 14.57$, $P < .001$. The Group × Hand interaction was not significant, $F(1,37) = 1.25$, $P = .27$.

Manual Dexterity and Imitation

In order to establish the extent to which differences between the groups in imitation performance might be accounted for by differences in motor skill, between-groups differences in imitation were examined using the dexterity measure as a covariate. First, an analysis of covariance was conducted on mean total posture imitation ratings between the two clinical groups (i.e. collapsed over Model Absent, Model Present, and No Visual Feedback conditions), with receptive language level (PPVT-R mental age) as the covariate. As expected, there was a significant Group difference, $F(1,36) = 4.54$, $P = .003$; Autistic $M = 299.5$, $SD = 51.1$; Language Impaired $M = 345.3$, $SD = 32.2$. Receptive language level accounted for 11% of the variance associated with the imitation ratings scores, $F(1,36) = 4.60$, $P = .04$. The posture imitation rating data were then reanalysed, with both the receptive language mental age (PPVT-R mental age) and dexterity (Grooved Pegboard dominant hand time-per-peg) measures as covariates. Regression analysis indicated that an additional 37% of the variance was accounted for by inclusion of the dexterity measure, $F(2,35) = 16.44$, $P < .001$. With both of these effects covaried, significant Group differences in Nonsymbolic Posture imitation scores remained, $F(1,34) = 4.81$, $P = .04$.

DISCUSSION

The children and adolescents with autism in this study performed relatively poorly on gesture imitation tasks, consistent with previous research. This study extended prior findings by clearly demonstrating that neither delayed receptive language skills nor memory deficits accounted for the autistic imitative impairment. It was also shown that significantly reduced fine motor skills contributed to, but did

Table 10. Mean Times-per-peg (SD) for Completion of the Grooved Pegboard

Hand	Autistic Mean (SD)	Language Impaired Mean (SD)
Dominant	4.9 (1.9)	3.5 (1.2)
Nondominant	5.5 (1.7)	3.9 (1.6)

not entirely account for, the relative praxic deficits of children with autism.

The finding that imitation of simple nonsymbolic gestures by children with autism was less accurate than that by controls confirms the earlier findings of DeMyer et al. (1972), Jones and Prior (1985), and Ohta (1987). Careful matching for intellectual and language levels has been incorporated into recent studies by Rogers et al. (1996) and by Stone, Ousley, and Littleford (in press), both of which also confirmed that imitation was specifically deficient in autism.

Information about the nature of the fundamental nonsymbolic imitative deficit in autism comes from the examination of the effects of manipulation of stimulus conditions and from analysis of the children's errors. Specifically, the role of visual information in the guidance of imitative movements was investigated. It has been hypothesised that the effects of lack of visual feedback during imitation tasks would be reduced for children with autism, based on analogous findings that have been reported for nonimitative visual-motor tasks (Frith & Hermelin, 1969; Hermelin & O'Connor, 1970; Masterton & Biederman, 1983). All three groups showed improved imitation in the second (Model Present) condition, presumably, as a joint function of practice and the availability of a model. Although definitive conclusions are precluded by the confound of order, visual memory differences could not account for the poorer gesture imitation by the Autistic group. Similarly, no group differences were observed when the model remained present but no view of the performing hand was available to the child, nor when sequences were performed out of sight (on the head). Instead, these conditions tended to be more difficult for all children, not differentially more or less so for those with autism. Inspection of the means in Fig. 1 suggests that the Language Impaired control group may have been more adversely affected by being unable to see their hands than the Autistic group (intermediate scores were obtained for the Normal group).

A notable finding of this study was that a particular error involving rotation of the hand by 180° (e.g. palm toward, rather than away from, the viewer) was more common for children with autism. Barresi and Moore (1996, commenting on data from Ohta, 1987) speculated that such errors reflect the child with autism's difficulty in translating his or her view of another's action into a matching action of the self (see also Whiten & Brown, 1998). This difficulty with appropriate spatial transformation of the gesture could be regarded as an action analogue of the pronoun reversal errors frequently observed in the speech of individuals with autism (Charney, 1980). Further exploration of the conditions under which this action error occurs in individuals with autism, and its relationship to other child characteristics (e.g. language and social skills) may provide important insights into the implications of imitation deficits for the cognitive and social development of these individuals. Smith (1996) noted anecdotally that there may be a period in normal development during which infants are prone to this error. It would be of interest to examine relationships among language, social functioning, and imitation in very young normal children and those presenting with possible autism, to explore the origins and functional significance of this behaviour.

Certainly the finding of more 180° Rotation errors by the Autistic group poses a challenge for any account based on diminished use of visual feedback in autism. Indeed, it appears more consistent with the contrary position, that the children with autism are responding on the basis of what is directly accessible to visual perception, that is, to their view of their own hands. This might imply increased, rather than decreased, dependence on visual feedback. These alternatives may be reconciled by the suggestion that it is the integration of information in perception, rather than simply the predominance of one modality over another, that is abnormal in autism, and that this may be fundamentally an attentional problem (Bryson, Landry, & Wainwright, 1997; Smith & Bryson, 1994). Barresi and Moore (1996) cited rotation errors as indirect support for their hypothesis that individuals with autism show a general failure to integrate information specifying first- and third-person perspectives, on the levels of action, affect, and cognition. They speculated that this might be attributable to an early developmental failure by infants with autism to process the intermodal information that is available in social interchanges, or perhaps to an even more general failure of intermodal integration (cf. Smith & Moore, 1997).

In the present study, the imitation performance of the Autistic may have been poor across all of the conditions relative to controls because of their diminished ability to integrate information from vision and kinaesthesis. Removal of visual feedback in the third imitation condition would thus have been irrelevant to their performance, which would already reflect poor integration of the two information sources (as opposed to lack of dependence on vision). In order to resolve this issue, it will be necessary to correct the confounding effects of practice. In addition, perhaps other tasks could address this issue more directly than in the context of praxis. Of particular interest is the issue of whether concomitant failures of intermodal integration would be observed on tasks that do and do not tap an interpersonal dimension (Smith & Moore, 1997; see Haviland, Walker-Andrews, Huffman, Toci, & Alton, 1996, and Loveland et al., 1995).

The significant decline in the quality of imitation in the No Visual Feedback condition for all participants was an expected finding (Bergès and Lézine, 1965). What was unexpected, however, was the significant increase in Left-Right Reversal errors that occurred in this condition. We suggest that this effect is a result of the children reverting under challenging conditions to a less mature imitation strategy, as though they were facing a mirror. These responses were found by Bergès and Lézine to constitute the majority in normally developing children up until the age of 6 years, and non-mirror responses predominated only in children over 10 years of age. Mirror responses might be interpreted as another form of visually driven matching, with the gesture reproduced relative to an external, rather than an egocentric (body-based) frame of reference. A similar argument for 180° Rotation errors would attribute their more frequent occurrence in the Autistic group and in young typically/developing children to a failure to encode the gestures of others accurately within a self-referenced framework. One would then expect an increased frequency of mirror (Left-Right Reversal) errors in the Autistic group. Although no significant preponderance was observed, the trend was consistent with somewhat more mirror (Left-Right Reversal) errors being produced by the children with autism when their hands were visible to them (see Table 7). Perhaps the 180° Rotations by the children with autism in the present study stand out because they are so developmentally inappropriate, given the age range of the sample, whereas the mirror responses are still relatively common in the other groups. One might predict that left-right reversals would persist in an older, higher-functioning autistic group.

The children with autism in this study performed as well as controls in their recognition of nonsymbolic postures. Three picture choices were used in the recognition task in the present study, in order not to tax too greatly the attentional capacities of the younger and lower-functioning participants. The results of the manipulation of the similarity of foils to target postures indicated that the children with autism were attentive to the features of the target gesture. That is, there was no indication of more impulsive responding by the children with autism, in that when errors were committed, the foils were not chosen randomly. Rogers et al. (1996) also found no problem with gesture recognition for their autistic group. They reported that their gesture recognition task used six foils, but did not describe how similar the targets and foils were, or report any analysis of errors.

Children in both control groups erred more often in recognising asymmetrical bimanual postures, whereas the errors of children with autism were evenly divided between symmetrical and asymmetrical postures. While this result might suggest some difference between the autistic and nonautistic groups in the encoding of these two types of gestures, it is difficult to interpret what that effect might be. For example, it did not appear to have implications for the production of the gestures. More detailed exploration of relationships between recognition and production errors, perhaps on an individual basis, might be warranted. It may be that neither the coding of quality of imitation nor of symmetry errors in this study was sufficiently sensitive to detect differences.

The present results indicate that the gesture imitation difficulties of children with autism are accounted for in part, but not entirely, by deficits in motor coordination. Manual dexterity was significantly impaired in this sample of children and adolescents with autistic spectrum disorders. That is, they completed the Grooved Pegboard more slowly than did controls, both with the dominant and nondominant hands (consistent with data from Szatmari, Tuff, Finlayson, & Bartolucci, 1990). The population of individuals who present clinically with autistic spectrum disorders is extremely variable with respect to performance on tasks assessing manual motor skill, but performance tends to be poorer than that predicted by language mental age. Where results are not consistent with this conclusion, both sample and task differences may be factors. An important issue is the extent to which the heterogeneity in motor skills within autistic spectrum disorders may be related to subgroup differences (Smith, in press).

One somewhat surprising aspect of the present findings was the unimpaired performance of the children with autism on the tasks assessing recognition and imitation of posture sequences. Whereas group differences were seen in the imitation of static postures, there were none for simple sequences (when the demands for postural accuracy were minimised). The failure to observe an increased number of perserverative responses is perhaps particularly surprising, given evidence of perseveration for individuals with autism on tasks such as the Wisconsin Card Sorting Test (Ozonoff et al., 1991; Szatmari et al., 1990), and others (Hughes & Russell, 1993; McEvoy, Rogers & Pennington, 1993). However, increased perseverative responding in autism is not a universal finding (Minshew, Goldstein, Muenz, & Payton, 1992). The range of language functioning in the present study was great, with some children obtaining mental age equivalents below 3 years. Multiple instances and forms of perseveration (e.g. of postures, of sequences) were observed for some children in all three groups, especially those with lower verbal mental ages. We collapsed categories of perseverative responding to improve the reliability of scoring these errors; it is possible that some alternative parsing of the data might reveal group differences. The present result (that is, no relative difficulty with gesture sequences) is not inconsistent with the sequencing deficit claimed by Rogers et al. (1996), as their autistic group also showed lower performance than controls in their *single* movement conditions (which were considerably more complex than the postures employed in this study), as well as in sequential movements. When single gestures cannot be accurately imitated, failure to imitate complex sequences cannot be attributed parsimoniously to a sequencing deficit. The assertion that "sequencing" is deficient in autism dates at least from the work of Hermelin and O'Connor (1970). On a variety of auditory tasks (both linguistic and nonlinguistic), these investigators demonstrated atypical patterns of both recall and production of ordered information. However, in a visual-spatial ordering task, children with autism were able to reproduce the order of a series of four pictures that had just been shown to them, and to use semantic associations between the pictures to guide their performance (Hermelin & O'Connor, 1970). This picture task resembles the sequence memory task in the present study, in that both perception and reproduction of a visually presented series were required. The Autistic group in the present study did not differ from controls even when they imitated the sequences of postures directly. This latter finding is challenging, given that Hermelin (1976) summarised years of careful studies by concluding that children with autism can deal effectively with information presented in a spatial framework, but that their processing of temporal sequences is impaired. Furthermore, she hypothesised that the autistic tendency is to process information in the modality in which it is presented, rather than recoding it to a more abstract (amodal) form. Nonetheless, in the present study, sequences were reproduced spatially (in the case of the memory task) and temporally (in the case of the imitation task). We can suggest only that these tasks were simple enough (both in terms of the postures and the brevity of the two- to three-item sequences) to be accommodated by the strategies available to the participants with autism. In any case, it is clear from the present results that it could not be merely the sequencing aspect of gesture production that underlies the pronounced difficulties in this domain consistently reported for people with autism. It would be of interest to document the performance of individuals with autism on alternative action-sequencing tasks such as those employed by Kimura (1977) or Roy (1981) in their studies of adults with acquired apraxia. In addition, it may be important to examine the relationships between the elements of sequences, whether these are semantic relationships, or whether the sequences of movements share "directedness" toward a goal.

In a recent report, Hughes (1996) provided clear evidence of motor planning impairments for a group of children with autism, and advocated additional work to tease apart sequencing versus prediction deficits in autism. The present study makes a contribution to this effort. The sequencing task used here reduced the demands for posture imitation by using very simple discrete postures.

Short sequences were also used, but more importantly, a manipulation of sequence length was introduced. The present results indicate that children with autism were not affected disproportionately by the requirement to imitate longer sequences. Therefore, a simple sequencing account for their failure to execute more complex series of movements can be rejected. The distinction between the necessity for serial ordering (simple sequencing, as in the present task) versus planning (movements directed toward a goal, as in Hughes, 1996) appears important.

Future more hypothesis-driven studies of imitative behaviour can build on the information provided by the present research. For example, it is now possible to construct a gesture series and tests of recognition that would maximise the likelihood of errors that could increase our understanding of the processes at work. These kinds of data require interpretation within a developmental context, as the behaviour of the Autistic group on the present tasks often resembled that of the youngest normally developing children (cf. Hughes, 1996). It is as though some aspects of information processing remain "stuck" in a developmentally early mode (although the mechanisms responsible in the case of autism may well differ from those in typical development). In order to provide an adequate account of what is typical in autism, a well-articulated neuropsychological account of the normal development of praxic abilities is required (cf. Cermak, 1985; Dewey, 1995; Roy et al., 1990). Further empirical evidence of the disruption of praxis in autism would, in turn, assist the effort of constructing useful developmental models, both normal and pathological.

This analysis of autistic deficits in the imitation of simple meaningless manual gestures lays the foundation for the second part of our study (Smith & Bryson, 1997). It is our contention that attempts to understand complex imitative phenomena without such a foundation have resulted in a lack of recognition of basic information processing deficits in autism. Our data on communicative gestures and actions with objects clarify the nature of the imitation problem seen here by suggesting a fundamental deficit in the representation of gestures and of actions more generally. We elaborate this working hypothesis in our companion paper.

REFERENCES

Abrahamsen, E.P., & Mitchell, J.R. (1990). Communication and sensorimotor functioning in children with autism. *Journal of Autism and Developmental Disorders, 20*, 75–86.

American Psychiatric Association. (1994). *Diagnostic and statistical manual of mental disorders* (4th edn.). Washington, DC: Author.

Bakeman, R., & Gottman, J.M. (1986). *Observing interaction: An introduction to sequential analysis*. Cambridge: Cambridge University Press.

Barresi, J., & Moore, C. (1996). Intentional relations and social understanding. *Brain and Behavioural Sciences, 19*, 107–122.

Bartak, L., Rutter, M., & Cox, A. (1975). Comparative study of infantile autism and specific developmental receptive language disorder: I. The children. *British Journal of Psychiatry, 126*, 127–145.

Bergès, J., & Lézine, I. (1965). The imitation of gestures. *Clinics in Developmental Medicine, No. 18*. London: Spastics Society/William Heinemann.

Bryson, S.E. (1990). Autism and anomalous handedness. In S. Coren (Ed.), *Left-handedness: Behavioural implications and anomalies* (pp. 441–456). Amsterdam: Elsevier North-Holland.

Bryson, S.E., Clark, B.S., & Smith, I.M. (1988). First report of a Canadian epidemiological study of autistic syndromes. *Journal of Child Psychology and Psychiatry, 29*, 433–445.

Bryson, S.E., Landry, R., & Wainwright, A. (1997). A componential view of executive dysfunction in autism: Review of recent evidence. In J.A. Burack & J.T. Enns (Eds.), *Attention: Development and psychopathology*. New York: Guilford Press.

Bryson, S.E., Wainwright-Sharp, J.A., & Smith, I.A. (1990). Autism: A developmental "spatial" neglect syndrome? In J. Enns (Ed.), *The development of attention: Research and theory*. (pp. 405–427). Amsterdam: Elsevier North-Holland.

Cermak, S. (1985). Developmental dyspraxia. In E.A. Roy (Ed.), *Neuropsychological studies of apraxia and related disorders (Advances in Psychology, Vol. 23)*, (pp. 225–248). Amsterdam: North Holland.

Charney, R. (1980). Pronoun errors in autistic children: Support for a social explanation. *British Journal of Disorders of Communication, 15*, 39–43.

Curcio, F. (1978). Sensorimotor functioning and communication in mute autistic children. *Journal of Autism and Childhood Schizophrenia, 8*, 281–292.

Dawson, G. (1988). Cerebral lateralization in autism: Clues to its role in language and affective development. In D.L. Molfese & S.J. Segalowitz (Eds.), *Brain lateralization in children: Developmental implications*. New York: Guilford Press.

Dawson, G. (1991). A psychological perspective on the early socio-emotional development of children with autism. In D. Cicchetti & S. Toth (Eds.), *Rochester Symposium on Developmental Psychopathology, Vol. 3*. Rochester, NY: University of Rochester Press.

Dawson, G., & Lewy, A. (1989a). Arousal, attention, and the socioemotional impairments of individuals with autism. In G. Dawson (Ed.), *Autism: Nature, diagnosis, and treatment* (pp. 49–74). New York: Guilford Press.

Dawson, G., & Lewy, A. (1989b). Reciprocal subcortical-cortical influences in autism: The role of attentional mechanisms. In G. Dawson (Ed.), *Autism: Nature, diagnosis, and treatment* (pp. 144–173). New York: Guilford Press.

Dawson, G., Warrenburg, S., & Fuller, P. (1983). Hemisphere functioning and motor imitation in autistic persons. *Brain and Cognition, 2*, 346–354.

DeMyer, M.K., Alpern, G.D., Barton, S., DeMyer, W.E., Churchill, D.W., Hingtgen, J.N., Bryson, C.Q., Pontius, W., & Kimberlin, C. (1972). Imitation in autistic, early schizophrenic, and nonpsychotic children. *Journal of Autism and Childhood Schizophrenia, 2*, 264–287.

Dewey, D. (1993). Error analysis of limb and orofacial praxis in children with developmental motor deficits. *Brain and Cognition, 23*, 203–221.

Dewey, D. (1995). What is developmental dyspraxia? *Brain and Cognition, 29*, 254–274.

Dunn, L.M., & Dunn, L.M. (1981). *Peabody Picture Vocabulary Test-Revised.* Circle Pines, MN: American Guidance Service.

Fleiss, J.L. (1981). *Statistics for rates and proportions.* New York: Wiley.

Frith, U., & Hermelin, B. (1969). The role of visual and motor cues for normal, subnormal and autistic children. *Journal of Child Psychology and Psychiatry, 10*, 153–163.

Harris, S.L., & Handleman, J.S. (Eds.) (1994). *Preschool education programs for children with autism.* Austin, TX: PRO-ED.

Haviland, J.M., Walker-Andrews, A.S., Huffman, L.R., Toci, L., & Alton, K. (1996). Intermodal perception of emotional expressions by children with autism. *Journal of Developmental and Physical Disabilities, 8*, 77–88.

Hermelin, B. (1976). Coding and the sense modalities. In L. Wing (Ed.), *Early childhood autism* (2nd ed.) (pp. 135–168). Oxford: Pergamon Press.

Hermelin, B., & O'Connor, N. (1970). *Psychological experiments with autistic children.* Oxford: Pergamon Press.

Hughes, C. (1996). Planning problems in autism at the level of motor control. *Journal of Autism and Developmental Disorders, 26*, 90–107.

Hughes, C., & Russell, J. (1993). Autistic children's difficulty with mental disengagement from an object: Its implications for theories of autism. *Developmental Psychology, 29*, 498–510.

Jones, V., & Prior, M.P. (1985). Motor imitation abilities and neurological signs in autistic children. *Journal of Autism and Developmental Disorders, 15*, 37–46.

Kanner, L. (1973). Autistic disturbances of affective contact. In *Childhood psychosis: Initial studies and new insights.* Washington, DC: V.H. Winston. (Original work published in 1943.)

Kimura, D. (1977). Acquisition of a motor skill after left hemisphere damage. *Brain, 100*, 527–542.

Krug, D.A., Arick, J.R., & Almond, P.J. (1980). Behavior checklist for identifying severely handicapped individuals with high levels of autistic behavior. *Journal of Child Psychology and Psychiatry, 21*, 221–229.

Landis, R.J., & Koch, G.G. (1977). The measurement of interobserver agreement for categorical data. *Biometrics, 33*, 159–174.

Loveland, K.A., Tunali-Kotoski, B., Chen, R., Brelsford, K.A., Ortegon, J., & Pearson, D.A. (1995). Intermodal perception of affect in persons with autism or Down syndrome. *Development and Psychopathology, 3*, 409–418.

Masterton, B.A., & Biederman, G.B. (1983). Proprioceptive versus visual control in autistic children. *Journal of Autism and Developmental Disorders, 13*, 141–152.

McEvoy, R.E., Rogers, S.J., & Pennington, B.F. (1993). Executive function and social communication deficits in young autistic children. *Journal of Child Psychology and Psychiatry, 34*, 563–578.

McManus, I.C., Sik, G., Cole, D.R., Mellon, A.F., Wong, J., & Kloss, J. (1988). The development of handedness in children. *British Journal of Developmental Psychology, 6*, 257–273.

Meltzoff, A.N., & Gopnik, A. (1993). The role of imitation in understanding persons and developing a theory of mind. In S. Baron-Cohen, H. Tager-Flusberg, & D.J. Cohen (Eds.), *Understanding other minds: Perspectives from autism* (pp. 335–366). Oxford: Oxford University Press.

Minshew, N.J., Goldstein, G., Muenz, L.R., & Peyton, L.R. (1992). Neuropsychological functioning in nonmentally retarded autistic individuals. *Journal of Clinical and Experimental Neuropsychology, 14*, 749–761.

Ohta, M. (1987). Cognitive disorders of infantile autism: A study employing the WISC, spatial relationship conceptualization and gesture imitations. *Journal of Autism and Developmental Disorders, 17*, 45–62.

Ozonoff, S., Pennington, B.F., & Rogers, S.J. (1991). Executive function deficits in high-functioning autistic individuals: Relationship to theory of mind. *Journal of Child Psychology and Psychiatry, 32*, 1081–1105.

Raade, A.S., Rothi, L.J.G., & Heilman, K. (1991). The relationship between buccofacial and limb apraxia. *Brain and Cognition, 16*, 130–146.

Rogers, S.J., Bennetto, L., McEvoy, R.E., & Pennington, B.F. (1996). Imitation and pantomime in high-functioning adolescents with autism spectrum disorders. *Child Development, 67*, 2060–2073.

Rogers, S., & Pennington, B.F. (1991). A theoretical approach to the deficits in infantile autism. *Development and Psychopathology, 3*, 137–162.

Rothi, L.J.G., Ochipa, C., & Heilman, K.M. (1991). A cognitive neuropsychological model of limb praxis. *Cognitive Neuropsychology, 8*, 443–458.

Roy, E.A. (1981). Action sequencing and lateralized cerebral damage: Evidence for asymmetries in control. In J. Long & A. Baddeley (Eds.), *Attention and performance, Vol. 9*, (pp. 487–500). Hillsdale, NJ: Lawrence Erlbaum Associates Inc.

Roy, E.A., Elliott, D., Dewey, D., & Square-Storer, P. (1990). Impairments to praxis and sequencing in adult and developmental disorders. In C. Bard, M. Fleury, & L. Hay (Eds.), *Development of eye-hand coordination across the life span* (pp. 358–384). Columbia, SC: University of South Carolina Press.

Smith, I.M. (1996). *Imitation and gesture representation in autism.* Unpublished doctoral dissertation, Dalhousie University, Halifax, Nova Scotia, Canada.

Smith, I.M. (in press). Motor functioning in Asperger syndrome. In F.R. Volkmar, A. Klin, & S. Sparrow (Eds.), *Asperger syndrome*. New York: Guilford Press.

Smith, I.M., & Bryson, S.E. (1994). Imitation and action in autism: A critical review. *Psychological Bulletin, 116*, 259–273.

Smith, I.M., & Bryson, S.E. (1997). *Gesture imitation in autism II: Symbolic gestures and pantomimed object use.* Manuscript in preparation.

Smith, I.M., & Moore, C. (1997). *Intersensory functioning in autism: A reconsideration.* Manuscript in preparation.

Soper, H.V., Satz, P., Orsini, D.L., Henry, R.R., Zvi, J.C., & Schulman, M. (1986). Handedness patterns in autism suggest subtypes. *Journal of Autism and Developmental Disorders, 16*, 155–167.

SPSS, Inc. (1988). *SPSS-X User's Guide* (3rd ed.). Chicago, IL: Author.

Stone, W., Ousley, O.Y., & Littleford, C.D. (in press). Motor imitation in young children with autism: What's the object? *Journal of Abnormal Child Psychology.*

Szatmari, P., Tuff, L., Finlayson, A.J., & Bartolucci, G. (1990). Asperger's syndrome and autism: Neurocognitive aspects. *Journal of the American Academy of Child and Adolescent Psychiatry, 29*, 130–136.

Tabachnick, B.G., & Fidell, L.S. (1989). *Using multivariate statistics* (2nd ed.). New York: Harper Collins.

Tallal, P., Miller, S., & Fitch, R.H. (1993). Neurobiological basis of speech: A case for the preeminence of temporal processing. *Annals of the New York Academy of Sciences, 682*, 27–47.

Trites, R.L. (1977). *Neuropsychological test manual.* Ottawa, ON: Royal Ontario Hospital.

Uzgiris, I.C., & Hunt, J.M. (1975). *Assessment in infancy.* Urbana, IL: University of Illinois Press.

Volkmar, F.R., Klin, A., & Cohen, D.J. (1997). Diagnosis and classification of autism and related conditions: Consensus and issues. In D.J. Cohen and F.R. Volkmar (Eds.), *Handbook of autism and pervasive developmental disorders* (2nd edn.) (pp. 5–40). New York: Wiley.

Wadden, N.P.K., Bryson, S.E., & Rodger, R.S. (1991). A closer look at the Autism Behavior Checklist: Discriminant validity and factor structure. *Journal of Autism and Developmental Disorders, 21*, 529–541.

Wainwright-Sharp, J.A., & Bryson, S.E. (1993). Visual orienting deficits in high-functioning people with autism. *Journal of Autism and Developmental Disorders, 23*, 1–14.

Whiten, A., & Brown, J.D. (1998). Imitation and the reading of other minds: Perspectives from the study of autism, normal children and non-human primates. In S. Bråten (Ed.), *Intersubjective communication and emotion in ontogeny: A sourcebook*. Cambridge: Cambridge University Press.

Young, J.M., Krantz, P.J., McClannahan, L.E., & Poulson, C.L. (1994). Generalized imitation and response-class formation in children with autism. *Journal of Applied Behavior Analysis, 27*, 685–697.

DISORDERED ACTION SCHEMA AND ACTION DISORGANISATION SYNDROME

G.W. Humphreys
University of Birmingham, UK

E.M.E. Forde
University of Aston, Birmingham, UK

We assessed the performance of four patients on a range of everyday tasks. Two patients were identified as having the symptoms of "Action Disorganisation Syndrome" (ADS) on the basis of their errors. The other two patients were matched to the ADS patients for short-term and longer-term episodic memory performance and for performance on the Stroop task. The ADS patients were selectively impaired both when carrying out the everyday tasks and when generating and ordering the necessary component subactions. These patients also had problems in following novel commands, though similar difficulties occurred in a patient with frontal lobe damage but without deficits on everyday tasks. In addition, different patterns of perseverative error arose in the two patients showing symptoms of ADS. We suggest that ADS can be caused by a disorder of action schema, and we discuss the different ways in which the retrieval of such schema can break down.

INTRODUCTION

The frontal lobes have long been regarded as being important for relatively "high-level" cognitive functions, and for determining successful and appropriate everyday human behaviour. For example, in the last century, Burdach (1819) referred to the frontal lobes as the "special workshop of the thinking process" (cited in Shallice, 1988) and, more recently, Duncan, Emslie, and Williams (1996) suggested that "general intelligence" (g) might be related to the activity of the frontal lobes. However, despite the general agreement that the frontal lobes are important in "high-level" control of human behaviour, there have been relatively few cognitive models developed to explain how they might be involved in controlling everyday actions, and what the nature of the control processes might be. Here we attempt to decompose some of the operations of these control processes, by assessing how the

---

Requests for reprints should be addressed to Professor Glyn Humphreys, Department of Psychology, University of Birmingham, Edgbaston, Birmingham, B15 2TT, UK (E-mail: G.W.Humphreys@bham.ac.uk).
This work was supported by a grant from the Medical Research Council awarded to the first author.

performance of everyday tasks is disrupted following damage involving the frontal lobes.

One of the first attempts to develop a cognitive model of the functions of the frontal lobes in everyday behaviour was outlined by Norman and Shallice (1986) (see Fig. 1). The first part of their model, the Contention Scheduling System, was designed to explain how we might execute routine tasks. They suggested that we store schemas for routine tasks and, when a triggering stimulus activates a schema above its threshold, that schema would remain active until the goal is attained or the schema is actively inhibited by competing schemas. However, human behaviour goes beyond the activation of routine actions from a triggering stimulus and Norman and Shallice proposed that a second system, the Supervisory Attentional System (SAS), might be involved in the control of non-routine "intentional" or "willed" actions that would require "higher-order" cognitive control. For example, dealing with novelty, or overcoming instinctive behaviour to act in a socially appropriate manner, would require involvement of the SAS.

On other accounts, the Contention Scheduling and Supervisory Attentional Systems are not conceptualised as functionally separate but rather they reflect the operation of knowledge sources at different levels of abstraction. For example, Grafman (1995) has argued that complex sequential everyday behaviours depend on the activity of stored memories, organised into "Structured Event Complexes"

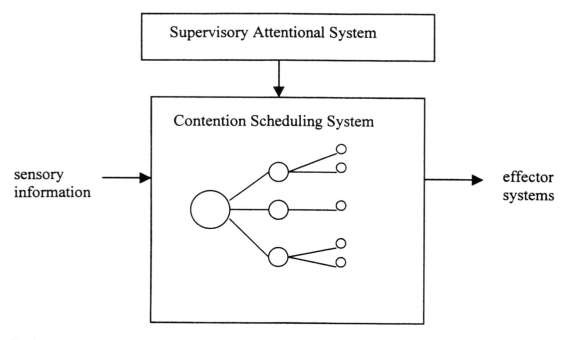

Fig. 1. Norman and Shallice's (1986) model of the control of action.

and "Managerial Knowledge Units". Structured Event Complexes are the more primitive units that store a sequence of events or actions, and different Structured Event Complexes store different types of information. For example, some may specify norms in social behaviour (e.g. waiting to be seated in a restaurant) whereas others underpin particular skills (e.g. the sequence of actions necessary to use chopsticks). A number of Structured Event Complexes can become associated to form "higher-level" Managerial Knowledge Units at varying levels of abstraction (see Fig. 2). According to this proposal, Norman and Shallice's Supervisory Attentional System is simply a set of context-free Managerial Knowledge Units that are relatively abstracted from specific behavioural instances and so can serve to guide performance in underspecified conditions. These high-level units are not functionally distinct from the lower-level managerial units and Structured Event Complexes that they modulate, but they are thought to be held in the frontal and prefrontal cortices. The idea that frontal lobe representations play a role in linking behaviours across time so that performance is not determined solely by the immediate stimulus is common to animal models of frontal lobe functioning, which hold that frontal lobe structures keep stimulus representations active over time (Goldman-Rakic, 1992) and that they are involved in representing the temporal order of behaviours (Fuster, 1991). We return to the importance of

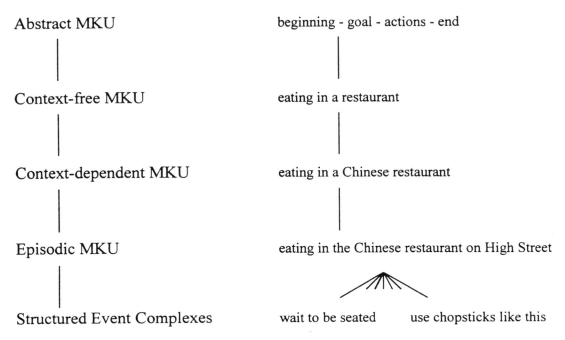

Fig. 2. The control of action using Managerial Knowledge Units and Structured Event Complexes (Grafman, 1995).

representing temporal structure when we discuss patients with "action disorganisation syndrome" (ADS) later.

There have been numerous studies illustrating how damage to the frontal lobes can impair performance on tasks that require "intentional" control over behaviour (Klosowska, 1976; Milner, 1963; Milner, Petrides, & Smith, 1985). However, patients with frontal lobe damage may not only have problems with novel tasks but also with routine everyday tasks, such as making a cup of tea or dressing themselves. For example, Luria (1966) described patients with frontal lobe lesions who were no longer able to perform familiar activities. One patient, when asked to light a candle, lit the candle correctly but then put it to his mouth in an attempt to smoke it. On another occasion, the patient lit the candle and then broke it in half and threw it away as if it was a match. In Norman and Shallice's model, everyday tasks should be accomplished without the involvement of the SAS, by routine activation of the Contention Scheduling System. In Grafman's view, however, even everyday actions may involve activation of lower-level Managerial Knowledge Units and so may be compromised to some degree by frontal lobe damage. Patients with impairments in routine everyday tasks can help shed light on what are conceived of as the Contention Scheduling System or low-level Managerial Knowledge Units in these models.

Early reports of frontal lobe patients who failed to perform routine activities were relatively anecdotal. Recently, however, Schwartz and her colleagues have documented systematic case studies of patients with so-called "action disorganisation syndrome" (ADS) following damage including the frontal lobes, and developed an "action coding system" to assess how efficiently and accurately patients perform routine tasks such as making a cup of coffee (Schwartz, 1995; Schwartz, Mayer, Fitzpatrick De Salme, & Montgomery, 1993; Schwartz, Montgomery, et al., 1995b; Schwartz, Reed, Montgomery, Palmer, & Mayer, 1991). We review their attempts to describe how performance on everyday tasks can be measured before describing our analysis of the factors that may contribute to success or failure on these tasks.

ACTION DISORGANISATION SYNDROME

Schwartz et al. (1991) developed the action coding system to provide a systematic way of measuring the ability of ADS patients to carry out everyday tasks. They started by dividing a given task into a number of smaller units of action which they called A1 steps and defined as "the smallest component of a behavioural sequence that achieves a concrete, functional result or transformation, describable as the movement of an object from one place to another or as a change in the state of an object". These A1 steps were grouped into A2 steps that reflected a higher-level representation in the action hierarchy that specified a particular subgoal. For example, "lift the cream, open the cream, move the cream to the cup" would be A1 steps in the A2 step "put cream in the coffee". Schwartz et al. (1991) used the proportion of independent A1s (ie. A1 steps that were not performed as part of an A2 step) as a measure of the coherence of the patients' performance. They also classified the errors made by the patients into six groups: (1) place substitutions, in which an object was moved to the wrong destination (e.g. putting orange juice in the coffee); (2) object substitutions, in which an object was used incorrectly (note that these are difficult to distinguish from place substitutions, since putting orange juice in the coffee might be a place substitution or an object substitution); (3) anticipations, which reflected the patient performing actions in the wrong sequence (e.g. drinking from the cup then adding coffee granules); (4) tool substitutions (e.g. stirring the coffee with a fork); (5) quality errors, in which the patient generally did the correct action but in a clumsy way (e.g. only partially opening a packet of sugar); and (6) omissions (e.g. failing to put any coffee granules in the cup).

Schwartz, Fitzpatrick-DeSalme, and Carew (1995a; Schwartz & Buxbaum, 1997) conducted a study of the performance of 14 controls and 15 patients on 6 everyday tasks: preparing a letter for mailing, preparing toast with butter and jam, wrapping a present, repotting a plant, preparing a pot of coffee, and packing a lunchbox. They found that control subjects tended to make sequence errors (15/35, or 43%, of their errors were in this category) whereas the patients predominately made omissions (61/204, 30%) and sequence errors (24%), with a few additions and semantic errors. Schwartz et al. found that patients made different types of errors on different tasks. For example, they made a larger proportion of semantic errors on the coffee task (25%) compared to the lunchbox task (11%), despite the fact that on overall difficulty (measured by the total number of errors) the tasks were equated. Interestingly, Schwartz et al. also found that individuals within the patient group performed relatively poorly on some tasks but at control levels on others. They suggested that the qualitative distribution of errors, and the individual inconsistency over tasks, may reflect differences in the number of steps in the task, in the sequencing constraints of the task or in the number of objects presented simultaneously in the array. For example, the number of objects could affect the number of competing "bottom-up" associations that may trigger action schemas. In the current paper, we were interested in assessing the influence of both the number of steps in the task and the number of objects in the array on the performance of ADS patients.

In a more detailed single case study, Schwartz et al. (1991) described an ADS patient, HH. For example, their analysis of him making a cup of coffee revealed a relatively large proportion of independent A1 steps (31% of his A1 actions were independent of any A2 goal). Over 28 test sessions in which he made a cup of coffee, he made 97 errors (a mean number of 3.5 per session); the majority were place and object substitutions (57/97, 59%), a few were anticipations (4/97, 4%), and the remainder were tool substitutions, quality errors, and omissions (in total 36/97, 37%). Schwartz et al. (1991) suggested that the large number of independent A1 actions arose because "intentions are pathologically weakened or poorly sustained". Consequently, HH was particularly vulnerable to the presence of salient objects in his visual array. This "bottom-up" or "data-driven" activation may result in the patient "toying" with objects as each capture his attention. Schwartz et al. defined "toying" as reaching towards or lifting the object without actually using it for any purpose. They distinguished this from capture errors/object substitutions, where the association with an irrelevant object was strong enough to trigger an alternative schema and cause the patient to pick up and use the object. Schwartz et al. suggested that the object substitutions made by HH (e.g. pouring coffee into the oats while making breakfast) did not reflect a degraded semantic system, but "substitutions in the argument structure of the action plan, akin to word substitutions in slips of the tongue", which arise from "the momentary heightened salience of competing affordances". They proposed that 'toying' and capture errors were on the same continuum and reflected differences in the degree to which the irrelevant object activated an alternative schema (see also Shallice, Burgess, Schon, & Baxter, 1989).

In terms of the Norman and Shallice (1986) model of the control of action, patients with disorders of everyday action may have an impairment to stored schemas for routine tasks in the Contention Scheduling System. These schemas may store both the correct actions and the correct temporal sequence for performing a task. Consequently, degradation of the component actions, or of the correct temporal sequence, could impair performance. Alternatively, damage to Managerial Knowledge Units could lead to failures in activating appropriate behaviours or to linking such behaviours correctly over time. The relations between accessing the component actions of a schema and sequencing those actions correctly has been examined in group studies of patients with frontal lobe damage. These studies have also compared patients' performance for routine and non-routine schemas. In general, the findings have pointed to the particular importance of the frontal lobes in non-routine or novel activities; with familiar actions, problems only emerge when distracting stimuli are present or when the event is not well defined. Sirigu et al. (1995) had patients with either frontal lobe damage (one unilateral left, four unilateral right and four with bilateral lesions) or posterior damage, and neurologically unimpaired controls, verbally generate a script which was "routine" (preparing to go to work), "non-routine" (going to Mexico) and "novel" (opening a beauty salon). They found no significant differences between the three groups in the total number of actions generated or in the mean evocation time per action for the three types of script (routine, non-routine and novel). Subjects in all three groups also tended to select the same actions for the three scripts. However, the frontal patients tended to close the scripts too early and made more sequence errors than normal controls, particularly in the nonroutine script (though, perhaps surprisingly, not with the novel script). In addition, when subsequently shown the scripts they had generated and asked to make sure they were in the correct sequence, frontal lobe patients made sequence errors on nonroutine and novel events (errors on routine sequences were made but to a lesser degree). When asked to rate the importance of various elements for a given script, frontal lobe patients also made deviant judgements compared to normal subjects and the posterior patient group. Sirigu et al. (1995) suggested that patients with frontal lobe damage can have relatively preserved representations of familiar scripts, "the finding that the same actions tended to fall at the same relative position ... among the three groups, is a further argument for the integrity of the script data base and of the semantic or associative links within that database following frontal lobe damage". Rather than there being a problem in retrieving action schemas, Sirigu et al. proposed that frontal lobe damage produces particular difficulties in sequencing actions when formulating a new plan of action (i.e. for nonroutine and novel actions). They distinguished between a "semantic" component that specifies the behaviours making up an action and a "syntactic" component that specifies the temporal relations between the behaviours. They suggested that deficits in the syntactic component are most strongly associated with damage to the frontal lobes. Since this syntactic component may be most stressed in the nonroutine or novel actions, when the temporal order of behaviours is not clearly established, behavioural problems arise most clearly in the nonroutine or novel tasks. However, the frontal patients in their study also showed some problems in assessing the importance of specific actions in fulfilling a goal, which might suggest that semantic components in the stored action schemas were not entirely intact. In addition, problems in retrieving the components of familiar action sequences can be observed in other conditions. Sirigu et al. (1996) asked frontal lobe patients (one unilateral left, five unilateral right and four with bilateral lesions), patients with posterior lesions, and neurologically unimpaired controls, to arrange a set of 20 commands into four coherent scripts (five in each), either with or without the appropriate headers as cues. In a further condition, related and unrelated distractor items were also included. The frontal lobe patients made significantly more "boundary violations" (i.e. incorporating an action from one script into another) and sequence errors than either control group. For example, one frontal lobe patient incorporated the action "switch the coffee maker on" into the "making a phone call" script; another placed "go to a pedestrian crossing" after the action "cross the first traffic light" in the "crossing the street" schema. In addition, the frontal lobe patients failed to detect distractor actions and included both related and unrelated distractors in their scripts. They were also significantly slower than the other two groups when distractors were included and when there were no appropriate headers to constrain the task. Sirigu et al. (1996) concluded that patients with frontal lobe damage not only have problems with the temporal ordering of action schemas, but also with stored representations of the correct components within a script. Perhaps, in severe cases, these problems in retrieving both the semantic and syntactic components of a schema, even for familiar actions, could contribute to ADS.

Indeed, Sirigu et al. suggested that an impairment in the script analysis task could serve as a predictor of deficits in everyday life, when patients actually perform such tasks. However, the schemas assessed by Sirigu et al. tended not to include practical everyday activities, so it is not clear how far the results can be generalised. One of the aims of the current paper was to assess the relationship between ADS, following frontal lobe damage, and performance on script analysis tasks, both in terms of generating the component actions of everyday tasks and in sequencing the actions into the correct temporal order.

In line with the distinction between semantic and syntactic components of action knowledge, recent functional anatomical studies (using PET scanning techniques) suggest that separate brain regions represent knowledge about the components of a script and knowledge of the temporal order of events. Partiot, Grafman, Sadato, Flitman, and Wild (1996) measured regional blood flow in normal volunteers in four different conditions: (1) font discrimination, (2) script event action verification (e.g. subjects would have to verify if "say hello" and "open the door" were events that belonged to the action script "talking"), (3) script event membership verification (e.g. subjects would be asked to verify if "go to church" and "put on your swimsuit" were events within the script for going to a wedding); and (4) script event order verification task (e.g. two events were presented and subjects had to indicate whether or not they occurred in the correct order: (pay the bill, receive the menu—"no"). They found that the left frontal lobe, left anterior cingulate, and the anterior part of the left superior temporal gyrus were involved when subjects had to decide if an event belonged to a particular script (a test of the semantic component). However, the right frontal lobe, left superior temporal gyrus, and left and right middle temporal gyri were more active when event order had to be verified (a test of the syntactic component). Partiot et al. suggested that the temporal lobe activation probably reflected semantic processing involved in retrieving word meanings, the left frontal activity may be involved in activating and maintaining the action schema, and the right frontal lobe was in retrieving order information.

These group studies suggest that patients with frontal lobe lesions tend to have particular difficulty with relatively unfamiliar and novel tasks, compared to familiar tasks. The left and right frontal lobes may also be involved in activating and maintaining different aspects of action schema: the left frontal lobe in generating the correct component actions and the right frontal lobe the correct temporal sequence. Following on from these studies, we attempted to assess in more detail the factors contributing to the behavioural problems found in ADS, following frontal lobe damage. The method adopted was to compare the performance of patients displaying the behavioural symptoms of ADS with that of other patients who had deficits on cognitive abilities that may comprise some of the component skills necessary for successful performance on everyday tasks (e.g. verbal short-term memory, recall from long-term memory). We compared patients matched on these component abilities on (1) the performance of everyday tasks, (2) long-term memory for action schema (including action and order knowledge), (3) general sequencing abilities, and (4) performance on nonroutine, novel tasks. We assessed whether the poor performance of the ADS patients on everyday tasks was related to (1) deficits in stored long-term knowledge, (2) sequencing abilities that could not be linked to poor short or long-term memory, or (3) to general problems in moderating overlearned responses.

CASE REPORTS

We assessed three patients whose neuropsychological damage included lesions to the frontal lobe (FK, HG and DS) on relatively simple, everyday tasks. Clinical observation indicated that FK and HG were impaired on everyday tasks and presented with the symptoms described in other ADS patients (e.g. Schwartz et al., 1995). DS did not perform nearly so poorly on everyday tasks in clinical testing, though his performance was equally impaired on tests assessing "executive" cognitive abilities (see following). These three patients all performed poorly on the General Memory score (Wechsler Memory Scale (WMS) and so we also tested one amnesic patient, FL (who scored similarly on the WMS), to assess the importance of long term episodic memory in performing routine tasks. All four patients had lower IQ scores on the Wechsler Adult Intelligence Scale (WAIS) than would be expected from their occupations, or from their predicted IQ on the National Adult Reading Test (NART), although FL and DS were still within 2 SDs of the control mean (see Table 1). More detailed background tests assessed cognitive skills that should contribute to good performance on everyday tasks, such as short-term memory, executive skills, and object recognition.

Patient FK

FK was 29 years old at the time of testing. He suffered carbon monoxide poisoning when he

Table 1. Summary of Cognitive Abilities

Task	FK	HG	DS	FL
General cognitive abilities				
WAIS				
Verbal	64	–	68	81
Performance	56	–	78	64
Full	58	–	72	73
NART predicted IQ	94	106	93	92
Executive skills				
Stroop (raw score)	27	0	0	11
Memory				
WMS				
Verbal	58	55	56	55
Visual	< 50	62	79	< 50
General	< 50	< 50	< 50	< 50
Concentration	63	65	63	64
Delayed memory	< 50	54	61	< 50
Digit span forwards	7	5	4	4
Digit span backwards	2	2	2	4
Language				
BPVS (raw score)	38	–	60	52
Visual perception				
Unusual views				
Minimal feature	24/25	–	25/25	22/25
Foreshortened	22/25	–	25/25	22/25
Object naming (from Expt 2)	41%	93%	86%	86%

was at university studying for an engineering degree (in 1989); he has subsequently lived at home with close family support. An MRI scan revealed bilateral damage to the superior and middle frontal gyri, superior and middle temporal gyri, and lateral occipital gyri (see Fig. 3).

Following his accident, FK's full IQ score on the WAIS was 58 and his General Memory Score on the WMS was < 50. However, in contrast to his generally poor performance on verbal and visual memory, concentration, and delayed recall, his forwards digit span (7) was

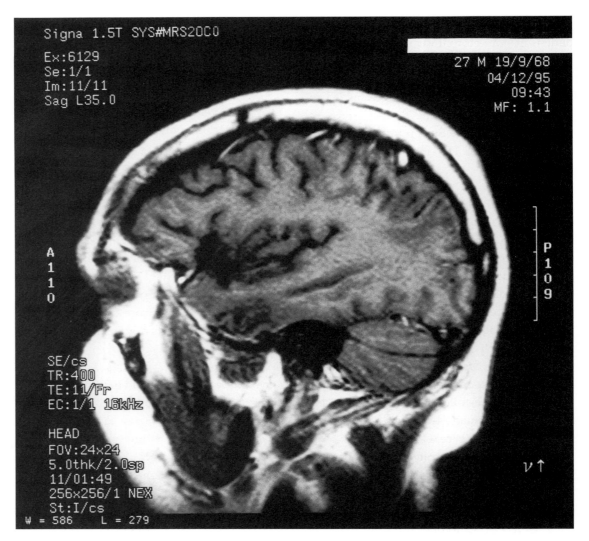

Fig. 3. MRI scan of patient FK.

normal, indicating that he had no impairment in verbal short-term memory. FK's low-level visual perception was normal (he scored 24/25 and 22/25 on unusual views matching from Birmingham Object Recognition Battery (BORB) (Riddoch & Humphreys, 1993) (control range 18.5–25 and 16.7–25, respectively). However, he did have severe problems in naming the objects that would be presented to him later in the everyday tasks (Experiment 2) (12/29, 41% correct) and the majority of his errors were close semantic neighbours. Consistent with his frontal lobe damage, FK also scored poorly on the Stroop test of executive function (see Table 1).

Patient HG

HG was a 78-year-old former railway driver who suffered a stroke in 1994. Following the stroke, HG lived in residential care and needed complete supervision. An MRI scan revealed damage to the right inferior, superior, and middle frontal gyri, precentral gyrus, inferior parietal lobule and angular gyrus (see Fig. 4). We were unable to complete a WAIS on HG, though he had a scaled score of 9 on Information, 9 on digit span, 7 on vocabulary and 3 on picture completion, compared to a mean and SD of 10 and 3, for 70–74-year-old controls. His General Memory Score on the WMS was < 50. HG performed poorly on verbal and visual memory, concentration, and delayed recall, although his forwards digit span (5) was just inside the normal range (7 ± 2). HG's low level visual perception and object recognition were good: He correctly named 27/29 (93%) of the objects that were presented in the everyday tasks. His two errors were to call a kettle a "jug" and a lunchbox a "container". There was no evidence of neglect or other forms of spatial deficit. Clinical symptoms of the parietal damage were not apparent.

HG was unable to name the colour of the ink on any of the colour words in the Stroop test, although he could when he was presented with noncolour words written in coloured ink (e.g. he could respond "black" to the written word "dog" but not to the written word "blue").

Patient DS

DS is a 64-year-old former train inspector who suffered a stroke in 1995. He lives at home with his wife and is able to function in a relatively self-sufficient manner. An MRI scan revealed damage to the left inferior, superior, and middle frontal gyri (see Fig. 5). Following his accident, DS's full IQ score on the WAIS was 72, his General Memory Score on the WMS was < 50, and his forwards digit span (4) was outside the control range. DS's low level visual perception and object naming was relatively normal: He scored 50/50 (100%) on unusual views matching and 25/29 (86%) on naming the objects in the everyday tasks. His errors were to call a bowl a "jug or cup", a knife a "fork", a lunchbox a "dolly", and a teapot a "kettle". Despite the presence of a few semantic errors in naming, DS used the objects appropriately. Consistent with his frontal lobe damage, DS performed poorly on the Stroop test (see Table 1).

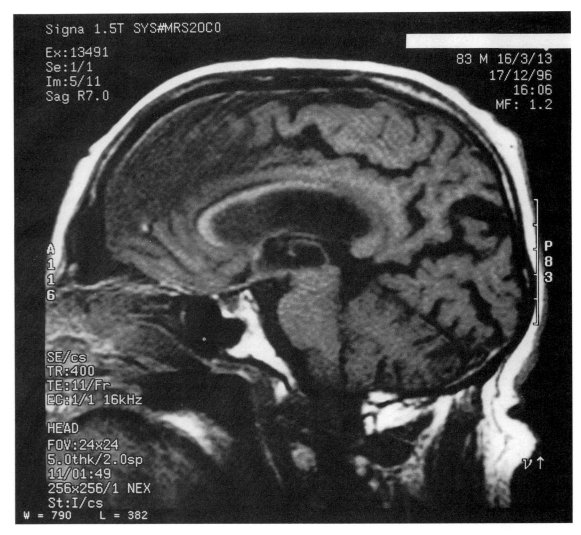

Fig. 4. MRI scan of patient HG.

Patient FL

FL is a 61-year-old former carpenter who suffered carbon monoxide poisoning in 1993. He lives at home with family support and attends day centres for occupational therapy. His main restriction on independent living is his amnesia. An MRI scan revealed bilateral globus pallidus infarctions and bilateral damage to superficial, lateral occipital gyri (see Fig. 6). Following his accident, FL's full IQ score on the WAIS was 73, his General Memory Score on the WMS was < 50, and his forwards digit span was 4. FL's visual perception of real objects

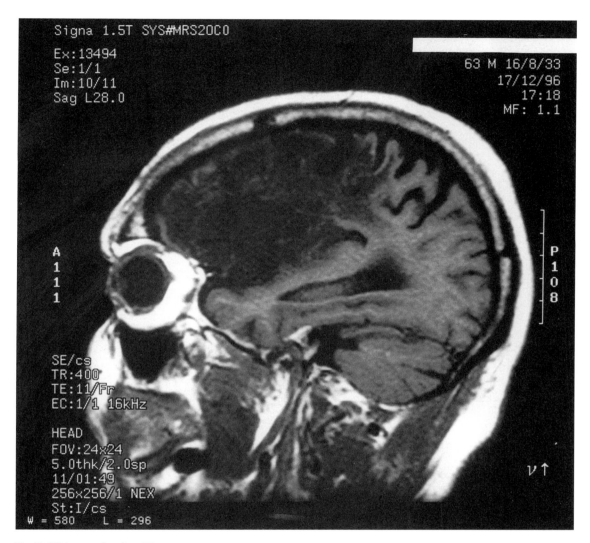

Fig. 5. MRI scan of patient DS.

was relatively good (he scored 22/25 and 22/25 on unusual views matching tasks taken from BORB, and he was able to name 25/29 (86%) of the objects used in the everyday tasks). His errors were to call a kettle a "jug", a lunchbox a "flask holder", a sandwich bag "Sellotape", and Sellotape "masking tape". FL also scored poorly on the Stroop test.

These background tests reveal that the patients were matched for predicted Full Scale IQ on the NART (HG scoring highest), on episodic memory recall (WMS), and on a measure of executive function associated with frontal damage (the Stroop test). The two patients

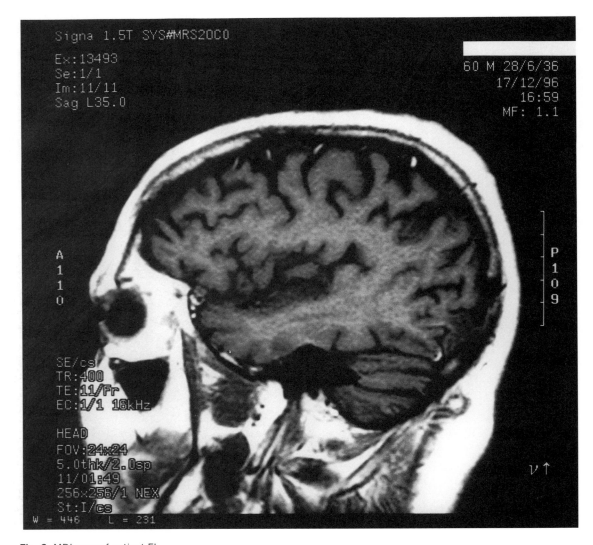

Fig. 6. MRI scan of patient FL.

with the behavioural symptoms of ADS, FK and HG, scored highest on the test of short-term verbal memory (digit span). The diagnosis of ADS for FK and HG is justified in the experimental tests that follow. It should also be noted that these two patients have brain damage extending beyond the frontal lobes. For this reason our emphasis is on providing a functional account of the symptoms associated with ADS rather than trying to link their behavioural problems closely with the site of their lesion.

EXPERIMENTAL INVESTIGATIONS

Generation of standardised scripts

Models of action generation typically propose that, for routine everyday tasks, we store the knowledge that allows us to accomplish these tasks successfully in hierarchically organised schemas (Grafman, 1995; Norman & Shallice, 1986). Indeed, Bower, Black, and Turner (1979) have suggested that other types of knowledge, such as socially acquired knowledge about how to behave, or what to expect in certain circumstances, may also be stored in schemas (or scripts) (see also Schank & Abelson, 1977). Bower et al. (1979) tested this hypothesis by asking a number of subjects to write down what they would expect to happen in five social situations (going to a restaurant, going to the doctor, attending a lecture, shopping at a grocery store, and getting up in the morning to get ready for school). They argued that if subjects tended to include the same component actions, in the same temporal sequence, this would provide empirical support for the "cultural uniformity assumption" inherent in script theories. Bower et al. (1979) found that subjects did tend to write the same component actions in the same temporal order, and they also tended to describe actions at the same level of specificity. For example, in the schema for visiting a restaurant, subjects might include "eating soup" as a component action, but would not write "pick up a spoon", "put it in the bowl of soup", "move the spoon towards your lips", and so on. Bower et al. identified the five "basic-level" scripts by including all components that were included by ≥ 25% of subjects. They suggested that these social schema might be hierarchically organised with more detailed "subordinate" actions (e.g. lift the spoon, put it in the soup, move the spoon towards your lips) represented at lower levels in the schema network.

In Experiment 1 here, we assessed whether routine everyday activities are also stored in fixed schema, using a similar methodology to Bower et al. (1979). We asked neurologically unimpaired subjects to write down how they would perform certain tasks to see if they would, in general, include the same components (at the same level of specificity) in the same temporal order.

Experiment 1: Generating Scripts for Action Tasks and Social Schemas

Method. Forty-five undergraduate students (mean age = 19.5 years) were asked to write descriptions of how they would accomplish nine tasks: (1) making a cup of tea with milk and sugar, (2) cleaning one's teeth, (3) writing a letter and getting it ready for the post, (4) preparing and eating breakfast cereal, (5) making a cheese sandwich and putting it in a lunchbox for a child to take to school, (6) painting a block of wood, (7) wrapping a gift, (8) making toast, and (9) shaving one's face. They were told to write the descriptions in a recipe style and provide enough information so that someone could subsequently follow their instructions and complete the task. A list of the items that should be used in each task were also given. We included this list so we would be able to compare the schema generated by the control subjects with the performance of the patients who were actually asked to carry out the task using a specified set of objects (in Experiment 2). The tasks were presented in a different random order for each subject.

Results. As shown in Table 2, the mean number of actions produced by the subjects varied for each task. However, in general we found that subjects tended to produce the same action components for each task. They also tended to use a similar level of specificity to describe an action in a sequence (e.g. put the butter on the bread), and only five subjects used very specific "low-level" actions, and only on a few occasions (e.g. take the lid off the butter, put the lid on the table, pick up the knife, put butter on the knife, spread the butter on the bread, put the knife down). We suggest that although these very detailed "subordinate" actions must be represented at some level within the script, the "basic-level" elements are those produced by the majority of the subjects. These basic-level elements may provide a schematic "frame" into which different subordinate actions may be slotted, according to the particular situation faced by the subject (e.g. whether there is a lid on the butter or not). Hence there are more likely to be individual differences in the frequency with which subordinate actions are elicited by subjects than the frequency with which basic-level actions are elicited. We designated the basic-level actions within the scripts to be those that were mentioned by 80%+ of the subjects. We also noted that the basic-level actions were almost all produced in the same temporal order, suggesting that script information is stored in a particular sequence. The only exception to this was for the tea sequence, in which 30 subjects put milk in the cup before tea and 15 put tea in before milk. The basic-level elements within each script and the order in which they were produced are shown in Appendix A.

It might be argued here that basic level actions are produced in a standard order not because they form part of our long-term knowledge but because of the physical constraints of the situations. For example, to make a cheese sandwich, the bread needs to be taken out of the bag before it can be buttered. However, although this is true for some steps in the tasks, it was by no means the rule. In the task "write and post a letter", there is nothing that prevents the address being written on the envelope before the letter is written; in making a cup of tea, milk could be placed in the cup prior to the kettle being poured into the teapot, etc. However, although these orderings of the

Table 2. Number of Actions Generated by Controls in Schema Generation Study (Experiment 1)

Task	Mean (SD)	Number of Actions Generated	
		By 80% of subjects	*Basic-level actions Used in Exp. 2*
Letter	10.5 (2.1)	8	8
Gift	8.4 (1.5)	8	8
Sandwich	11.0 (2.4)	7	7
Tea	11.3 (2.8)	9	6
Toast	10.1 (2.1)	7	6
Paint	8.6 (1.8)	5	5
Cereal	5.4 (1.6)	3	3
Cigarette	10.2 (1.9)	8	8
Teeth	9.1 (2.2)	6	6
Shave	5.1 (2.4)	3	3

component actions are possible, they did not occur. The consistency in the proposed actions, and in their ordering, suggested that these basic-level steps are stored in a temporally ordered schema in long-term memory. The basic-level action schemas, common to the majority of the control subjects (80% +), were then used in the following section to provide an objective measure for the everyday actions performed by the neuropsychological patients.

Neuropsychological Study

Experiment 2: Performing Everyday Tasks

In their investigations of ADS, Schwartz et al. (1991, 1995b) have typically used natural settings in which the objects required for one task (e.g. preparing a cup of coffee) were randomly mixed with others in the environment (e.g. a breakfast tray with food and utensils). In this context, the other objects present may have activated competing schemas, which lead to confused behaviour—especially if the content and structure of individual schemas were weakly activated (Schwartz et al., 1991). In Experiment 2a, we attempted to provide a purer measure of patients' performance on individual tasks by only giving them task-relevant objects. In Experiment 2b, the patients were presented with the target objects and three distractor objects that were task-irrelevant but semantically related to three of the target objects.

Method. Each patient was placed in front of a table and asked to perform a particular task. Performance was videoed for later analysis. The tasks were a subset of seven of those that had been standardised in Experiment 1: making a cup of tea, wrapping a gift, posting a letter, making toast, making a cheese sandwich, preparing a cereal, and painting a block of wood. In Experiment 2a, the patient was only given objects that would be needed to complete the task (e.g. for wrapping the gift he was given wrapping paper, the gift, scissors, sellotape, and a bow). In Experiment 2b the patients had three additional items that were semantically related to three of the task-relevant items (e.g. a newspaper, masking tape, and a knife). The tasks were presented to each patient in a different random order. Six elderly control subjects (five male, one female; age range 57–74) also performed the same set of tasks under the same conditions as the patients.

The videos were transcribed to record every action made by each patient. This was done initially by one of the investigators (EF) but a subset of the actions were also verified by a naïve rater. The number of independent A1 steps was used as a measure of how efficiently and coherently the patient performed the task. These independent A1 steps corresponded to "subordinate" actions that were not part of a complete A2 step, such as lifting a knife and then putting it down. In addition to this measure of performance, the patients' errors were also classified into a number of different categories:

1. Semantic: When a semantically related object was used in place of the target

object (e.g. pouring the milk into the teapot rather than the cup).
2. Sequence: When an action was performed at the wrong time within the overall sequence of A2 steps, according to the norms collected in Experiment 1 (e.g. putting the sugar in the cup before the tea).
3. Perseveration: When an action was repeated within the sequence (e.g. pouring the milk into the cup twice). In a first analysis, each repetition was counted as an additional perseverative error if an action was repeated a number of times (e.g. pouring the milk into the cup four times would be counted as three perseverations). In a second analysis, this type of error was counted as only one perseveration (with the milk).
4. Step omission: When the patient omitted one of the basic-level steps (e.g. failing to put any milk in the tea).
5. Tool omission: When the patient failed to use the correct object to accomplish a goal (e.g. stirring the tea with a finger).
6. Addition: When the patient added an action that was outside the range of actions produced by the normal group (e.g. ripping a teabag open and pouring loose tea into the teapot).
7. Quality: When the patient misjudged the appropriate amount of something (e.g. filling the cup with more milk than tea).
8. Spatial: When the patient misjudged the spatial orientation of the objects (e.g. missing the cup and pouring the tea onto the table).

Experiment 2a: Results and Discussion

Total number of errors. The total numbers of errors made by each patient on each task are shown in Table 3. The data were analysed by counting each error made by each patient (including every repetition, when a particular action was repeated a number of times), and entering each task as a subject, and each patient as a level, in a repeated measures ANOVA. This revealed a significant main effect of patient, $F(3,18) = 6.7$, $P = .003$. A post hoc Newman-Keuls comparison ($P = .05$) showed that there was no significant difference in the number of errors made by FK (38) or HG (49). However, there were significant differences between these two patients and FL (9) and DS (9).

A multiple regression analysis with number of steps and number of objects as predictors showed that number of steps ($r = .87, P = .004$), but not number of objects ($r = -.14, P = .08$), was a significant predictor of performance for HG. For FK, both the number of steps ($r = .79$,

Table 3. Number of Errors Made by Each Patient on Each Task (Experiment 2a)

Task	Number of steps	FK	HG	DS	FL
Letter	8	8	14(11)	1	2
Gift	8	9(8)	18(10)	2	3
Sandwich	7	7	5	2	1
Tea	6	8(7)	4	2	1
Toast	6	5	5	2	1
Paint	5	1	3	0	1
Cereal	3	0	0	0	0
Total		38(36)	49(38)	9	9

Numbers in parenthesis refer to the total number of errors when repeated perseverations of the same error are counted as one perseverative error.

$P = .01$) and the number of objects ($r = .73$, $P = .02$) were significant individual predictors, although only number of steps was reliable in a stepwise regression, with the variance due to the other factor take out. FL and DS made too few errors to provide a meaningful analysis.

When only new perseverative responses were included, a one way ANOVA showed the same pattern. There was a significant main effect of patient, $F(3,18) = 10.7$, $P = .003$, and HG (38) and FK (36) made more errors than FL (9) or DS (9).

Types of errors. The total numbers and proportions of each type of error, made by each patient, are shown in Table 4. FK made a relatively large proportion of step omissions (37% of his errors) but also tended to perseverate (18%), make sequence errors (16%), and add inappropriate actions (16%). The majority of HG's errors were step omissions (35%) and perseverative errors (31%), although he also made some semantic (10%) and sequence errors (10%). DS and FK only made nine errors each, but when they did make an error they were most likely to omit a step, make a sequence error, or make a quality/spatial error.

Although there was no significant difference in the total number of errors made by HG and FK, there seemed to be some differences in the types of errors made. For example, FK made a larger number of task additions in which he either did something bizarre (e.g. ripping the teabag open and pouring loose tea into the cup of tea) or became distracted and began an irrelevant task (e.g. starting to read all the writing on the lid of the butter). Interestingly, HG and FK made the same number of new perseverative errors (HG made 4 and FK made 5): However, when all perseverative errors were included, FK's total only increased by 2 (to 7 perseverative errors) but HG's total increased by 11 (to 15 perseverative errors). There was also a difference in the nature of the perseverative responses: FK tended to repeat an earlier action later in the task (e.g. he put the milk in the cup, poured water into the cup, and then poured more milk into the cup); in con-

Table 4. Types of Errors Made by Each Patient in Experiment 2a

Task		Sem	Seq	P(all)	Step Omit	Tool Omit	Add	Quality & Spatial	Total
FK	No.	3	6	7	14	0	6	2	38
	%	8	16	18	37	0	16	5	
HG	No.	5	5	15	17	2	1	4	49
	&	10	10	31	35	4	2	8	
DS	No.	0	3	1	3	1	0	1	9
	%	0	33	11	33	11	0	11	
FL	No.	1	2	0	3	0	1	2	9
	%	11	22	0	33	0	11	22	

trast, HG would perform the same action over and over again (e.g. when wrapping the gift he kept cutting more and more sellotape until he had nine extra pieces). FK only repeated the same action consecutively in the sequence on one occasion, compared to HG who repeated the same action consecutively on nine occasions. This suggests that FK failed to maintain a record of actions already completed and consequently repeated actions later in the task. However, if so, this does not simply reflect a problem in verbal short-term memory since FK had a normal forward digit span (of 7). Thus, the repetitions he made in the everyday tasks did not reflect a general inability to maintain a record of events over the short-term. Instead we propose that, for FK, once an action had been activated it was likely to be reactivated in the context of the related task. In contrast, HG's perseverative errors seem to reflect a problem in inhibiting an action once it has been performed. This is consistent with there being separate processes that constrain the repetition of actions immediately and over the intermediate term (at least a few actions back). Immediate repetition of actions may be prevented by an error-monitoring process or perhaps more automatically by a form of "rebound" inhibition immediately after selection of an action. Note that such rebound effects are incorporated into recent connectionist models of serial behaviour (Cooper, Shallice, & Farrington, 1995; Houghton, 1990, 1994; Houghton, Tipper, Weaver, & Shore, 1996). The repetition of actions over the intermediate term may be prevented by the imposition of temporal order constraints in long-term memory representations for actions. For example, memory for the sequence of actions in a task may take the form of a gradient of activation over the representation of each subaction (Houghton, 1990). Once one subaction has been selected it is inhibited immediately after completion. This leads to the representation of the next subaction in the sequence being the most strongly activated, so that it wins the competition for selection. This gradient of activation should normally enable late actions in a sequence still to be selected even if representations of earlier actions return from their rebound inhibitory state. However, if the activation gradient is evenly distributed, earlier actions may return to compete, as in FK here. We discuss this point in more detail in the General Discussion.

Number of Independent A1 Steps. FK was the only patient to make a substantial number of independent A-1 steps that did not link together within an A2 action (16 in total). FL, DS, and HG made only 1, 3 and 2, respectively. FK, in particular, tended to lift an object and then set it down without using it. Schwartz et al. (1991) suggested that this "toying" reflected a minimal elaboration of the "bottom-up" activation from the object. We will return to the theoretical implications of this in the General Discussion.

Neurologically Unimpaired Controls. The six neurologically unimpaired subjects made an average of 3 (range 2-4) errors on the set of everyday tasks. It is interesting that their results when performing the actions were very close to the basic-level action scripts generated by the younger subjects (in Experiment 1), indicating that these scripts are appropriate for subjects of different ages and backgrounds. The errors made by these controls were predominantly sequence errors (50%), with a few step omissions (22%), tool omissions (11%), spatial/quality errors (11%), and one semantic error (6%). They made an average of 1.5 independent A-1 steps. The two "control" patients, FL and DS, did make more errors than the non-brain-damaged controls. However, this increase was relatively slight when compared with FK and HG, who made 12 times the number of errors made by the neurologically intact subjects.

Experiment 2a: Conclusions

The errors made by FK and HG placed them well outside the normal range and are characteristic of the symptoms of ADS. On the other hand, DS and FL made considerably fewer errors and reliably less than the other two patients (although still more than the control subjects). This basic finding is of some interest because the four patients were matched in terms of some of the component processes that could underlie performance in everyday actions. In particular, the patients were matched for their immediate short-term memory (digit span), their general memory recall (in the WMS), and their ability to moderate over-learned responses to stimuli (in the Stroop test). These results show that relatively poor immediate memory performance (for FL and DS) is not sufficient to generate ADS, nor is it necessary, since FK in particular had a normal digit span. In addition, poor general recall from long-term episodic memory (in the WMS) or an impaired ability to moderate overlearned responses (in the Stroop test) are not sufficient. All of the patients showed impairments in episodic memory and in the Stroop task, but the ADS patients (FK and HG) were certainly no worse than the control patients (DS and FL). We suggest that neither immediate and longer-term episodic memory nor moderation of overlearned responses are, on their own, critical factors in ADS. Finally, the fact that all the patients were somewhat worse than the controls suggests that there may be some problems associated with the performance of these relatively complex tasks even in patients without the clinical symptoms of ADS. For instance, there may be lowered monitoring and error correction on some occasions. However, such problems are slight relative to the impairment on everyday tasks shown by in ADS. There are also some qualitative shifts in performance in the control patients relative to the ADS patients, since only the ADS patients made substantial numbers of perseverative errors (see Tables 4 and 6).

Experiment 2b: Method

Experiment 2b examined the effects of distractor objects on routine everyday actions. The tasks and objects were the same as Experiment 2a except for the additional three distractor items per task. Only the four patients were tested.

Experiment 2b: Results and Discussion

Total number of errors. The total number of errors made by each patient on each task is shown on Table 5. In the first analysis (with every repetition of an action counted as an error), a one-way ANOVA revealed a significant main effect of patient, $F(3,18) = 4.5$, $P = .01$. Again, HG (47) and FK (39) made more errors than FL (17) and DS (18), although on a Newman-Keuls analysis, only HG was reliably worse than FL or DS ($P < .05$). When only new perseverative responses were included, a one-way ANOVA showed the same pattern ($F(3,18) = 3.3$, $P = .05$).

A multiple regression analysis with number of steps and number of objects as predictors, showed that number of steps ($r = 0.82$, $P = .02$) but not number of objects ($r = .62$, $P = .09$) was a significant predictor for HG. For FK, there was a trend for an effect of the number of steps ($r = .76$, $P = .09$) but no effect of the number of objects ($r = .42$, $P = .48$) on his performance.

Table 5. Number of Errors Made by Each Patient on Each Task (Experiment 2b)

Task	FK	HG	DS	FL
Letter	10(7)	6	1	2
Gift	7	11(4)	5	4
Sandwich	14(13)	10	4	2
Tea	5	7	1	3
Toast	2	9	6	3
Paint	1	4	1	3
Cereal	0	0	0	0
Total	39(35)	47(40)	18	17

Numbers in parentheses refer to the total number of errors when repeated perseverations of the same error are counted as one perseverative error.

Types of errors. The total numbers and proportion of each type of error, made by each patient, are shown in Table 6. FK made the same types of errors as he had in the previous condition: step omissions (33%), perseverations (26%), sequence errors (23%), and additions (10%). The majority of HG's errors were again step omissions (32%) and perseverative errors (21%), although he also made a number of semantic (17%) and sequence errors (17%). DS made a higher number of semantic errors in this condition (44% of his errors compared to 0% in Experiment 2a) and also made a number of tool omissions (22%), and additions (17%). FL's errors were predominantly omissions (41%), although he did make a few semantic (18%) and sequence errors (18%).

A comparison of the type of perseverative errors made by FK and HG revealed a similar pattern to Experiment 2a. Both FK and HG made a total of 10 perseverative errors but when only new perseverative errors were included, FK made 6 new perseverations and HG made only 3. Combining across Experiments 2a and 2b, the difference between the patients in the proportion of new to repeated perseverations was reliable [$\chi^2(1) = 5.6$, $P = .02$]. Again FK tended to repeat an action later in the sequence but HG perseverated with the same immediate action. For example, in this experiment, HG perseverated with cutting paper eight times when he was supposed to be wrapping a gift. He perseverated with cutting despite the fact that the paper was far too small and, interestingly, despite the fact that he verbally acknowledged that the paper was too small. For example, on one occasion he put the

Table 6. Types of Errors Made by Each Patient in Experiment 2b

				Type of error				
Patient		Sem	Seq	P(all)	Step Omit	Tool Omit	Add	Quality & Spatial	Total
FK	No.	1	9	10	13	0	4	2	39
	%	3	23	26	33	0	10	5	
HG	No.	8	8	10	15	0	3	3	47
	%	17	17	21	32	0	6	6	
DS	No.	8	0	1	2	3	3	0	18
	%	44	0	6	11	22	17	0	
FL	No.	3	3	0	7	2	1	1	17
	%	18	18	0	41	12	6	6	

scissors down and said "the paper is not big enough is it?". We said "no, it isn't", yet despite this, he immediately picked up the scissors and started cutting the paper again. He then put the scissors down and said "it needs to be bigger than that", then once again picked up the scissors and started cutting the paper.

Number of Independent A1 Steps. FK, HG and DS made a number of independent A1 steps (17, 15 and 14, respectively). FL only made three independent A1 steps.

Beginning vs. End of Sequence: Overall Analysis. To assess the patients' performance at the beginning and the end of an action sequence we compared the number of step omissions made during the first and second half of the six- (tea and toast) and eight- (gift and letter) step sequences. Since the presence of semantic distractors did not strongly affect the performance of the ADS patients, we combined the results from Experiment 2a and 2b to maximise the data available. FK included significantly more steps (i.e. made significantly fewer step omissions) during the first half of the sequences (23/28) compared to the second half (14/28), $\chi^2(1) = 6.5$, $P = .01$. However, for HG, DS, and FL there were no significant differences in the number of steps completed in the first and second half of the sequences (see Table 7). We also compared the number of "overt" errors made to the first and second half of the sequence (in terms of the first and last three [in the six-step actions] or four [in the 8-step actions] steps of the standardised schema). By "overt" we mean errors that were new perseverations, sequence, semantic, or quality/spatial errors, and not omissions or additions. Note that this analysis was performed according to the position of the action in the correct sequence, not its position in the patient's response. For example, if a patient put a teabag in the teapot at the very end of his attempt to make a cup of tea, this would be scored as a sequence error in the first half of the schema, since it is step 1 in our standardised set of basic-level actions. Using these criteria,

Table 7. Analysis of the First and Second Half of Action Sequences

	FK		HG		DS		FL		
Task	1st	2nd	1st	2nd	1st	2nd	1st	2nd	
Steps completed[a]	23	14**	14	19	27	27	27	25	(max=28)
Overt errors	13	10	8	22*	3	10	4	8	
Errors/step[b]	0.61	0.67	0.57	1.16	0.11	0.41	0.19	0.32	

[a] Maximum = 28.
[b] Errors/step is a measure of the number of errors per step the patient performed and not per number of steps in the standardised schema.
** indicates a significant difference ($P = .01$) and * indicates a trend ($P = .06$).

the distribution of FK's errors (13 in the first and 10 in the second half of the action schema) conformed to that expected if his errors were equally distributed across the first and second half of the action sequences [$\chi^2(1) < 1.0$]. However, HG tended to make more overt errors in the second (22) compared to the first half (8) of the schema, $\chi^2(1) = 3.46, P = .06$. For DS and FL there were no significant differences between the first (3 and 4, respectively) and the second half (11 and 8, respectively) of the action schema [$\chi^2(1) = 2.5, P = .1; \chi^2(1) < 1.0$, respectively]. Note also that, in terms of the number of overt errors per step performed by that patient, FK and HG were again worse than DS and FL (see Table 7).

Experiment 2b: Conclusions

In terms of both accuracy and error type, the data from Experiment 2b replicated the pattern in Experiment 2a. The presence of the semantic distractors did not further impair the performance of the ADS patients or lead to a new pattern of errors. There was no evidence that semantic distractors, which may have activated alternative schemas/actions, produced additional deficits for these patients. There was some suggestion of an effect on the control patient, DS, who previously made few action errors. DS made relatively high numbers of semantic errors and independent A1 steps when distractors were present. We return to this point in the General Discussion, noting now only that distractor effects are not necessarily marked in patients showing the behavioural symptoms of ADS. From this we conclude that the ADS patients had detailed enough action schemas to activate the appropriate objects to achieve subgoals of the task. For instance, in wrapping a gift both HG and FK used the sellotape, rather than masking tape (one of the semantic distractors), which would also have been sufficient to achieve the goal but was less appropriate than the sellotape. These data also demonstrate that ADS patients may not all show the same pattern. FK tended to be better at accessing the initial steps in a sequence, and made omissions of the final steps, whereas HG tended to make more overt errors per step for the steps in the second half of the sequence. In addition, HG had difficulty in inhibiting an action immediately after he had selected it, whereas although FK also perseverated he tended to repeat an earlier action later in the sequence. We discuss how more detailed models of routine action generation may explain different patterns of ADS in the General Discussion.

Experiment 3: Generating Action Scripts

Experiment 2a showed that two of our patients (FK and HG) had severe problems in executing familiar everyday tasks. Schwartz et al. (1991, 1995b), using the Norman and Shallice (1986) framework, suggested that disorganised behaviour in everyday tasks may arise because the patients have degraded schemas in the contention scheduling system or, alternatively, because they have difficulty in inhibiting overlearned responses to objects in the visual array. If ADS reflects an impairment in inhibiting actions associated to objects in the environment, we would predict that patients should perform better when asked to describe verbally how a task should be carried out, since in this condition there are no objects in the visual field to drive "bottom-up" activation of alternative schemas. On the other hand, Sirigu et al. (1996) suggested that verbal generation tasks, tapping the patient's knowledge about particular schemas, would be valid predictors of actual performance. We tested these competing ideas in the next experiment that required subjects to generate verbal schemas for the tasks that had been presented in Experiment 2.

Method. The four patients were asked to give verbal descriptions of how a number of tasks should be completed. The tasks and the instructions were the same as in Experiment 1 except that the patients gave spoken descriptions, which were written down by the experimenter.

Results and discussion. The numbers of basic and subordinate level actions generated by the patients are shown on Table 8. There was a significant difference in the number of basic-level actions produced by the four patients, $\chi^2(3) = 12.6$, $P = .006$. Again the four patients divided into two distinct pairs: there was no difference between DS (who produced 56% of the basic-level actions generated by controls in Experiment 1) and FL (60%) or between FK (37%) and HG (31%), but there was a significant difference between DS and FK (the lowest and the highest in the relatively good and relatively poor pairs, respectively), $\chi^2(1) = 3.87$, $P = .05$. However, both DS and FL scored below control performance (see Experiment 1). This task probably underestimated DS's knowledge about the content of action schemas since his dysphasia made this task tiring and he often struggled to find the right words. For FL, performance may also have been limited by his amnesia, which could impair recall for some tasks. Nevertheless, the scores for DS and FL were still substantially higher than for FK or HG. The data are consistent with FK and HG having some deficit in their stored schemas for everyday actions.

Experiment 4: Ordering Action Schemas

In Experiment 3 the patients had to generate both the component actions and the temporal

Table 8. Number of Actions Generated by Patients in Schema Generation Study (Experiment 3)

	Basic level					Subordinate level			
Task	Controls	FK	HG	DS	FL	FK	HG	DS	FL
Letter	8	3	2(1)	3	5	0	0	0	1
Gift	8	1	2	3	4	0	1	0	0
Sandwich	7	2	0	3	3	2	3	2	1
Tea	6	3	4	4	7(6)	2	0	2	2
Toast	6	5	2	4	5	0	0	0	0
Paint	5	1	1	2	1	0	0	0	3
Cereal	3	2	2	2	2	2	0	0	1
Teeth	6	1	2	5(4)	2	1	0	0	2
Shave	3	1	1	3	2	1	2	1	1
Total	52	19	16	29	31	8	6	5	11

The numbers in parentheses refer to the number of actions given in the correct sequence if all were not generated in the correct sequence.

sequences of everyday tasks. The deficits apparent with FK and HG could reflect poor memory for the component actions, their order, or poor retrieval of both types of information[1]. FK and HG often only produced one or two basic-level actions for tasks in Experiment 3, making it difficult to evaluate order knowledge. To assess this in more detail, and to try to reduce retrieval demands, we gave the patients written labels for the basic actions in each schema and asked them to arrange the actions into the correct temporal sequence.

Method. Basic-level actions from each schema were written on separate cards[2]. The actions from a particular schema were presented to each patient in a jumbled array and their task was to arrange the cards into the correct temporal sequence. The tasks were presented to each patient in a different random order. There was no time limit imposed and the patients were encouraged to check the order of the actions when they had finished. The numbers of correct consecutive actions were recorded. For example, if the patients ordered the cards 1-2-4-3-5 they would score 1, since there was

---

[1] In other work, we have found that FK is impaired even at discriminating correct from incorrect instructions, when given written steps for a given action (an incorrect step here involved inappropriate use of two objects from the task set; for instance, "pour the milk into the teapot" for the task "make a cup of tea"). Order information is minimised under these circumstances. The remaining deficit suggests a problem in memory for the component actions in addition to any other deficit in order information.

[2] This experiment was carried out before Experiment 1, so the "basic level" actions are not exactly the same as those generated in Experiment 1.

only one pair of consecutive correct responses (1-2). There were 46 possible correct consecutive pairs.

Results and Discussion. There was a significant difference in the number of correct consecutive responses between the four patients (see Table 9), $\chi^2(3) = 8.1$, $P = .04$. More detailed analysis showed that the only significant differences were between FL and the two ADS patients (FK and HG), $\chi^2(1) = 5.7$, $P = .02$.

The results from Experiment 4 are generally consistent with those from Experiment 3. FK and HG were impaired at ordering the basic component actions in familiar tasks relative to FL. Nevertheless, in this experiment, as in Experiment 3, the two control patients (FL and DS) scored some way below ceiling. DS, in fact, did not differ reliably from the patients showing ADS symptoms [$\chi^2(1) = 1.8$, $P = .2$]. Here DS's relatively low performance cannot simply be attributed to his dysphasia and it appears that he has some problems in ordering

Table 9. Number of Correctly Ordered Consecutive Responses (Experiment 4)

Task	Max	FK	HG	DS	FL
Letter	6	2	0	4	1
Gift	6	1	4	2	0
Sandwich	6	0	2	2	4
Tea	7	0	3	4	3
Toast	6	2	1	2	3
Paint	3	3	0	0	3
Cereal	3	3	1	0	3
Teeth	4	1	1	0	2
Shave	4	0	0	4	4
Total	45	12	12	18	23
% of maximum		27	27	40	51

component actions in the correct temporal sequence. This may be part of a general problem in ordering information, associated with frontal lobe damage (Lepage & Richer, 1996; Petrides & Milner, 1988), and we examine this possibility in the next experiment. FL's relatively poor performance (compared to ceiling) may again be attributable to his severe memory problems which prevent him holding a number of items in working memory. Yet the ADS patients, FK and HG, still performed worse than FL or DS. Poor memory alone cannot account for their problems. Rather, it appears that they have a problem in accessing long-term knowledge about the sequential order of actions in everyday tasks, which may be part of a more general problem in ordering information in the correct temporal sequence.

In Experiment 5 we assess whether the problem with ordering actions was part of a more general difficulty, affecting the patients' ability to sequence. Patients were required to sequence letters and numbers. Letter sequencing was contrasted with number sequencing because letter sequences are arbitrary and correct ordering depends on retrieving long-term knowledge about the alphabet; knowing the order of letters in the alphabet is not fundamentally important for understanding the role of letters in our language. In contrast, it can be argued that order (or magnitude information) is inherent in each number, and so number ordering can be more stimulus driven and less reliant on long-term memory retrieval. In addition, we varied the number of items in each list. In Experiment 2a, FK and HG performed better on the tasks that had fewer steps than on tasks with a larger number of subactions. It may be that the retrieval of sequence information is more difficult when a larger working memory load is imposed; if so, then sequencing numbers and letters should be worse when there are more items in the array.

Experiment 5: Sequencing Letters and Numbers

Method. The patients were given a set of written letters or numbers (one per card) in a random order, and asked to order the items from early in the alphabet to late in the alphabet (for letters) or from small to big (for numbers). There were 10 three-item arrays and 10 six-item arrays for both letters and numbers. The test was repeated for FK on a separate occasion with different stimuli. One patient, FL, was only tested with numbers. FL is an attentional dyslexic (Mayall & Humphreys, 1998) and has problems in processing multi-letter arrays. Any problem he showed in letter ordering may be due to this rather than a sequencing problem per se. Sequencing performance was only measured when patients could verbally identify all the items in an array.

Results and Discussion. The numbers of correctly ordered arrays are shown in Table 10. The control patients, DS and FL, performed at ceiling on this test (though DS made some errors with the longer number sequence). FK was impaired with both types of stimuli, although, summing across conditions, his performance was better with numbers [$\chi^2(1) = 15.6, P < .0001$] and with shorter arrays [$\chi^2(1) = 15.6, P < .0001$]. HG was rather better than FK

Table 10. Sequencing Letters and Numbers (Experiment 5)

Condition	No. of items	FK	HG	DS	FL
Letters	3	6/20	7/10	10/10	–
	6	0/20	3/10	10/10	–
Numbers	3	17/20	10/10	10/10	10/10
	6	6/20	10/10	7/10	10/10

on these tasks: he was at ceiling with numbers and scored 10/20 with letters. He scored 7/10 and 3/10 on three and six letter sequences, and made 15/20 (75%) and 26/50 (52%) correct consecutive responses in each case. Relative to DS, FK and HG were worse at the letter sequencing tasks [$\chi^2(1) = 39.2, P < .0001$ and Fisher exact probability = .0004, respectively] and FK was worse at the number sequencing task [$\chi^2(1) = 4.5, P = .03$].

Interestingly, both HG and FK performed better with numbers compared to letters, despite having no problems in letter identification. Both patients could read all the letters with ease. We suggest that the difficulty in sequencing letters may reflect the fact that there is less intrinsic order in letters compared to numbers, so that letter ordering requires that long-term alphabetic knowledge is drawn upon. For numbers, magnitude information is inherent in their semantic representations. The performance of both patients seems to be worse when long-term knowledge must be accessed for sequencing and when there is a greater working memory load. A contributing factor in ADS may be a deficit in retrieving long-term schemas for sequential order and in maintaining the to-be-ordered elements in working memory during task performance.

Experiment 6: Following Commands (Utilisation Behaviour).

In Experiment 2, we assessed how ADS patients perform in routine everyday tasks, and we showed that, in addition to omission and sequence errors, FK and HG also sometimes used objects inappropriately (e.g. HG put milk in the teapot and FK ripped open the teabag instead of simply placing it in the teapot). Under the circumstances of measuring behaviour in everyday tasks, it is difficult to assess the factors that determine such actions. It might be that the patients failed to recognise the objects present, and so used them inappropriately, or alternatively it might be that other objects activated other action schemas that may have lead to inappropriate object use. In fact, we suggest that it is unlikely that a failure to recognise the objects was crucial here. For example, on other tests we have shown that HG's object recognition is extremely good (see also Buxbaum, Schwartz, & Carew, 1997; Forde & Humphreys, 1998). To provide a more analytic assessment of inappropriate action errors, we devised an experimental test of object use. The patients were given a simple command to follow, which had to be carried out on two target objects presented alongside two distractor objects. The four items were chosen to be objects that would be used together in an action schema (e.g. teapot, cup, milk, and spoon; stamp, pen, notepaper, and envelope). The idea was that these objects would activate a particular schema (make a cup of tea, writing a letter), which would have a top-down influence on a number of basic-level actions. These basic-level actions would also receive bottom-up activation from the objects in the visual array. Consequently, a number of basic-level actions would be activated relatively highly. We wanted to assess if the patient could inhibit the irrelevant competing actions and selectively follow a single command. The commands were either conventional or unconventional actions for the target pair. For instance, given the objects teapot, cup, milk, and spoon, a conventional action command (for the target objects teapot and cup) would be "pour from the teapot into the cup" and an unconventional command would be "pour from the cup into the teapot". We were particularly interested to see if the patients with frontal lobe damage could override the conventional actions (e.g. pour from the teapot into the cup) and use the objects in a novel way (e.g. pour from the cup into the teapot). The model of frontal lobe function outlined by Norman and Shallice (1986) would predict that the patients with an impairment to the SAS would find this novel condition particularly difficult.

Method. On each trial, a patient was given a written command and asked to read it aloud. If he did not read it correctly he was asked to read it again until it was read correctly. The commands involved using two objects in a conventional way (e.g. pour from the teapot into the cup, sign your name on the notepaper) or in a novel way (e.g. pour from the cup into the teapot, sign your name on the stamp). The patient was given two target objects and two related objects on each trial. There were two trials for each of the seven everyday schema that were used in Experiment 2. The patients were told that they would not need to use all the objects and were asked to follow the command as precisely as possible.

Results and Discussion. The number of trials on which the patients followed the command correctly (correct only) or followed the command correctly but also did something additional with the objects (correct + incorrect) is shown in Table 11.

An analysis of the correct only trials showed that there was a significant difference in the number of commands followed correctly between the standard and novel conditions for FK (57% and 0% correct, respectively, Fisher exact probability = .002) and DS [79% and 36% correct, respectively, $\chi^2(1) = 5.3$, $P = .02$]. HG showed a nonsignificant trend in the same direction [71% and 43% correct, respectively, $\chi^2(1) = 2.3$, $P = .1$]. However, for FL there was no difference between the standard and novel conditions [79% and 71% correct, respectively, $\chi^2(1) < 1.0$].

The errors made by the patients were divided into schema-congruent actions, novel actions, and others. An error was classified as a schema-congruent action when the patient used the objects in a task-congruent way; for example, if they were asked to pour from the cup into the teapot, but actually did the reverse. A novel action error occurred when the patient used the objects in an unconventional way; for example, such an error occurred when patients were asked to pour from the kettle into the teapot and they poured from the teapot into the kettle instead.

In the standard condition, FK and FL's errors were predominantly schema congruent actions, whereas HG and DS made more novel errors (although note that the total number of errors was small). However, in the novel condition all the patients tended to make schema congruent errors (see Table 12). In terms of the Norman and Shallice (1986) model, it seems that HG and FK have problems both in the Contention Scheduling System and in the SAS. DS appears to have no impairment in the Contention Scheduling System, but his poor performance in the novel condition may reflect an executive impairment within the SAS that is needed to overcome routine conventional behaviour. In terms of Grafman's (1995) account, the patients may differ in the level of any defi-

Table 11. Using Objects in Standard and Novel Ways (Experiment 6)

Condition	FK	HG	DS	FL
Correct + incorrect				
Standard	12	10	12	14
Novel	4	7	7	11
Correct only				
Standard	8	10	11	11
Novel	0	6	5	10

Table 12. Types of Errors Made by Each Patient in Experiment 6

Condition	FK	HG	DS	FL
Standard				
Schema congruent	5	0	0	2
Related novel	0	3	3	1
Other	1	1	0	0
Novel				
Schema congruent	13	6	6	2
Related novel	0	2	0	1
Other	1	0	3	1

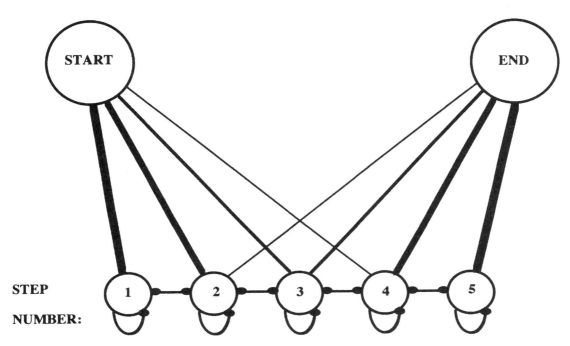

Fig. 8. A schema for task ordering based on an activation gradient. Units correspond to time signals to start and end the task and the component actions within the task. The activation assigned to the action units is a function of the size of the weight connecting each action unit to each time signal (stronger weights = thicker lines). Action units are mutually inhibitory and compete to fire. However, after firing each action unit self-inhibits itself enabling the next action to be made (rebound inhibition) (after Houghton, 1990).

bers of independent A1 actions, not produced in the appropriate temporal order. Evidence for there being specific problems in maintaining an activation gradient as sub-actions are produced is provided by the findings that (1) FK's performance was related to the number of basic-level actions in the schema and (2) he accomplished fewer basic-level actions at the end compared to beginning of the action sequences (Experiment 2). FK was also affected by the length of the array when sequencing letters and numbers. Consistent with our general proposal, we have some preliminary data showing that FK has problems on other tasks that require monitoring of serial order behaviour. For example, FK is significantly worse at spelling aloud (101/170 words, 59%) compared to written spelling (123/170 words, 72%), $\chi^2(1) = 7.2$, $P < .01$. We propose that this is because spelling aloud requires more internal monitoring of the sequence of letters compared to written spelling, where there is a visual record of the letters already produced. FK also tended to make more repetitions/perseverations with letters in spoken spelling (on 16/69, 23% of his errors) compared to written spelling [5/47 (11%), $\chi^2(1) = 2.8, P = .09$].

HG, like FK, may also have an impairment in maintaining an activation gradient, since he too was affected by the number of steps in a sequence (with everyday actions and with letters); however, any such problems seemed reduced for him relative to FK and, unlike FK, HG completed as many end as beginning steps in the action sequences. Consistent with both patients having problems in retrieving and maintaining stored knowledge of action sequences, we found that the patients were poor at recounting the action sequences verbally and in ordering the events when given the component actions on cards (Experiments 3 and 4).

Impairments to the Supervisory Attentional System (SAS) after Frontal Lobe Damage

It is interesting that neither the patient with left frontal lobe damage (DS) nor the amnesic control (FL) made many errors on our routine tasks, indicating that neither frontal lobe damage per se nor poor short or longer-term episodic memory are *sufficient* to cause significant problems on everyday tasks. Schwartz et al. (1995b) also found no correlation between performance on routine tasks and standard measures of frontal lobe skills (Trails test, Wisconsin Card Sorting test, Tower of London). Nevertheless, our frontal control patient DS performed as poorly as the ADS patients when required to carry out novel commands using objects that would normally be used together in a standard way (Experiment 6). This is exactly the pattern of performance that might be expected if there were damage to a specific system required for the performance of non-routine actions, much as envisaged in Norman and Shallice's (1986) proposal for a SAS separated from a Contention Scheduling System that deals with everyday action. Here it seems that damage to the SAS does not produce large numbers of ADS symptoms in everyday tasks. Indeed, the main problems with everyday tasks experienced by our frontal control, DS, occurred only when semantic distractors were present (Experiment 2b), when a number of semantic errors and independent A1 actions were induced. Such errors may reflect competition in selection for action between objects that can each be used for the task. In performing everyday tasks, the SAS may only be called upon when such conflicts arise.

However, although damage to the SAS does not seem sufficient, it may still be necessary for ADS to occur. Schwartz and Buxbaum (1997), in their "unifying account" of ADS, argued that problems in performing routine actions result from the combination of two impairments: (1) to the SAS and (2) to stored knowledge in routine action schemas. An impairment to the SAS alone would not lead to problems in activities that are automatic/routine and consequently can be performed with little "attentional control". An impairment to stored schemas for routine actions may also on its own not lead to difficulties on everyday tasks, since an intact SAS could develop strategies for accomplishing the particular goal. However, this compensatory system could not be used if the SAS was also impaired, and so performance on routine everyday actions would be pathologically impaired. Whether an impairment to the SAS is necessary to cause ADS cannot be gauged from the present data, since we report an association between "supervisory deficits" and ADS. Thus, in addition to the problems on everyday tasks, both HG and FK had deficits on the novel action task (Experiment 6) and on other tasks (e.g. the Stroop) where overlearned responses to stimuli would need to be modified. More important for our present purposes is our finding that deficits in retrieving routine action schemas are impaired in such cases, over and above the "dysexecutive" problems apparent in DS. Also we propose that such deficits can themselves be fractionated, with different forms of impairment apparent in maintaining an appropriate temporal order for the output of sequential actions (impaired rebound inhibition and impaired temporal ordering across the sub-actions within the task). Note that, even if action schemas are impaired (e.g. due to a loss in the temporal gradient necessary to retrieve actions over time), they may still be activated to some degree in a bottom-up manner by related objects in the novel action task. This may be sufficient to elicit utilisation errors when supervisory/monitoring processes are also impaired.

Everyday Actions, Bottom-up Activation and Physical Constraints on Performance

One final point to note concerns the performance of the amnesic control subject, FL, who made few errors in everyday tasks and so showed no major signs of ADS. He also performed better in the everyday tasks than in the tasks requiring him either to recall or to order the steps (from written labels) (Experiments 3 and 4). We have suggested that recall and ordering were hampered by FL's amnesia. What is interesting is that, when asked to perform the everyday tasks, these problems could be reduced. This is likely because real objects provide stronger "bottom-up" signals to help access stored knowledge about the tasks, and also some of the physical constraints on task performance reduce errors (e.g. bread cannot be placed in a toaster until it has been removed from the bag; paint cannot be applied to wood until the lid has been taken from the paint tin). Though these factors can improve performance for an amnesic patient, they did not seem to for our ADS patients. We propose that this is at least partly due to the impairment of stored knowledge of the basic-level actions, and their order, in such patients.

Manuscript received 12 November 1997
Revised manuscript received 15 May 1998
Manuscript accepted 17 September 1998

REFERENCES

Bower, G.H., Black, J.B., & Turner, T.J. (1979). Scripts in memory for text. *Cognitive Psychology, 11*, 177–220.

Burgess, P., & Hitch, G.L. (1992). Towards a network model of the articulatory loop. *Journal of Memory and Language, 31*, 429–460.

Buxbaum, L.J, Schwartz, M.F., & Carew, T.G. (1997) The role of semantic memory in object use. *Cognitive Neuropsychology, 14*, 219–254.

Cooper, R., Shallice, T., & Farrington, J. (1995). Symbolic and continuous processes in the automatic selection of actions. In J. Hallam et al. (Eds.), *Hybrid problems, hybrid solutions*. London: IOS Press.

Duncan, J., Emslie, H., & Williams, P. (1996). Intelligence and the frontal lobe: The organisation of goal-directed behaviour. *Cognitive Psychology, 30*, 257–303.

Forde, E.M.E., & Humphreys, G.W. (1998). The importance of semantic memory in performing routine everyday actions. Manuscript in preparation.

Fuster, J.M. (1991). The prefrontal cortex and its relation to behaviour. *Progress in Brain Research, 87*, 201–211.

Goldman-Rakic, P.S. (1992). Working memory and the mind. *Scientific American, 267*, 110–117.

Grafman, J. (1995). Similarities and distinctions among current models of prefrontal cortical functions. *Annals of the New York Academy of Sciences, 769*, 337–368.

Houghton, G. (1990). The problem of serial order: A neural network model of sequence learning and recall. In R. Dale, C. Mellish, & M. Zock (Eds.), *Current research in natural language generation*. London: Academic Press.

Houghton, G. (1994). Inhibitory control of neurodynamics: Opponent mechanisms in sequencing and selective attention. In M. Oaksford & G.D.A. Brown (Eds.), *Neurodynamics and psychology*. London: Academic Press.

Houghton, G., Tipper, S., Weaver, B., & Shore, D.I. (1996). Inhibition and interference in selective attention: Some tests of a neural network model. *Visual Cognition, 3*(2), 119–164.

Humphreys, G.W., Forde, E.M.E., & Francis, D. (1998). The organisation of sequential action. In S. Monsell & J. Driver (Eds.), *Attention and performance XVIII*. Cambridge, MA: MIT Press.

Klosowska, D. (1976). Relation between ability to program actions and location of brain damage. *Polish Psychological Bulletin, 7*, 245–255.

Lepage, M., & Richer, F. (1996). Inter-response interference contributes to the sequencing deficit in frontal lobe lesions. *Brain, 119*, 1289–1295.

Luria, A.R. (1966). *Higher cortical functions in man*. New York: Basic Books.

Mayall, K.A., & Humphreys, G.W. (1998). An early processing locus in a case of attentional dyslexia. Manuscript submitted for publication.

Milner, B. (1963). Effects of different brain lesions on card sorting. *Archives of Neurology, 9*, 90–100.

Milner, B., Petrides, M., & Smith, M.L. (1985). Frontal lobes and the temporal organization of memory. *Human Neurobiology, 4*, 137–142.

Norman, D.A., & Shallice, T. (1986). Attention to action: Willed and automatic control of behaviour. In R.J. Davidson, G.E. Schwartz & D. Shapiro (Eds.), *Consciousness and self regulation*. New York: Plenum Press.

Partiot, A., Grafman, J., Sadato, N., Flitman, S., & Wild, K. (1996). Brain activation during script event processing. *Cognitive Neuroscience, 7*, 761–766.

Petrides, M., & Milner, B. (1982). Deficits on subject-ordered tasks after frontal- and temporal-lobe lesions in man. *Neuropsychologia, 20*, 249–262.

Riddoch, M.J., & Humphreys, G.W. (1993). *BORB: The Birmingham Object Recognition Battery*. Hove, UK: Lawrence Erlbaum Associates Inc.

Schank, R.C., & Abelson, R. (1977). *Scripts, plans, goals and understanding*. Hillsdale, N.J.: Lawrence Erlbaum Associates Inc.

Schwartz, M.F. (1995). Re-examining the role of executive functions in routine action production. *Annals of the New York Academy of Sciences, 769*, 321–335.

Schwartz, M.F., & Buxbaum, L.J. (1997). Naturalistic action. In L. Rothi & K. Heilman (Eds.), *Apraxia: The Neuropsychology of Action*. Hove, UK: Psychology Press.

Schwartz, M.F., Fitzpatrick DeSalme, E.J., & Carew, T.G. (1995a). The multiple objects test for ideational apraxia: Aetiology and task effects on error profiles. *Journal of the International Neuropsychological Society, 1*, 149.

Schwartz, M.F., Mayer, N.H., Fitzpatrick De Salme, E.J., & Montgomery, M.W. (1993). Cognitive theory and the study of everyday action disorders after brain damage. *Journal of Head Trama and Rehabilitation, 8* (1), 59–72.

Schwartz, M.F., Montgomery, M., Fitzpatrick De-Salme, E.J., Ochipa, C., Coslett, H.B., & Mayer, N.H. (1995b). Analysis of a disorder of everyday action. *Cognitive Neuropsychology, 12(8),* 863–892.

Schwartz, M.F., Reed, E.S., Montgomery, M., Palmer, C., & Mayer, N.H. (1991). The quantitative description of action disorganisation after brain damage: A case study. *Cognitive Neuropsychology, 8(5),* 381–414.

Shallice, T. (1988). *From neuropsychology to mental structure.* Cambridge: Cambridge University Press.

Shallice, T., Burgess, P.W., Schon, F., & Baxter, D.M. (1989). The origins of utilization behaviour. *Brain, 112,* 1587–1598.

Sirigu, A., Zalla, T., Pillon, B., Grafman, J., Agid, Y., & Dubois, B. (1995). Selective impairments in managerial knowledge following pre-frontal cortex damage. *Cortex, 31,* 301–316.

Sirigu, A., Zalla, T., Pillon, B., Grafman, J., Agid, Y., & Dubois, B. (1996). Encoding of sequence and boundaries of scripts following prefrontal lesions. *Cortex, 32,* 297–310.

Vitkovitch, M., & Humphreys, G.W. (1991). Perseverant responding in speeded picture naming: It's in the links. *Journal of Experimental Psychology: Learning, Memory and Cognition, 17,* 664–680.

Vitkovitch, M., Humphreys, G.W., & Lloyd-Jones, T.J. (1993). On naming a giraffe a zebra: picture naming errors across different object categories. *Journal of Experimental Psychology: Learning, Memory and Cognition, 19,* 243–259.

Vitkovitch, M., Kirby, A., & Tyrell, I. (1996). Patterns of excitation and inhibition in picture naming. *Visual Cognition, 3(1),* 61–80.

APPENDIX A

Norms for Action Schemas (Experiment 1)

A. write and post a letter (paper, pen, stamp, envelope)	**B. wrap a gift** (bow, wrapping paper, selotape, gift, scissors)
1. write the letter 2. sign the letter 3. fold the letter 4. put the letter in the envelope 5. seal by pressing down 6. write the address on the envelope 7. lick the stamp 8. stick the stamp on the envelope	1. unfold the paper 2. put the gift in the centre 3. cut the paper 4. fold the paper over the gift 5. secure with selotape 6. fix one end 7. fix the other end 8. stick the bow on top
C. make a cheese sandwich and put it in a lunchbox for a child to take to school (cheese, sandwich bags, bread, plate, knife, butter, lunchbox)	**D. make a cup of tea with milk and sugar** (teapot, spoon, teabags, cup, milk, sugar, kettle)
1. put bread on the plate 2. put butter on the bread 3. put cheese on the bread 4. put the other slice of bread on top 5. cut the sandwich in half 6. put it in a sandwich bag 7. put it in a lunchbox	1. *put water in the kettle* 2. *let the kettle boil* 3. put a teabag in the teapot 4. pour hot water into the teapot 5. *wait a few minutes* 6. put milk in the cup 7. pour tea into the cup 8. put sugar in the tea 9. stir the tea
E. make toast (toaster, bread, plate, butter, jam, knife)	**F. paint a block of wood** (paint, wood, paintbrush, stirrer)
1. get some bread 2. put it in the toaster 3. switch the toaster on (press the lever to lower the bread) 4. *wait for the bread to toast* 5. put toast on the plate 6. butter the bread 7. put jam on the bread	1. take the lid of the paint 2. stir the paint 3. put the brush in the paint 4. paint the wood 5. continue painting until the wood is covered
G. prepare a cereal (spoon, bowl, milk, cereal)	**H. smoke a cigarette** (cigarette, ashtray, matches)
1. pour cereal into the bowl 2. pour milk into the bowl 3. eat the cereal with the spoon	1. get the cigarette out of the packet 2. put the cigarette in your mouth 3. strike the match 4. light the cigarette 5. inhale 6. exhale 7. flick ash into the ashtray 8. stub out the cigarette

| **I. clean your teeth** | **J. shave your face** |
(toothbrush, toothpaste, glass)	(shaving brush, razor, soap)
1. put the toothbrush under running water	1. brush the soap onto your face
2. squeeze toothpaste onto the toothbrush	2. shave with the razor
3. put toothbrush in your mouth	3. rinse the razor after each shave
4. brush	
5. fill the glass with water	
6. rinse your mouth	

Actions in italics were not scored as a basic-level action step in the schema.

SUBJECT INDEX

Abulia 515–17
Action
  see also Movement
  imagined 604–5
  imitation 549, 553, 554
  inappropriate object utilisation 799–801
  meaningless 559–60, 561, 565, 569, 573–77
  naturalistic 617–43
  perception 553–82
  willed 483–533, 772
Action disorganisation syndrome 771–811
Agnosia, visual 705–22
Akinesia 484, 502–4, 505, 509
'Alien hand' phenomenon 515, 646–47
'Anarchic hand' syndrome 645–83
Apathy 515–17
Apraxia 557–58, 617–43, 685–703
Ataxia, optic 708
Attention 490–91
Autism 747–70

Bradykinesia 502–4, 505, 509

Cognitive tasks, willed actions 494–95, 500–501, 506, 512, 517
Contention Scheduling System 772, 774, 806
Coordinated movements 501–2
Cortical potential 498–99, 502, 508, 514, 518
Corticobasal ganglionic degeneration 645–83
Counting 496, 501

Decision-making 499
Deep brain stimulation 509–10

Disordered action schema 771–811
Dopamine 504–6, 508, 521, 724–25, 741

Everyday tasks 787–95, 801, 807
External cues 508, 700, 733

Familiarity, visual 645, 674–78
Frontal apraxia 619
Frontal cortex 486–89, 500–501, 700
Frontal lobe 518, 771, 779, 801
Frontostriatal circuits 486–88, 500, 511, 517, 519–20
Functional imaging 490–96, 504–6, 511–12

Gesture imitation 747–70
Grasping movements
  brain regions involved 558–59, 573, 576, 597
  manual interference 660–64, 667–78
  visual agnosia 707

Handedness
  gesture imitation 752, 760–61
  left-right judgements 583, 584–86, 600, 608
  transcranial magnetic stimulation 496
Hands
  dyssynchronous apraxia 685–703
  gesture imitation 762–64
  manual interference 645–83
  movement perception 553, 558–79
  movement and shape 583–615
  naturalistic actions 617–43
Human locomotion 535–52, 554, 558

SUBJECT INDEX

Ideational apraxia 617–43
Ideo-motor action 484, 557
Imagined movement 604–5
Imitation of actions
   autism 747–70
   movement perception 549, 553, 554, 558, 559, 563, 565–67, 574–77
Initiation of movement 491–94, 498–99, 501–2, 506–8, 512–14, 728, 731, 734

Left hemisphere, apraxia 617–43, 678–79
Limb movements
   dyssynchronous apraxia 685–703
   left-right judgements 584–88, 599–601, 603
Locomotion 535–52, 554, 558

Macaque monkey 556–57
Manual interference effects 645–83
Meaningless actions 559–60, 561, 565, 569, 573–77
Monkeys 489, 556–57
Motor function
   autism 748–50, 761–62
   dyssynchronous apraxia 685–703
   hand movements 597–608
   manual interference effects 647–48, 653–54
   preprogrammed movements 723–45
   spatial perception 709
   willed actions 483–533
Movement
   see also Action
   brain regions involved 556–59, 564–67, 591–97
   dyssynchronous apraxia 685–703
   human locomotion 535–52, 554, 558
   preprogrammed 685, 687, 693–96, 698–99, 723–45

   willed actions 483–533
Movement-related cortical potential 498–99, 502, 508–9, 514

Naturalistic action 617–43
Numbers 495, 496, 501, 512, 518, 798

Object naming 804
Object recognition 584, 599
Object use 799–801
Optic ataxia 708

Pallidotomy 509–10
Parietal cortex 557, 558, 565, 567, 605
Parkinson's disease 484, 502–10, 516, 518, 520–21, 723–45
Pointing tasks
   manual interference effects 645, 655–60, 665–67, 674
   spatial perception 711–12, 714–15, 718
Positron emission tomography 490, 553, 558–79, 586, 590–97
Posture 535–36, 747, 752–60, 764–65
Preparation for actions 493, 501, 508, 514, 576, 602–4, 608
Preprogrammed movements 685, 687, 693–96, 698–99, 723–45
Primary obsessional slowness 514–15

Random number generation 495, 501, 512, 518
Reaching tasks 645, 660–64, 667–78
Reaction-time
   Parkinson's disease 726–28, 731, 733
   willed actions 485, 493–94, 501, 507–8, 512–13, 518

Schizophrenia 484, 511–14, 518, 520–21
Sequencing tasks 765–66, 772–79, 785–98, 801, 805–7
Shape, hands 583–615
Simultaneous movement 691–93, 698
Spatial perception 705–22
Speech production, object naming 804
Supervisory Attentional System 485–86, 772, 773, 806–7

Timing of actions 491–93, 498–99, 507, 734–35, 801–2
Transcranial magnetic stimulation 496–97

Visual affordances 645–83
Visual agnosia 705–22
Visual feedback 749
Visual perception
  hand shape 583–615
  human biological motion 553–82
  human locomotion 535–52
Visuospatial perception 705–22

Walking 535–52, 554
Willed action 483–533, 772
Word generation task 494–95, 501, 512